An Introduction to
INDUSTRIAL ORGANIC CHEMISTRY

An Introduction to

INDUSTRIAL ORGANIC
CHEMISTRY

PETER WISEMAN

Lecturer in Chemistry
University of Manchester Institute of Science and Technology

WILEY INTERSCIENCE
A DIVISION OF JOHN WILEY & SONS INC.
NEW YORK–LONDON–SYDNEY–TORONTO

© 1972 APPLIED SCIENCE PUBLISHERS LTD
RIPPLE ROAD, BARKING, ESSEX, ENGLAND

First published in Great Britain
1972 by Applied Science Publishers Ltd

Published in the U.S.A. 1972 by
Wiley Interscience Division, John Wiley
& Sons Inc., 605 Third Avenue, New
York, N.Y. 10016

ISBN : 0 471 95620 1

Printed in Great Britain by
Alden & Mowbray Ltd
at the Alden Press, Oxford

Preface

THE purpose of this book is to provide an introduction to the way in which organic chemistry is applied in industry, in terms which will be appreciated by modern students of chemistry and chemical engineering. While there are already available a number of very useful books on the organic chemical industry, and on industrial organic chemistry, in the main these consider the chemistry at a purely descriptive level, and to a reader accustomed to the discussion of reactions in mechanistic terms this tends to suggest that industrial organic chemistry is either simple and uninteresting, or is incapable of rationalisation. In fact, neither of these is generally true. Much industrial organic chemistry is far from simple, as may become clear in the following chapters, and although there are a number of areas where present knowledge of the mechanisms involved is meagre, there are many others where a substantial amount is known about the chemistry. However, the main interest in the study of industrial chemistry does not lie merely in a consideration of the mechanisms of the reactions involved, but in the relationships between chemistry, technology and economics, and this book attempts to bring out such relationships.

In writing a book of this type, there are two major ways of organising the material, on the basis of the technological structure of the industry, and on a chemical basis. The latter approach has been used, since this seems appropriate to a treatment in which the emphasis is on chemistry. As is often the case, a gain in one direction requires a sacrifice in another, and this arrangement is not ideal for demonstrating the overall technological structure of the organic chemical industry. Thus the steps in process chains from basic raw material to the final product which leaves the chemical industry will often be found to be distributed through a number of chapters. For example, steps involved in the manufacture of poly(ethylene terephthalate) from naphtha are discussed in Chapters 2, 3, 5, 6 and 7. It will be instructive for the student to follow through such process chains by use of the index and the table of contents. Cross-referencing of key points is also provided. To help provide an overall appreciation of the main transformations brought about in the industry, the main raw materials used for organic chemical manufacture are discussed in Chapter 1.

For many products, figures have been given to indicate the scale of operation. It will be found that these figures are for 'production', 'consumption' or 'capacity'. The first two are self-explanatory: capacity, or production capacity, is the amount which could be produced given that the demand existed. Production of a chemical by a company, or in a country, is often below the capacity for that chemical, and production and consumption often differ, because of the existence of import and export trading. The figures given generally refer to the United Kingdom, because it is with the UK chemical industry that the author is most familiar, but in some cases where data for the UK could not be found, US data are given. (Statistics for the American chemical industry are more readily available than for the chemical industries of many other countries, and it is useful to remember that the American chemical industry is about the same size as the West European chemical industry, and about six times the size of the U.K. chemical industry.) The statistics given have been taken from a number of sources and are of varying reliability. They are provided merely in order to give perspective to the discussion, and no attempt has been made to provide such data for all products discussed.

SI units have been used except where the use of other units seemed particularly desirable. For instance, pressures have been expressed in atmospheres since this clearly indicates the relationship between process pressure and atmospheric pressure.

I take great pleasure in expressing my thanks to my colleagues, Dr. R.E. Banks, Mr. F.R. Benn, Dr. J.M. Birchall, Dr. B.L. Booth, Mr. J. Dwyer and Dr. D.R. Taylor, who each read, and made valuable comments on, one or more chapters, and to Dr. A.R. Thompson for helpful discussions on some aspects of dyestuffs chemistry. I gratefully acknowledge the provision of various data on production and production capacities by ECN Chemical Data Services, London and by Mr. P.G. Godfrey of Chem Systems International Inc., London. Finally, I would like to acknowledge the contribution that has been made by my wife, who has considerably facilitated the progress of this book by her help and understanding.

PETER WISEMAN

Manchester
October 1971

Table of Contents

TABLE OF CONTENTS

Chapter 7 POLYMERIZATION
Stereochemistry of

Chapter 1

Introduction

THE chemical industry is an industry of major importance in all highly developed countries; in the UK, for example, its output at the present time is about 9% of total manufacturing industrial output. The products of the chemical industry make a major contribution to present-day standards of living, and life as we know it today would be impossible without these products. Some of the more obvious examples of materials which make life more pleasant and convenient for us are plastics, synthetic fibres, pharmaceuticals and dyes, but the products of the industry cover a much wider range than this. It is, in fact, difficult to think of many goods in modern society, in the manufacture of which products of the chemical industry have played no part.

In the application of chemistry in industry, factors are involved which the student generally does not meet in his chemistry courses, i.e. economic factors. Chemistry in industry is always eventually judged in economic terms, i.e. the benefits which it may produce are balanced against the resources which will be necessary to produce them. These economic judgements are applied over all aspects of an organisation's activities, but they apply with particular directness to manufacturing processes.

ECONOMICS OF CHEMICAL PROCESSES

Making a chemical involves the deployment of a number of resources by the organisation involved. These are of three types:

(1) Resources involved in acquiring the technological capability to design and operate the plant, i.e. in acquiring 'process know-how'.

(2) Resources involved in building the plant.

(3) Resources involved in operating the plant.

It is convenient, in capitalist and communist societies alike, to express the expenditure of resources of this type in terms of money. Thus, the resources involved in building the plant—materials, labour, energy and so on—are purchased by the organisation, and the sum of their costs is called the *capital cost* of the plant. To operate the plant requires raw materials, labour and supervision, services, i.e. electricity, fuel, steam, water, etc. and provision of various other items and facilities. The sum of the costs of these is the *production cost* of the product; it is normally expressed as cost per weight of product, e.g. pence/kg.

Acquiring the process knowhow involves the organisation in carrying out research and development work, involving costs of salaries, equipment, etc. Alternatively, the knowhow may be purchased from some other organisation.

It is not difficult to see that for a particular product, the less expenditure of resources its manufacture requires the better. In other words, it is desirable that chemical processes should require cheap plants and have low production costs. Although the relationship between capital and production costs and process chemistry is a complex one, and cannot be discussed fully here, some simplified observations will be of help.

Capital Cost

To a first approximation, the more simple a process, the cheaper will be the plant required to run it. Features of processes which tend to increase capital costs are:

(1) *Multiple reaction stages.*
In a large plant, each separate reaction carried out requires a separate reactor; a three-stage process requiring three reactors is evidently likely to have a higher capital cost than a single-stage process requiring one reactor.

(2) *Extensive separation.*
Separation of product from by-product, unreacted raw materials, etc. involves the provision of separation equipment, e.g. distillation columns, filters, etc. The more separation has to be done, the more equipment is required, and the higher is the capital cost.

(3) *Use of extremes of temperature and pressure.*
In very general terms, the further removed are the operating temperature and pressure from ambient conditions, the higher the capital cost of the necessary plant will tend to be. The use of

temperatures above about 1 200 K has special engineering problems associated with it which lead to high capital costs. Temperatures below ambient temperature require the use of refrigeration equipment, which is expensive. Use of increased pressure requires the use of compressors or pumps, and of pipes, reactors and other equipment designed to operate under pressure.

(4) *Use of corrosive materials.*

This will involve the use of corrosion-resistant materials, which in some cases are very expensive.

(5) *Solids handling.*

Handling solids mechanically is much more expensive in terms of capital expenditure than handling liquids or gases. For example, continuous addition of a liquid to a reaction vessel is very readily achieved by means of a pump and a pipeline; continuous addition of a solid, with the use of the same amount of labour, requires much more complex and expensive mechanical equipment.

Production Costs

For most chemical processes the *raw material cost* is by far the most important of the production costs, and it is therefore highly desirable that processes should use cheap raw materials, and should convert them efficiently into product. The latter means not only that high yields are desirable, but also that the use of materials which do not appear in the product should be avoided if possible. There are many processes in which such ancillary chemicals are used. For example, the monochlorobenzene process for phenol (see page 183) gives quite a high yield of phenol, but is inefficient from the point of view of raw materials usage in that, in addition to producing phenol, it converts chlorine and sodium hydroxide into low value sodium chloride:

$$C_6H_6 \xrightarrow{\text{Cl}_2} C_6H_5Cl \xrightarrow{\text{NaOH}} C_6H_5ONa \longrightarrow C_6H_5OH + NaCl$$
$$\begin{array}{ccc} + & & + \\ HCl & & NaCl \end{array}$$

It is not possible to summarise in simple terms the relationship between the other production costs and process chemistry. However, one or two generalisations can be made. Labour costs are likely to be higher for multistage processes than for single-stage processes, for obvious reasons, and are also generally increased when solids have to be handled. Compression of gases to high pressures and heating reactants to high temperatures consume large amounts of energy and give rise to high service costs. Strongly endothermic reactions also give rise to high energy consumption. Extensive separation operations tend to result in high service costs.

There is a component of the production costs, not discussed here, which is proportional to the capital cost of the plant.

Research and Development Costs

As we have seen, in addition to the capital and production costs, resources have to be expended in developing, or less often in buying, a process. There tends to be an inverse relationship between research and development costs and capital and production costs, in that the more is spent on research and development, the better process one is likely to achieve, and therefore the less the capital and/or production costs will be. In any particular situation, there is an optimum amount of research and development spending, after which further expenditure will not bring sufficient savings in costs to make it worthwhile. (It should be pointed out that it is not possible to know what this optimum amount is: this is one of the things which makes research management difficult.)

An important consequence of this is that different types of processes are developed to different degrees of technical sophistication. If one proposes the manufacture of acetic acid at the rate of 75 000 tons per annum, which may involve a capital investment of the order of £5 million and an annual production cost of the order of £2½ million, then the savings which would result from even minor improvements to the process are very great. In such a situation extensive research and development work is justified. A process to manufacture 100 tons per annum of butyric acid would evidently warrant much less expenditure on development. Consequently, small-tonnage processes are generally less highly developed than large-tonnage processes.

Effect of Scale on Costs

A very important factor in chemical process economics, which does not greatly affect the discussions in this book but which should be noted is that chemical processes are very susceptible to *economies of scale.* That is to say, for most processes the capital cost per annual ton of product and the production cost decrease as the scale of operation is increased.

Because of this effect, there is a very considerable incentive to build large plants, and one of the important developments in the chemical industry in recent years has been a great increase in scale of operation in the manufacture of a number of chemicals. In ethylene manufacture the size of plants has increased by about ten times in the past twenty years. Because of the economic advantages of building large plants there is a tendency for there to be only a small number of producers of a chemical in a country the size of the United Kingdom.

BASIC RAW MATERIALS FOR ORGANIC CHEMICALS

There are a number of possible sources of carbon available in large quantities, any of which could, in principle, form the basis for the manufacture of all organic chemicals, except for the relatively small number of naturally occurring products which cannot be synthesised:

	animal materials	e.g. fats and oils
	vegetable materials	e.g. oils, carbohydrates
CARBON SOURCES	*coal*	
	petroleum	
	natural gas	
	carbonates	e.g. calcium carbonate

What is in fact used is determined by price and availability of the material, and the ease or otherwise with which it can be converted into useful chemicals using currently available technology. Of the sources listed, all except metallic carbonates have been used for the manufacture of organic chemicals. The relative importance of the different sources has changed substantially during the history of the chemical industry. The most important change, which has occurred during the past forty years in the USA, and since World War II in most other highly industrialised countries, has been the emergence of petroleum as the major raw material for organic chemical manufacture (see Table 1-1).

We shall briefly discuss the major methods that have been used for the manufacture of organic chemicals from the various raw materials, and will consider the main reasons for the changes that have occurred in the raw material basis of the industry.

TABLE 1-1

PRODUCTION OF ORGANIC CHEMICALS IN THE UNITED KINGDOM BY SOURCE

Source	Thousand metric tons									
	1949	1953	1955	1959	1962	1964	1965	1966	1967	1968
Coal	300	380	465	595	700	705	775	815	825	830
Carbohydrates	160	170	140	80	30	30	30	30	30	30
Petroleum	45	195	290	595	1 400	1 705	2 390	2 850	3 260	3 730
Total All Sources	505	745	895	1 270	2 130	2 440	3 195	3 695	4 115	4 590

Source: Estimates by Esso Chemical Ltd. staff.

Fats and Oils
These are to be found in both animal and vegetable organisms. For example, tallow and whale oil are obtained from animal sources and palm oil, cottonseed oil and castor oil from vegetable sources. They consist of esters of glycerol and long-chain fatty acids; these esters are known as 'triglycerides':

$$CH_2O_2CR_1$$
$$CHO_2CR_2 \qquad triglyceride$$
$$CH_2O_2CR_3$$

R_1, R_2 and R_3, which may be all the same, are straight-chain aliphatic groups which may be unsaturated and/or carry hydroxyl groups.

Because these materials are obtained from plants, which have to be harvested, or animals, which have to be reared or hunted, they are fairly expensive. Palm oil, for instance, costs about £110/ton in the UK (1970). They have therefore never been used as general purpose carbon sources, but rather have been, and are still, used for applications where their structure makes them particularly convenient raw materials. Examples of such uses are shown below.

SOAP MANUFACTURE
Hydrolysis of triglycerides with sodium hydroxide gives glycerol and the sodium salt of the fatty acid, i.e. soap:

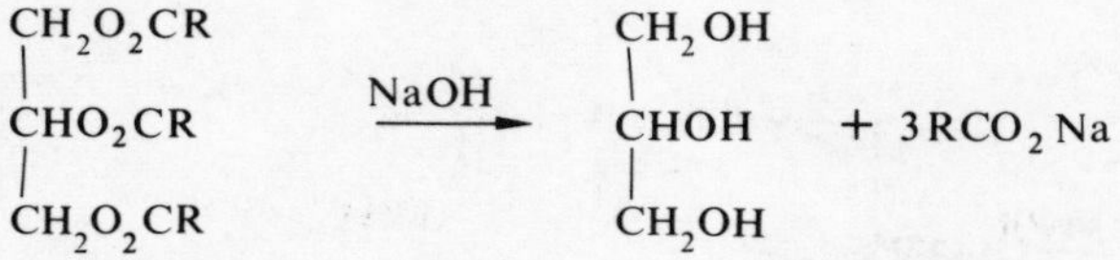

$$\begin{array}{l} CH_2O_2CR \\ | \\ CHO_2CR \\ | \\ CH_2O_2CR \end{array} \xrightarrow{\text{NaOH}} \begin{array}{l} CH_2OH \\ | \\ CHOH \\ | \\ CH_2OH \end{array} + 3\,RCO_2Na$$

The importance of soap does not need to be emphasised; this process is carried out on a large scale. The process also provides a major source of glycerol. For example, all glycerol production in the UK is from this source.

LONG-CHAIN FATTY ACIDS

Hydrolysis of triglycerides with steam gives the free acids, e.g.:

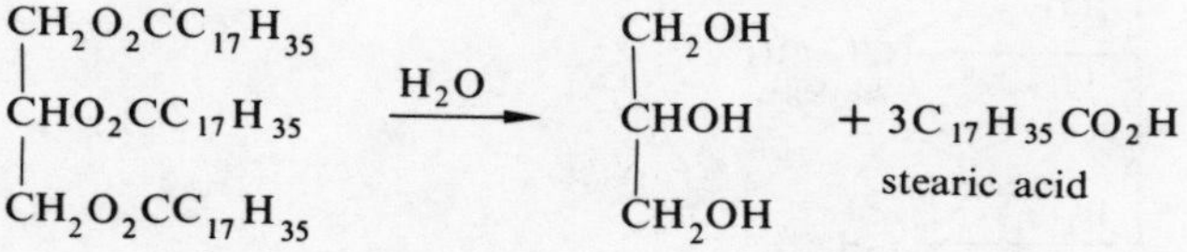

$$\begin{array}{l} CH_2O_2CC_{17}H_{35} \\ | \\ CHO_2CC_{17}H_{35} \\ | \\ CH_2O_2CC_{17}H_{35} \end{array} \xrightarrow{\text{H}_2\text{O}} \begin{array}{l} CH_2OH \\ | \\ CHOH \\ | \\ CH_2OH \end{array} \begin{array}{l} + 3\,C_{17}H_{35}CO_2H \\ \text{stearic acid} \end{array}$$

Long-chain acids of this type are difficult and expensive to make by synthetic methods, and hydrolysis of fats and oils is currently the preferred method of manufacture.

LONG-CHAIN ALCOHOLS

A number of long-chain alcohols are made by hydrogenolysis of triglycerides, or of fatty acid methyl esters prepared from triglycerides (see page 209), e.g.:

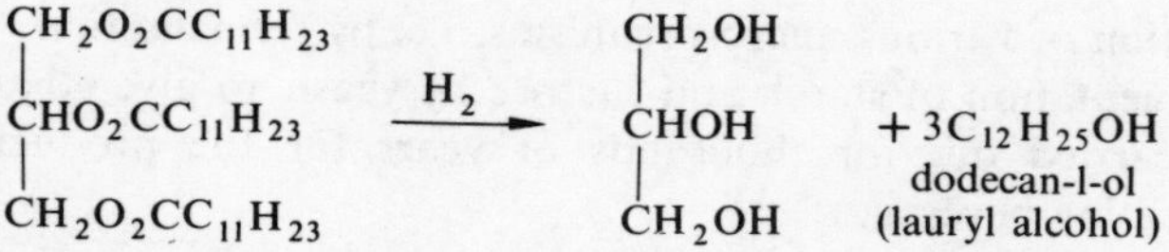

$$\begin{array}{l} CH_2O_2CC_{11}H_{23} \\ | \\ CHO_2CC_{11}H_{23} \\ | \\ CH_2O_2CC_{11}H_{23} \end{array} \xrightarrow{\text{H}_2} \begin{array}{l} CH_2OH \\ | \\ CHOH \\ | \\ CH_2OH \end{array} \begin{array}{l} + 3\,C_{12}H_{25}OH \\ \text{dodecan-1-ol} \\ \text{(lauryl alcohol)} \end{array}$$

Although synthetic methods are available for the manufacture of long-straight-chain primary alcohols (see page 245), they are relatively expensive, and manufacture from fats and oils remains competitive at present.

Carbohydrates

Carbohydrates are obtained from plants as, for example, sucrose $C_{12}H_{22}O_{11}$ and starch $(C_6H_{10}O_5)_n$. They are polyhydroxy compounds

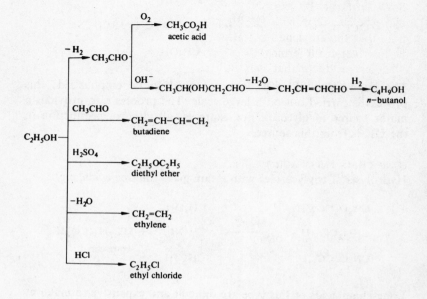

Fig. 1-1. *Some major products from ethanol.*

whose structures are not particularly relevant to the present discussion. The potential of these materials as raw materials for general organic chemical manufacture stems from the fact that they can be converted into a number of simple aliphatic chemicals by the action of various micro-organisms, i.e. by fermentation.

Fermentation of starch and sucrose by yeasts to give ethanol has been carried out for thousands of years for the production of intoxicating beverages:

$$C_{12}H_{22}O_{11} \longrightarrow 4C_2H_5OH + 4CO_2$$
sucrose

$$(C_6H_{10}O_5)_n \longrightarrow 2n\,C_2H_5OH + 2n\,CO_2$$
starch

Such fermentation also provided a major route to a number of aliphatic chemicals for many years (see Fig. 1-1). It is interesting to note that the ethylene from which the first polyethylene was made was derived from fermentation ethanol.

Other products that can be made by fermentation of carbohydrates are butanol, acetone and citric acid:

$$C_{12}H_{22}O_{11} \xrightarrow[\text{36-48 hrs.}]{\textit{Clostridium acetobutylicum}} \begin{array}{l}\text{n-butanol} \\ \text{acetone} \\ \text{ethanol}\end{array}$$

$$C_{12}H_{22}O_{11} \xrightarrow[\text{6-12 days}]{\textit{Aspergillus niger}} \begin{array}{l}CH_2CO_2H \\ | \\ HOCCO_2H \\ | \\ CH_2CO_2H \\ \text{citric acid}\end{array}$$

The manufacture of butanol and acetone by this route is no longer important, but fermentation is still the major route to citric acid, the only other source being extraction from lemons and pineapples. Ethanol is now mainly made from petroleum. Fermentation ethanol accounted for 9% of total production in the USA in 1963, and about 12% of total production in the UK in 1965. The decline in importance of fermentation of carbohydrates as a source of chemicals is illustrated by Table 1-1.

A number of factors have been responsible for this decline in importance. Carbohydrates are agricultural products and their production requires a large amount of labour. Consequently they are relatively expensive; sucrose in bulk, for instance, costs about £40/ton (UK, 1970). Further, fermentation is wasteful of raw materials in that a considerable proportion of the raw material does not appear in the product. Thus, in the ethanol fermentation, if one assumes 100% yield according to the equation shown, 1 kg sucrose will give 0·54 kg ethanol. The actual yield is less. Fermentations are slow and have to be carried out in dilute solutions. An ethanol fermentation, for instance, takes about two days, and gives a solution containing about 9% ethanol; the final concentration of products in the butanol–acetone process is 2%. This means that a large amount of reactor volume is required, and that separation costs are high. These factors have combined to render fermentation routes to organic chemicals unattractive in highly developed countries, compared to alternative routes from other raw materials, except where, as in the case of citric acid, synthesis is difficult.

Coal

The structure of coal is extremely complex, and is by no means fully elucidated. It is fairly clear, however, that coals (which vary widely in constitution) are made up of large molecules consisting of fused aromatic and hydro-aromatic rings, and a small proportion of nitrogen-containing rings. Oxygen is present as hydroxyl and carbonyl groups. A typical approximate constitution is $(C_{22}H_{20}O_3)_n$ +small amounts of nitrogen and sulphur.

It is fairly clear from this that coal is far from ideal as a raw material for general organic chemical manufacture. It contains a much smaller proportion of hydrogen than most technically important organic chemicals, and its complex, fused-ring structure is far removed from their relatively simple structures. It would seem likely that drastic treatment would be required to convert coal into useful chemicals. A further disadvantage is that it is a solid and, as has been pointed out, handling solids on the large scale is difficult and leads to high capital and processing costs. On the other hand, coal is cheap. In the UK it costs about £5/ton (1969) in large quantities; in the USA it is much cheaper.

Despite its disadvantages, coal was for many years the major source of organic chemicals in the UK and in many other countries *via* the routes discussed below.

COAL CARBONISATION BY-PRODUCTS

If coal is heated to above about 570 K in the absence of air, decomposition occurs. As would be expected, the higher the temperature, the greater the extent of the breakdown. At 1 200 K the products are carbon, in the form of coke, and various gaseous and liquid decomposition products. This process is operated to make coke for use in steel manufacture, in 'coke ovens', or to make coal gas (town gas), in 'retorts'. In both cases, apart from the gas, a mixture of organic chemicals is obtained in the form of what are known as crude benzole and coal tar:

$$\text{coal} \xrightarrow{\text{1 200–1 400 K}} \text{coke} + \text{gas} + \text{crude benzole} + \text{tar}$$

By relatively simple physical and chemical treatments it is possible to separate a number of individual compounds from these products:

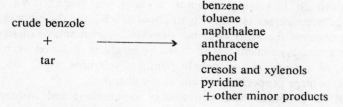

It is important to note that these are obtained as by-products of coke and gas in relatively minor quantities. Thus, under one set of typical conditions, carbonisation of 1 000 kg of coal yields, for example, about 5·3 kg benzene, 2·9 kg naphthalene and 0·4 kg phenol. There is thus no question of carrying out coal carbonisation for the primary purpose of making chemicals.

This being so, the amount of these products available from coal carbonisation is determined by the demand for the primary products, coke and gas; for most of them the supply is not keeping pace with the demands of the chemical industry. Manufacture of town gas by coal carbonisation has been rendered obsolescent in the UK and many other countries by the development of methods of producing gas from petroleum fractions and by the advent of cheap natural gas, and the amount of coal carbonised for this purpose is rapidly diminishing. In the steel industry, increased efficiency of blast furnace operation has led to a decrease in usage of coke per ton of steel. Overall, the amount of coal subjected to high-temperature carbonisation in the UK has decreased substantially during recent years (see Table 1-2).

TABLE 1-2

COKE PRODUCTION IN THE UNITED KINGDOM
(Million tons)

	1957	1959	1961	1963	1965	1966	1967	1968
Gas Works	12·03	9·97	9·82	9·72	7·77	7·20	6·23	4·60
Coke Ovens	20·43	17·00	17·78	15·48	17·07	16·12	15·32	16·25
Total	32·46	26·97	27·60	25·20	24·84	23·32	21·55	20·85

Source: *Annual Abstracts of Statistics Nos.* 105 *and* 106, Her Majesty's Stationery Office, London (1968 and 1969) (by permission of the Controller of Her Majesty's Stationery Office).

Although increasing amounts of coal are subjected to low-temperature carbonisation (e.g. at about 800 K), for the preparation of smokeless fuel, this does not provide a useful source of significant amounts of chemicals since the by-products obtained from this type of process are extremely complex mixtures which do not contain large proportions of any single chemical.

Of the products listed above, only naphthalene and anthracene are now obtained solely from coal in the UK. Benzene and toluene are still obtained in significant amounts from this source, and so, on a much smaller scale, are cresols, xylenols and pyridine, though these are all also made by other routes. Only a small proportion of total phenol manufacture comes from this source.

The other main way of making organic chemicals from coal is through synthetic processes based on coke. The coal is used only as a source of carbon, and the hydrogen is derived from water.

CARBIDE-BASED PROCESSES
Coke and calcium oxide are heated to about 2 300 K to give calcium carbide, and this is reacted with water to give acetylene:

$$CaO + 3C \longrightarrow CaC_2 + CO \qquad \Delta H = +460 \text{ kJ/mol}$$

$$CaC_2 + 2H_2O \longrightarrow CH{\equiv}CH + Ca(OH)_2$$

Acetylene can be readily converted into a variety of chemicals, and this provided a major source of aliphatic chemicals for many years (see Fig. 1-2).

There are a number of features of the carbide process which make it inherently expensive to operate. The most important factor is that, because of the very high reaction temperature and the high endothermic heat of reaction, the process consumes very large amounts of energy. Further, the only practicable way of obtaining the temperatures required is to use electrical heating, and electricity is, in most countries, a particularly expensive way of supplying energy. The electricity consumption is about 3·5 kWh/kg carbide or about 9·2 kWh/kg acetylene. At 0·33 p/kWh, which is as low a price for electricity as could reasonably be assumed in the UK, this gives a cost for electricity alone of about 3 p/kg acetylene. It is worth noting that this is more than the total production cost of ethylene in a modern plant.

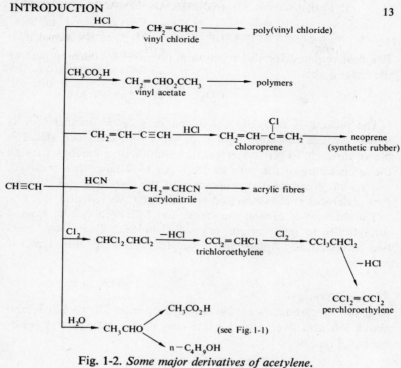

Fig. 1-2. *Some major derivatives of acetylene.*

The process requires two stages from carbon to acetylene, both involving solids handling, which leads to high capital and processing costs. It is wasteful of raw materials, in that one third of the carbon is converted to carbon monoxide, and calcium oxide is converted to impure calcium hydroxide.

The high cost of acetylene from carbide provided a considerable incentive to develop more economically attractive methods of manufacture of the products shown in Fig. 1-2. This has led to the development of routes not involving acetylene as an intermediate, and also to the development of processes for the manufacture of acetylene from natural gas and petroleum fractions. Consequently, the carbide route to organic chemicals is now obsolescent in many countries.

WATER GAS-BASED ROUTES

Water gas is a mixture of carbon monoxide and hydrogen, made by the reaction of white hot coke with steam:

$$C + H_2O \longrightarrow CO + H_2 \qquad \Delta H = +130 \text{ kJ/mol}$$

The heat required for this reaction is provided by burning part of the coke in air:

$$C + O_2 \longrightarrow CO_2 \qquad \Delta H = -390 \text{ kJ/mol}$$

The process is normally conducted in a cyclic manner. Air is blown through the coke until it is white hot, and then steam is blown through, to give water gas, the endothermic reaction causing the temperature of the coke to drop. Air blowing is then resumed, and so on. A substantial proportion of the carbon supplied therefore undergoes combustion and not the water gas reaction.

The mixture of carbon monoxide and hydrogen can be reacted catalytically to give various organic chemicals. There have been two commercially important processes involving this type of reaction.

Methanol Synthesis
Reaction of carbon monoxide and hydrogen at 570 to 670 K and about 300 atm over a zinc oxide–chromium oxide catalyst gives methanol (see page 71):

$$CO + 2H_2 \longrightarrow CH_3OH$$

It can be seen that whereas water gas would be expected to consist of approximately equimolar proportions of carbon monoxide and hydrogen, the proportions required for this reaction are two moles of hydrogen per mole of carbon monoxide. The proportion is adjusted by reacting the water gas with further steam over an iron oxide catalyst at 570 to 670 K, when the so-called 'shift reaction' occurs:

$$CO + H_2O \longrightarrow CO_2 + H_2$$

The extent of this reaction is adjusted so as to give the desired proportion of hydrogen to carbon monoxide. Carbon dioxide is subsequently removed from the gas stream.

This route was the major source of methanol in many countries, including the UK, until recently. It has now largely been replaced by processes in which the synthesis gas is obtained by reaction of natural gas or petroleum fractions with steam. The overriding disadvantage of the coke-based process compared with these processes is that it involves a solid raw material which leads to high capital costs, high labour costs and inefficient use of energy. In the UK, manufacture of methanol from coke ceased in 1962.

Fischer–Tropsch Process

Reaction of synthesis gas over iron or cobalt catalysts at about 450 K and pressures near atmospheric gives a complex mixture of products, mainly alkanes and alkenes with a wide range of molecular weights, together with much smaller quantities of oxygen-containing compounds, e.g. alcohols and acids. This process was used on a very large scale in Germany during World War II in order to provide an indigenous source of liquid fuels, and hydrocarbons for chemical synthesis.

At the present time, the Fischer–Tropsch process is not economically viable in most countries, petroleum being a substantially cheaper source of liquid fuels and hydrocarbons. It is, however, operated in South Africa. It is conceivable that processes of this type may become important as oil reserves become exhausted.

Petroleum

Petroleums, or crude oils, are complex mixtures, largely of hydrocarbons, which vary widely in composition depending on origin. The main components are alkanes, cyclopentane and cyclohexane derivatives (called 'naphthenes' in petroleum technology) and aromatic hydrocarbons, the latter generally being present in the smallest proportions. A very wide range of molecular weights from methane upwards is covered. Oxygen-, nitrogen- and sulphur-containing compounds are also present.

The major use of petroleum is for the manufacture of liquid fuels for use, for example, in internal combustion and jet engines, as boiler fuels in power stations, ships, etc. and for domestic heating. Very large quantities are consumed in these applications. For example, in 1969 UK consumption of petroleum products was 89·3 million tons of which about 80 million tons was for fuel uses.

Production of these liquid fuels involves a combination of physical and chemical processing known as *refining* (see Chapter 2). In addition to liquid fuels, this produces low molecular weight alkanes and alkenes, and in the USA petrochemical manufacture has tended to be based on gaseous by-products from oil refining, together with ethane, propane and butanes from natural gas. In Europe and Japan where ethane, propane and butanes are not available from natural gas, and where oil refining is carried out on a relatively smaller scale and has a different product pattern (in

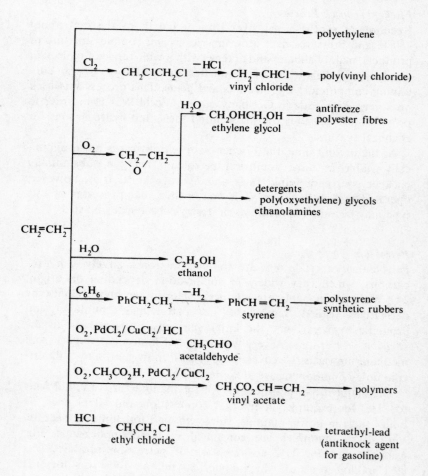

Fig. 1-3. *Some major derivatives of ethylene.*

particular, the proportion of gasoline required is much lower), most organic chemical manufacture has been based on *naphtha*, one of the liquid fractions obtained by distillation of petroleum (see page 29).

Although naphtha is a complex mixture, its components have relatively simple structures and can be envisaged as being fairly easily converted into useful products, in contrast with coal with its extremely complex structure. Being a liquid, it is easy to transport

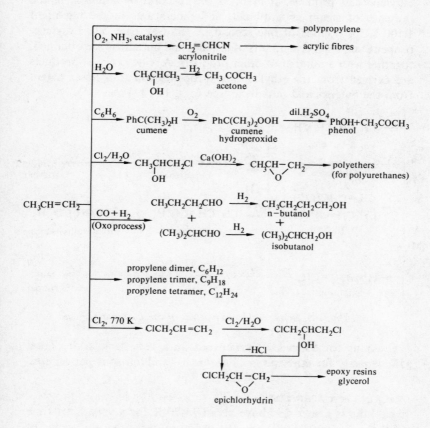

Fig. 1-4. *Some major derivatives of propene.*

and handle, and can be reacted in the liquid and the vapour phase. Further, it is fairly cheap [UK (1970) about £10/ton]. It is used for the manufacture of organic chemicals *via* five main routes. Although these routes are discussed in detail in subsequent chapters, they are described in outline here in order that a coherent overall picture may be obtained.

CRACKING FOR ALKENES

This process is of very great importance for the manufacture of ethylene and propene. It involves the reaction of naphtha, in the presence of steam as a diluent, at temperatures in the region of 1 100 K for periods of one second or less (see page 52). Ethylene, propene and smaller amounts of butenes and butadiene are obtained, together with a number of other products. A wide range of products are derived from the ethylene and propene, and to a lesser extent from the butenes and butadiene (see Figs. 1-3, 1-4 and 1-5).

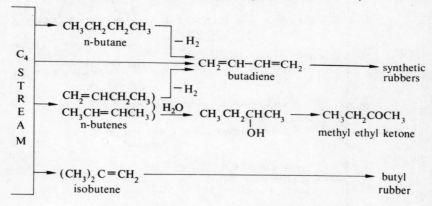

Fig. 1-5. *Some major derivatives of the C₄ stream.*

Cracking for alkenes is carried out on a very large scale. Total UK capacity for ethylene in 1970 was 1·7 million tons per annum.

CRACKING FOR ACETYLENE

If naphtha is heated to above about 1 470 K for a very short time (of the order of 0·01 sec) and the products quenched, fair yields of acetylene can be obtained (see page 60). It will be appreciated that there are major engineering problems involved in achieving these reaction conditions.

A number of processes were developed for the manufacture of acetylene by this route, and were expected to show economic advantage over the carbide route. However, unexpected difficulties in full scale operation of some of these processes, mainly associated with the formation of carbon and tar, which cause blockages, have led to them being much less economically attractive than expected. At least one large plant has been shut down without ever coming into full production, and this does not now seem to be a favoured method of making acetylene. The failure of these processes has, in the UK at least, hastened the replacement of acetylene-based routes by routes based on other intermediates. This applies with particular force to vinyl chloride manufacture.

STEAM REFORMING

Reaction of naphtha vapour with steam at about 1 100 K over a nickel catalyst gives a mixture of carbon monoxide and hydrogen (see page 65). This is thus a direct alternative to the water gas reaction as a source of synthesis gas for methanol and ammonia synthesis. The use of a liquid raw material, which can be reacted in the vapour phase, leads to a very much simpler and cheaper plant, which requires less labour and is much more thermally efficient than a coke-based plant. As a consequence, steam reforming of naphtha shows very clear economic advantage over coke-based manufacture of synthesis gas.

The process has been adapted for the manufacture of town gas, and shows economic advantage over manufacture by coal carbonisation.

CATALYTIC REFORMING

Catalytic reforming involves the reaction of naphtha vapour over a platinum-on-alumina catalyst at about 770 K (see page 36). A mixture of aromatic hydrocarbons is obtained. This process now provides the major source of benzene, toluene and xylenes.

DIRECT OXIDATION

Naphtha can be oxidised to acetic acid in reasonable yield (see page 90), and this forms the basis of a process operated in the UK at a scale of 90 000 tons per annum.

Natural Gas

The major constituent of natural gas is normally methane; North Sea gas, for instance, contains about 94% methane, the other two major constituents being, ethane (ca. 3%) and nitrogen (ca. 1%). Some natural gas, particularly that associated with oilfields, contains substantial proportions of C_2 to C_4 alkanes and, as we have seen, these alkanes form convenient raw materials for chemical manufacture, e.g. by cracking to alkenes. Methane can be used as a raw material for steam reforming and for cracking for acetylene.

Methane has technical advantages over naphtha in steam reforming (see page 70). Further, in countries with substantial deposits it is cheaper than naphtha. In such circumstances it is the preferred raw material for steam reforming.

The manufacture of acetylene from methane presents less difficulties than from naphtha, mainly because less carbon and tar are formed. However, pyrolysis plants for acetylene manufacture are complex and expensive, and consume large amounts of energy. At the present time, acetylene-based routes do not appear to compare well with alternative routes through ethylene, propene or other alkenes, even in locations where methane is very cheap. Certainly in the UK there is no sign of manufacture of acetylene from methane becoming important at the time of writing.

CHEMICAL PROCESSING TECHNOLOGY

It is not proposed to discuss in detail the equipment used for large-scale chemical processing: the design of such equipment forms part of the discipline of chemical engineering, and is outside the scope of this book.

Many of the operations in chemical manufacture, e.g. refluxing, distillation, filtration, are basically the same as those carried out in the laboratory, although the problems involved in carrying them out are of an entirely different order of magnitude. There are, however, some aspects of large-scale processing which require a brief discussion as a background to material in subsequent chapters.

Batch and Continuous Processing

There are two broad ways of carrying out chemical processes, i.e. by *batch processing* and by *continuous processing*. In batch processing, which is the type most commonly met when carrying out laboratory preparations, the reactants are put into a reactor, allowed to react for a specified time and are discharged from the reactor, possibly after being subjected to some separation process. The operation of the equipment is intermittent, and the activity of a particular portion of the equipment varies with time. In continuous, or flow processing, the raw materials are continuously fed to, and the reaction products are continuously removed from, the reactor, and are usually subjected to continuous separation operations in further items of equipment.

Two main types of continuous reactor are in use for liquid-phase reactions, i.e. *tubular reactors* and *stirred flow reactors*. Tubular reactors, as the name suggests, consist essentially of a tube through which reactants are passed. The tube is usually surrounded by a jacket through which is circulated a heat-transfer medium: for convenience of construction the tube is often coiled, or doubled back on itself so that a long tube may be accommodated in a jacket of reasonable dimensions (see Fig. 1-6a). There is little back or forward mixing along the length of the tube, so that elements of feed all spend approximately the same time in the reactor. The composition of the reaction mix changes along the length of the tube. The kinetic characteristics of a tubular reactor are the same as those of a batch reactor.

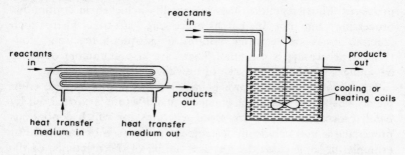

(a) Tubular reactor   (b) Stirred flow reactor

Fig. 1-6. *Continuous reactors for liquid-phase reactions.*

A stirred flow reactor consists essentially of a stirred reactor with provision for continuous feed of reactants and continuous removal of products (see Fig. 1-6b). Heat transfer may be provided for by a coil in the reactor or by a jacket. It is fairly evident that the kinetic characteristics of this type of reactor are different from those of tubular flow or batch reactors. The composition of the reaction mix is constant throughout the reactor, and there is a wide spread of the times spent by particular elements of feed in the reactor. Often a number of stirred flow reactors are used in series, the product from one forming the feed to the next one in the series. In such an arrangement, the kinetic characteristics approach those of a tubular reactor as the number of stirred flow reactors increases.

For non-catalytic gas-phase reactions, tubular reactors are usually used. Reactors for heterogeneous-catalysed gas-phase reactions are discussed in the next section.

Continuous processing has a number of advantages over batch processing for manufacturing operations. It is relatively easy to arrange for a high degree of automatic control of a continuous process, whereas automatic control of the intermittent operations of a batch process is considerably more difficult and expensive. In consequence, batch processes tend to require more labour to operate them than continuous processes, and therefore to have higher labour costs. Energy conservation is much easier with a continuous process than with a batch process. For example, hot process streams can be cooled by transferring the heat to the raw materials, to water for steam generation, etc. Integration of the intermittent operations of a batch process into an energy-conserving system is generally not feasible, and service costs therefore tend to be higher in batch processes. Short reaction times are readily achieved in continuous processing, but not in batch processing, so that where such reaction times are required, notably in gas-phase reactions, continuous processing is essential. There are also a number of other advantages which, for the sake of brevity, we shall not discuss.

The situation is not, however, as is sometimes supposed, completely one-sided, and much chemical manufacture is carried out by batch processing. There are some reactions for which continuous processing is not technically feasible for one reason or another. For example, in some cases the physical nature of the reactants, or the reaction conditions, are such that continuous processing is difficult to arrange. Suspension polymerisation (see page 237) is one example of such a process.

The development of a continuous process is often considerably more difficult than the development of a batch process, so that in any particular case the benefits which are likely to be gained from developing a continuous process, e.g. savings in labour and energy costs, have to be weighed against the extra research and development costs which will have to be incurred in order to generate them. The savings are proportional to the scale on which the process is to be operated; the research and development costs are not. Consequently, for small-scale manufacture it is often not worthwhile developing a continuous process. Again there are various other advantages of batch processes over continuous processes which it is not proposed to discuss.

In general and somewhat over-simplified terms, continuous processing is used for large-scale manufacture, where technically feasible, and for gas-phase reactions on any scale, and batch processing is used for small-scale manufacture except for gas-phase reactions, and for large-scale manufacture by processes which it has not been found technically feasible to carry out continuously.

Catalytic Reactors

Many commercial processes involve reactions carried out in the gas-phase over solid catalysts. A variety of types of reactor is used for these reactions, and major classes of these will be briefly discussed. Heat transfer is a major consideration in heterogeneous-catalysed gas-phase reactions, and catalytic reactors may be divided into classes on the basis of the type of provision that they make for heat transfer during the reaction.

The simplest type of catalytic reactor is one in which no provision is made for either adding or removing heat, called an *adiabatic reactor*. This consists essentially of a vessel in which the catalyst is held, usually as a single bed, with provision for feeding reactants to one side of the bed, and removing products from the other side (see Fig. 1-7a). This type of reactor is relatively cheap to construct. Evidently the temperature will vary through the bed, and will increase from inlet to outlet when an exothermic reaction is involved, and decrease from inlet to outlet when an endothermic reaction is involved. This characteristic limits its applications in that for satisfactory operation, many processes need to be carried out over a narrower range of temperature than would be produced in an adiabatic reactor. The use of a number of adiabatic reactors in series, with intermediate heating or cooling stages, increases the flexibility of the system, but also reduces the simplicity which is the main attraction of this type of reactor.

When an adiabatic reactor is unsuitable, an *isothermal reactor*, in which provision is made for adding or removing heat so as to control the temperature, is used. The most common type of isothermal reactor is one in which the catalyst is held in a number of tubes which are surrounded by a heating or cooling medium. A *tubular reactor* for use for an exothermic reaction is shown in schematic form in Fig. 1-7b. When an endothermic reaction is being carried out then, depending on the temperature required, a liquid or vapour heat transfer medium may be used, or the tubes may be heated direct by burning oil or gas.

The diameter of the tubes varies, depending, for example, on the rate of heat transfer required and on the temperature gradient that can be tolerated from the centre to the outside of the catalyst bed. For oxidations, which are highly exothermic and temperature

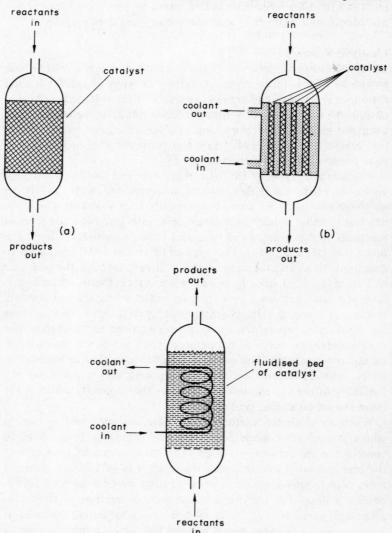

Fig. 1-7. *Catalytic reactors: (a) adiabatic reactor; (b) tubular reactor (with cooling); (c) fluidised-bed reactor (with cooling).*

sensitive, tubes of about 2·5 cm diameter are typically used, and a large number of tubes, commonly thousands, have to be used in a reactor. In steam reforming, an endothermic reaction, tubes of diameter 20 cm or more have been used. A tubular reactor is evidently more complicated, and therefore more expensive, than an adiabatic reactor of comparable size.

Another type of reactor in which heat transfer can be provided is the *fluidised-bed reactor*. This depends on the fact that if a gas is passed upwards through a bed of fine particles, then above a certain rate of gas flow the bed becomes fluidised and behaves in many respects like a liquid. In such a reactor heating or cooling can be carried out by means of coils or tubes immersed in the fluidised bed (see Fig. 1-7c), and very good temperature control is possible. The engineering of fluidised-bed reactors is somewhat more complicated than that of fixed-bed reactors, and their kinetic characteristics are different to those of fixed-bed reactors. Their major use is in catalytic cracking (see page 42) where they have the particular advantage of allowing continual regeneration of the catalyst.

Yield and Conversion in Chemical Processes

In industrial chemistry the term 'yield' often has a somewhat different meaning to that which it has when used to describe a laboratory experiment. Also, the term 'conversion' is often used in discussing processes. To avoid any confusion, we shall define the terms yield and conversion as used in this book.

Conversion generally refers to a single reaction stage of a process; it is the percentage of raw material which is consumed in such a stage.

Yield can refer to either a single stage or a complete process. It is the percentage theory yield as normally calculated, but based on material consumed, discarded or otherwise lost, and not necessarily on the amount of raw material originally added to the reactor.

As an example, let us assume that we are oxidising toluene to benzoic acid, and that the products from the reaction stage are as shown below:

$$PhCH_3 \longrightarrow PhCO_2H + PhCH_3 + \text{other products}$$

1 mole   0·7 mole   0·2 mole

In this reaction, the conversion of toluene, i.e. the percentage consumed in the reaction, is 80%. In the laboratory we would probably discard the unreacted toluene while working up, and we would say that we had obtained a yield of 70% of theory. In commercial operations, recovery of unreacted toluene might well be worthwhile. In this case, assuming that we obtained 100% recovery of the toluene (which is unlikely), we have obtained 0·7 mole benzoic acid from 0·8 mole toluene and the yield is therefore 87·5% of theory. We would say that the reaction was operated at 80% conversion and gave an 87·5% yield.

Many industrial processes operate at partial conversion of one or more reactants. Processes involving reversible reactions in which equilibrium is reached at only partial conversion are one obvious class of such processes, and processes in which one reactant is used in excess, and is inevitably only partially converted, are another.

One other very important class of processes which are run at partial conversion are those in which the product is subject to further reaction, i.e. processes involving reactions of the general type:

$$A \longrightarrow B \longrightarrow C$$

where B is the required product. In this case working at low conversions ensures that the concentration of product remains low, so that the rate of its further reaction is kept down.

When a process is based on a reaction which for one reason or another has to be operated at partial conversion, the decision has to be made whether unconverted reactant(s) will be recovered and recycled, or will be discarded. The basis for this decision is a comparison of the value of the material which could be recovered with the capital and operating costs which would be involved in recovery. Evidently the higher the proportion of unconverted material, and the higher its unit value, the more likely it is that recovery and recycling will be economically attractive.

Chapter 2

Reactions of Alkanes and Cycloalkanes

ALKANES were for many years regarded as being unreactive, and received little attention from early organic chemists. Fortunately, this initial assessment of their reactivity proved to be ill-founded, and reactions of alkanes and cycloalkanes are now of extreme importance in petroleum refining and in organic chemical manu-facture, and are carried out on a larger scale than any other organic chemical reactions. The reactions discussed in this chapter fall fairly clearly into two groups, petroleum refining reactions and petro-chemical reactions, although there are some interconnections between the two.

PETROLEUM REFINING REACTIONS

Petroleum refining is the combination of physical and chemical pro-cessing by which crude oil is converted into various grades of liquid fuel. It is carried out on an extremely large scale; thus, in 1969 the amount of crude oil refined in the UK was 90 million tons. The fuels produced by refining are used, for example, as gasoline or motor spirit, diesel fuel, jet fuel, domestic heating fuel, fuel oil for power stations, ships, etc. Fairly evidently, the properties required for these various applications differ, and refining operations are designed to produce fuels with the correct properties for the par-ticular applications.

Two of the important characteristics of gasoline, for instance, are its volatility and its burning properties. It has to be volatile enough to give efficient carburetion, but not so volatile that vapour locking in fuel lines, or excessive losses from fuel tanks, occur. Gasolines normally have a boiling range of about 305 to 460 K. It is of the utmost importance that gasoline should burn smoothly in the engine and should not detonate or 'knock', since knocking reduces engine power, and can cause damage to the engine. The tendency

27

of a fuel to cause knocking is expressed in terms of an *octane number*, the higher the value of the octane number, the greater being the resistance to knocking. Most modern automobile engines require a fuel with an octane number between 90 and 100, the higher the compression ratio, the higher the octane number required. The octane numbers of various individual hydrocarbons in

TABLE 2-1

OCTANE NUMBERS OF SOME HYDROCARBONS

	Hydrocarbon	Octane number	b.p. (K)
C5	n-pentane	62	309
	2-methylbutane	90	301
	cyclopentane	85	322
	n-hexane	26	342
	2-methylpentane	73	333
	2,2-dimethylbutane	93	323
C6	hex-1-ene	63	337
	hex-2-ene	81	341
	benzene	>100	353
	cyclohexane	77	354
	n-octane	0	399
	2-methylheptane	13	391
	2,2,4-trimethylpentane (iso-octane)	100	372
	oct-1-ene	35	395
	oct-2-ene	56	398
C8	oct-3-ene	68	396
	o-, m-, and p-xylenes	>100	—
	ethylbenzene	98	410
	1,2-dimethylcyclohexane	79	403
	ethylcyclohexane	41	403

Note: Octane number can be determined in different ways, which often give somewhat different results. The above values are determined by the 'motor method'.

Source of data: *Modern Petroleum Technology*, 3rd ed., The Institute of Petroleum, London, 1962. By permission of The Institute of Petroleum.

the gasoline boiling range are shown in Table 2-1. It will be seen that straight-chain alkanes with six carbon atoms and over have very low octane numbers; it would be impossible to run a modern automobile engine on a fuel composed solely of such materials.

Desirable constituents of gasoline from the point of view of producing a high octane number material are branched-chain alkanes, alkenes, and especially aromatics and, as we will see below, the major chemical processes carried out in refining operations are designed to produce materials containing compounds of these types.

At the present time, small quantities of tetraethyl-lead and, to a lesser extent, tetramethyl-lead are added to gasolines. This addition produces an increase in octane number ranging from 3 to 20 units depending on the quantity used and on the type of gasoline.

Other fuels have their own quality requirements of varying degrees of stringency. Domestic heating oil ('paraffin' in the UK), for instance, must for safety reasons have a low enough volatility that vapour over the liquid will not catch fire at ambient temperature, i.e. it must have a reasonably high flash point. Further, if it is to be used for lighting, or in portable heating appliances, it should not contain a high proportion of aromatic hydrocarbons, which give smoky flames, and produce smells.

The first operation in refining is fractional distillation of the crude oil. This splits it into a number of fractions with boiling point ranges appropriate to various fuel applications. Figure 2-1 indicates major fractions which are often taken in this initial distillation. It should be noted that the exact mode of operation varies considerably depending on circumstances; additional fractions may be taken, and the boiling ranges may differ from those in the figure.

The material in the gasoline boiling range is normally divided into two or more fractions, and of these only the lower boiling components, 'straight run gasoline', can be used directly in motor spirit, and then only for blending with high octane number material. The higher boiling material ('naphtha' in Fig. 2-1) has to be chemically processed to increase its octane number before it can be used in motor spirit.

The terminology of material in the gasoline boiling range is slightly confusing in that in the chemical industry the term 'naphtha' is extended to cover all the material in this range. Thus, in chemical operations, 'light naphtha' typically indicates a material with a boiling range of up to about 390 to 420 K, containing compounds (mainly alkanes) with carbon numbers in the range C_4 to C_8 or C_9. 'Full range naphtha', with a boiling range extending to about 460 K, contains materials with carbon numbers up to about C_{12}. In this present discussion of refining operations and processes. the term 'naphtha' refers to materials of the type indicated in Fig. 2-1; in other parts of this book it will be used in the sense indicated above.

The proportions of the various fractions obtained from crude oil distillation, and their properties as fuels, depend on the crude oil being used. Not surprisingly, they are seldom, if ever, in balance with the requirements of the market, and consequently further chemical and physical operations are carried out on the fractions obtained from the distillation unit. The most important of these are processes which produce high octane number gasoline. These are of three types:

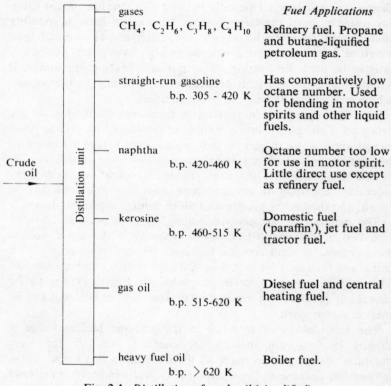

		Fuel Applications
gases	CH_4, C_2H_6, C_3H_8, C_4H_{10}	Refinery fuel. Propane and butane-liquified petroleum gas.
straight-run gasoline	b.p. 305 - 420 K	Has comparatively low octane number. Used for blending in motor spirits and other liquid fuels.
naphtha	b.p. 420-460 K	Octane number too low for use in motor spirit. Little direct use except as refinery fuel.
kerosine	b.p. 460-515 K	Domestic fuel ('paraffin'), jet fuel and tractor fuel.
gas oil	b.p. 515-620 K	Diesel fuel and central heating fuel.
heavy fuel oil	b.p. > 620 K	Boiler fuel.

Crude oil → Distillation unit

Fig. 2-1. *Distillation of crude oil (simplified).*

(i) Processes which produce gasoline from lower boiling feed-stocks.

(ii) Processes which lead to an improvement in the octane number of a material in the gasoline boiling range.

(iii) Processes which produce gasoline from a material in a higher boiling range.

Some important processes of these types are discussed below. It will be noted that they all involve carbonium ion intermediates.

Catalytic Alkylation

Catalytic alkylation provides a method of producing high octane fuel components from some of the gaseous products of refining operations. It involves the reaction of isobutane with alkenes, normally propene or butenes, in the presence of an acidic catalyst, usually 96 to 98% sulphuric acid or anhydrous hydrogen fluoride. Isobutane, propene and butenes are by-products of another refinery process, catalytic cracking, and isobutane is also present in the gases from primary crude oil distillation. The process of alkylation was discovered in the mid 1930s, and was of great importance in the production of aviation gasoline in World War II.

Alkylation gives fairly complex mixtures of products. The products given by alkylation of isobutane with propene under one set of conditions are shown below:

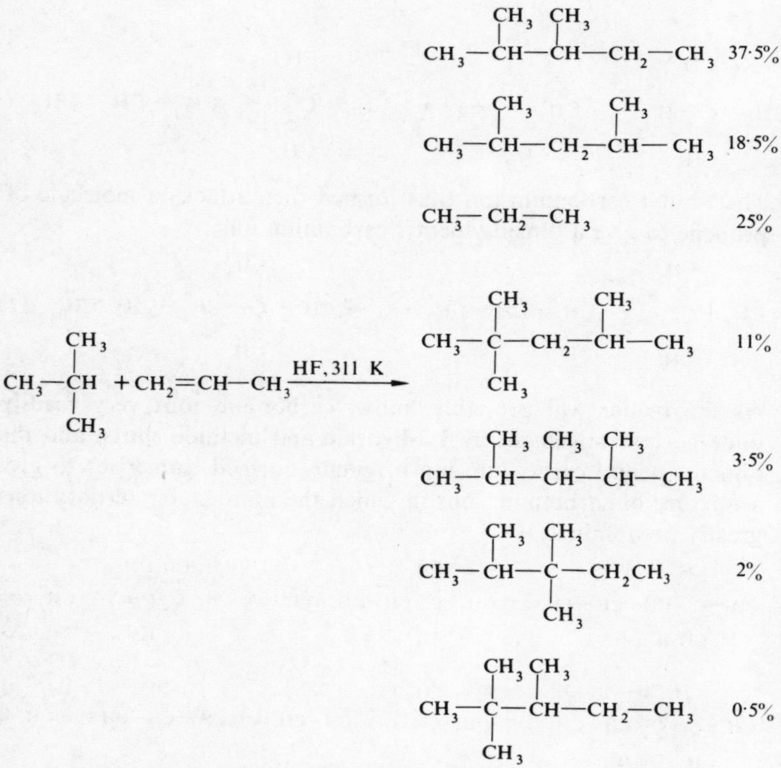

$$CH_3-\overset{\overset{\displaystyle CH_3}{|}}{CH}-\overset{\overset{\displaystyle CH_3}{|}}{CH}-CH_2-CH_3 \quad 37\cdot5\%$$

$$CH_3-\overset{\overset{\displaystyle CH_3}{|}}{CH}-CH_2-\overset{\overset{\displaystyle CH_3}{|}}{CH}-CH_3 \quad 18\cdot5\%$$

$$CH_3-CH_2-CH_3 \qquad\qquad 25\%$$

$$CH_3-\overset{\overset{\displaystyle CH_3}{|}}{\underset{\underset{\displaystyle CH_3}{|}}{C}}-CH_2-\overset{\overset{\displaystyle CH_3}{|}}{CH}-CH_3 \quad 11\%$$

$$CH_3-\overset{\overset{\displaystyle CH_3}{|}}{CH} + CH_2{=}CH-CH_3 \xrightarrow{\text{HF, 311 K}}$$

$$CH_3-\overset{\overset{\displaystyle CH_3}{|}}{CH}-\overset{\overset{\displaystyle CH_3}{|}}{CH}-\overset{\overset{\displaystyle CH_3}{|}}{CH}-CH_3 \quad 3\cdot5\%$$

$$CH_3-\overset{\overset{\displaystyle CH_3}{|}}{CH}-\overset{\overset{\displaystyle CH_3}{|}}{\underset{\underset{\displaystyle CH_3}{|}}{C}}-CH_2CH_3 \quad 2\%$$

$$CH_3-\overset{\overset{\displaystyle CH_3}{|}}{\underset{\underset{\displaystyle CH_3}{|}}{C}}-\overset{\overset{\displaystyle CH_3}{|}}{CH}-CH_2-CH_3 \quad 0\cdot5\%$$

It can be seen that in addition to compounds containing seven carbon atoms, propane and compounds containing eight carbon atoms are obtained. A further point of interest is that the alkylation products are not those which would be produced by simple addition of isobutane across the double bond.

MECHANISM

The steps which are thought to be involved in the main alkylation reaction are discussed below, using the reaction of isobutane and propene as an example. (In the following reaction schemes, the anions are omitted for simplicity.)

The reaction is a chain reaction in which initiation involves protonation of propene, followed by abstraction of a hydride ion from isobutane by the isopropyl carbonium ion:

$$CH_3CH{=}CH_2 + HF \rightleftharpoons CH_3\overset{+}{C}HCH_3 + F^-$$

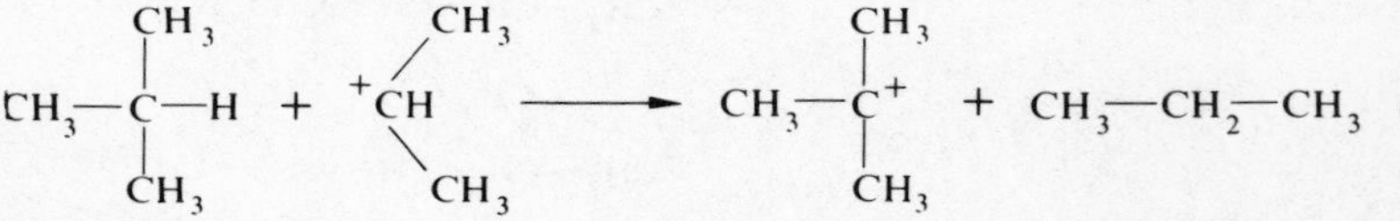

$$CH_3{-}\underset{\underset{CH_3}{|}}{\overset{\overset{CH_3}{|}}{C}}{-}H + {}^+CH\overset{CH_3}{\underset{CH_3}{\diagdown}} \longrightarrow CH_3{-}\underset{\underset{CH_3}{|}}{\overset{\overset{CH_3}{|}}{C}}{}^+ + CH_3{-}CH_2{-}CH_3 \quad (1)$$

The t-butyl carbonium ion thus formed then attacks a molecule of propene to give a dimethylpentyl carbonium ion:

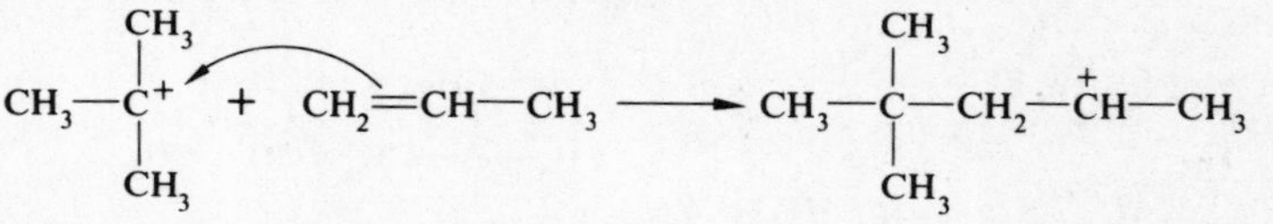

$$CH_3{-}\underset{\underset{CH_3}{|}}{\overset{\overset{CH_3}{|}}{C}}{}^+ + CH_2{=}CH{-}CH_3 \longrightarrow CH_3{-}\underset{\underset{CH_3}{|}}{\overset{\overset{CH_3}{|}}{C}}{-}CH_2{-}\overset{+}{C}H{-}CH_3 \quad (2)$$

As the reader will probably know, carbonium ions very readily undergo rearrangement by 1,2-hydride and methide shifts, and the dimethylpentyl carbonium ion originally formed isomerises to give a mixture of carbonium ions in which the more stable tertiary ions greatly predominate:

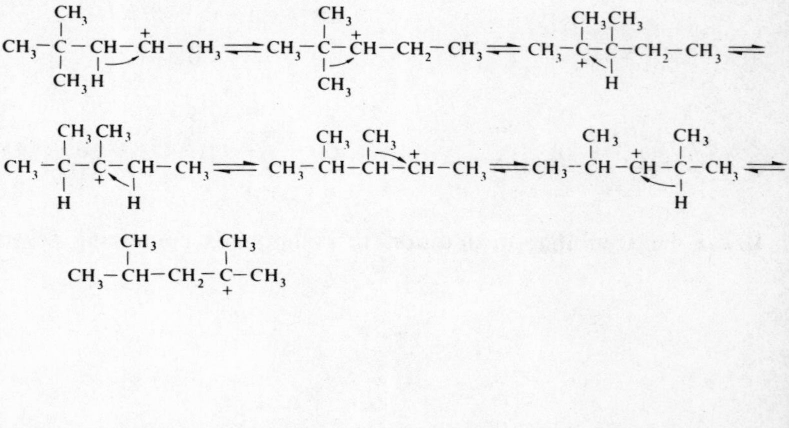

$$CH_3{-}\underset{\underset{CH_3}{|}\,\underset{H}{|}}{\overset{\overset{CH_3}{|}}{C}}{-}CH{-}\overset{+}{C}H{-}CH_3 \rightleftharpoons CH_3{-}\underset{\underset{CH_3}{|}}{\overset{\overset{CH_3}{|}}{C}}{-}\overset{+}{C}H{-}CH_2{-}CH_3 \rightleftharpoons CH_3{-}\underset{\underset{H}{|}}{\overset{\overset{CH_3CH_3}{|\,\,\,|}}{C{-}C}}\!\!{}^+{-}CH_2{-}CH_3 \rightleftharpoons$$

$$CH_3{-}\underset{\underset{H}{|}}{\overset{\overset{CH_3CH_3}{|\,\,\,|}}{C}}{-}\overset{+}{\underset{}{C}}{-}CH{-}CH_3 \rightleftharpoons CH_3{-}\underset{}{\overset{\overset{CH_3}{|}}{C}}H{-}\overset{+}{\underset{}{\overset{\overset{CH_3}{|}}{C}}}H{-}\overset{+}{C}H{-}CH_3 \rightleftharpoons CH_3{-}\underset{}{\overset{\overset{CH_3}{|}}{C}}H{-}\overset{+}{\underset{}{\overset{\overset{CH_3}{|}}{C}}}H{-}\overset{}{\overset{\overset{CH_3}{|}}{C}}{-}CH_3 \rightleftharpoons$$

$$CH_3{-}\overset{\overset{CH_3}{|}}{C}H{-}CH_2{-}\underset{+}{\overset{\overset{CH_3}{|}}{C}}{-}CH_3$$

Abstraction of a hydride ion from a molecule of isobutane by a tertiary dimethylpentyl carbonium ion gives a dimethylpentane and a new t-butyl carbonium ion:

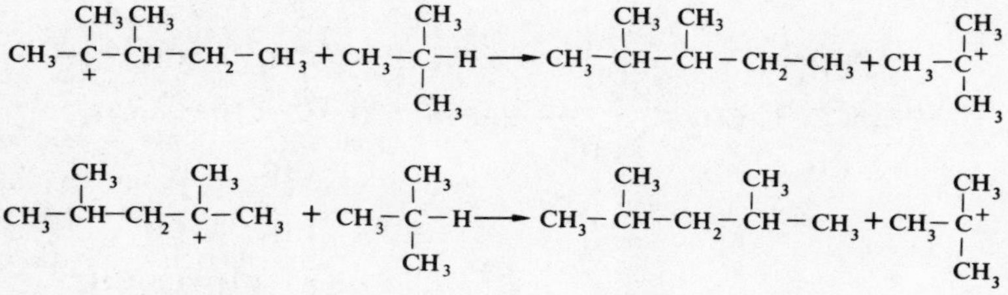

$$CH_3-\overset{CH_3}{\underset{+}{\overset{|}{C}}}\overset{CH_3}{\overset{|}{CH}}-CH_2-CH_3 + CH_3-\overset{CH_3}{\underset{\underset{CH_3}{|}}{\overset{|}{C}}}-H \longrightarrow CH_3-\overset{CH_3}{\overset{|}{CH}}-\overset{CH_3}{\overset{|}{CH}}-CH_2-CH_3 + CH_3-\overset{CH_3}{\underset{\underset{CH_3}{|}}{\overset{|}{C}}}{}^+$$

$$CH_3-\overset{CH_3}{\overset{|}{CH}}-CH_2-\overset{CH_3}{\underset{+}{\overset{|}{C}}}-CH_3 + CH_3-\overset{CH_3}{\underset{\underset{CH_3}{|}}{\overset{|}{C}}}-H \longrightarrow CH_3-\overset{CH_3}{\overset{|}{CH}}-CH_2-\overset{CH_3}{\overset{|}{CH}}-CH_3 + CH_3-\overset{CH_3}{\underset{\underset{CH_3}{|}}{\overset{|}{C}}}{}^+$$

The t-butyl carbonium ion thus generated attacks a further molecule of propene [reaction (2)], and the reaction chain is continued.

Transfer of a proton from t-butyl carbonium ions to propene can occur, and leads to the formation of isobutene and isopropyl carbonium ion:

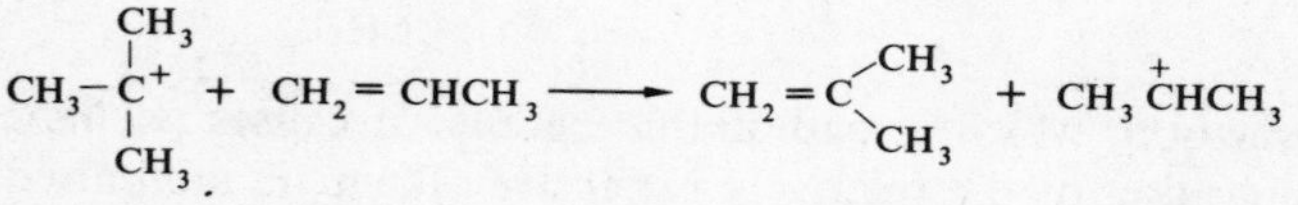

$$CH_3-\overset{CH_3}{\underset{\underset{CH_3}{|}}{\overset{|}{C}}}{}^+ + CH_2=CHCH_3 \longrightarrow CH_2=C\overset{\diagup CH_3}{\diagdown CH_3} + CH_3\overset{+}{C}HCH_3 \quad (3)$$

This reaction is undesirable, since the isopropyl carbonium ion thus formed is converted into propane by reaction (1), and the yield of material in the gasoline boiling range is reduced.

The isobutene formed in reaction (3) undergoes attack by t-butyl carbonium ion to give a C$_8$ carbonium ion, which can carry out a hydride ion abstraction to give a C$_8$ alkane:

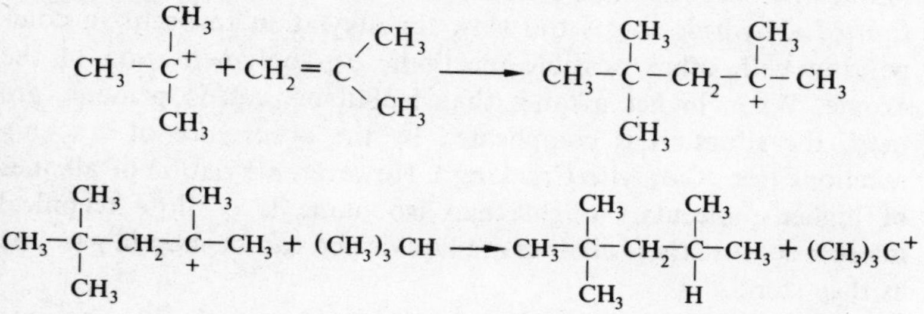

$$CH_3-\overset{CH_3}{\underset{\underset{CH_3}{|}}{\overset{|}{C}}}{}^+ + CH_2=C\overset{\diagup CH_3}{\diagdown CH_3} \longrightarrow CH_3-\overset{CH_3}{\underset{\underset{CH_3}{|}}{\overset{|}{C}}}-CH_2-\overset{CH_3}{\underset{+}{\overset{|}{C}}}-CH_3$$

$$CH_3-\overset{CH_3}{\underset{\underset{CH_3}{|}}{\overset{|}{C}}}-CH_2-\overset{CH_3}{\underset{+}{\overset{|}{C}}}-CH_3 + (CH_3)_3CH \longrightarrow CH_3-\overset{CH_3}{\underset{\underset{CH_3}{|}}{\overset{|}{C}}}-CH_2-\overset{CH_3}{\underset{\underset{H}{|}}{\overset{|}{C}}}-CH_3 + (CH_3)_3C^+$$

The C$_8$ carbonium ion may of course rearrange as discussed above before abstracting a hydride ion.

Polymerisation of the alkene can occur. For example, in alkylations with isobutene, the following reactions can take place:

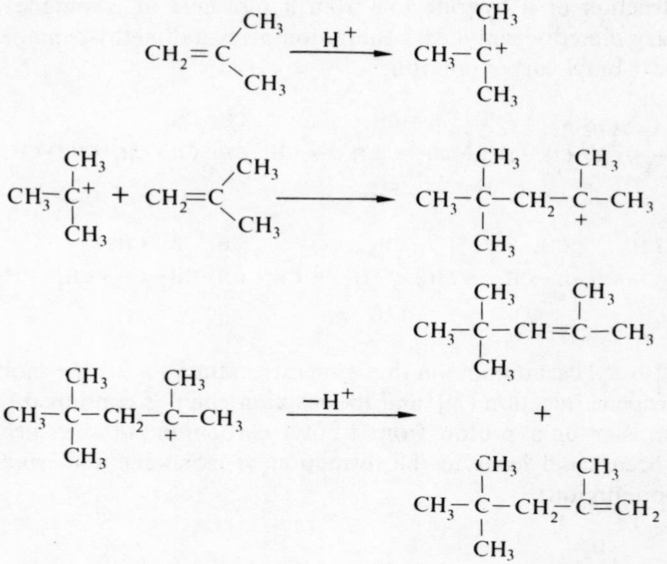

$$CH_2=C\begin{smallmatrix}CH_3\\\\CH_3\end{smallmatrix} \xrightarrow{H^+} CH_3-\overset{CH_3}{\underset{CH_3}{C^+}}$$

$$CH_3-\overset{CH_3}{\underset{CH_3}{C^+}} + CH_2=C\begin{smallmatrix}CH_3\\\\CH_3\end{smallmatrix} \longrightarrow CH_3-\overset{CH_3}{\underset{CH_3}{C}}-CH_2-\overset{CH_3}{\underset{+}{C}}-CH_3$$

$$CH_3-\overset{CH_3}{\underset{CH_3}{C}}-CH_2-\overset{CH_3}{\underset{+}{C}}-CH_3 \xrightarrow{-H^+}$$

$$CH_3-\overset{CH_3}{\underset{CH_3}{C}}-CH=\overset{CH_3}{C}-CH_3$$

+

$$CH_3-\overset{CH_3}{\underset{CH_3}{C}}-CH_2-\overset{CH_3}{C}=CH_2$$

Polymerisation, which is undesirable because it causes an increase in consumption of the relatively expensive alkene, is minimised by keeping the alkene concentration low.

There are various other side reactions which can be undergone by the intermediate carbonium ions in this type of system, and alkylation in high yield can only be achieved under rather specific conditions. Only alkanes with tertiary hydrogen atoms, and which therefore give a relatively stable tertiary carbonium ion on hydride ion abstraction, undergo the reaction. With other alkanes, abstraction of a hydride ion is too slow for alkylation to occur in competition with other possible reactions, e.g. polymerisation of the alkene. When higher alkanes than isobutane, e.g. isopentane, are used, the situation is complicated by the occurrence of cracking reactions (see 'Catalytic Cracking'). However, alkylation of alkanes of higher molecular weight than isobutane is of little technical interest because these alkanes are liquids and can be used in gasoline as they stand.

PROCESS OPERATION

Alkylation is brought about by contacting isobutane and the alkene, in the liquid phase, with the acid catalyst. For reasons already indicated, the isobutane is used in substantial excess, and is recycled. With sulphuric acid as catalyst, temperatures in the range 270 to 300 K are usually used, since at higher temperatures oxidation of the alkene occurs. Refrigeration is necessary in order to maintain these low reaction temperatures, and this adds to the capital and operating costs of the process. With hydrogen fluoride, temperatures up to about 320 K can be used, so there is no need for refrigerated cooling. The reaction is carried out under sufficient pressure to keep the alkene and the isobutane in the liquid phase.

The products from alkylation are mixtures of highly branched alkanes, and have octane numbers of about 90 (unleaded). Alkylation capacity in the USA in 1970 was about 800 000 barrels per day (1 barrel = 35 imperial gallons or 0·16 m^3).

Catalytic Isomerisation

The main use of the process of catalytic isomerisation has been the conversion of n-butane to isobutane, which, as we have seen, is required as a feedstock for alkylation. The reaction can conveniently be brought about by contacting n-butane with aluminium chloride, either in the gas phase, when the aluminium chloride is used on a solid support, e.g. bauxite, or in the liquid phase, under pressure, when the aluminium chloride is present as a liquid 'catalyst complex'. Temperatures in the range 350 to 420 K are used. The reaction is generally agreed to involve a carbonium ion mechanism.

The main difficulty in formulating such a mechanism is in accounting for the formation of carbonium ions from the alkane, and there has been much discussion of this point. One suggestion is that protonation of small amounts of butene in the butane gives rise to carbonium ions which act as initiators for a chain reaction:

$$C_4H_8 + HCl + AlCl_3 \longrightarrow C_4H_9^+ + AlCl_4^-$$

$$C_4H_9^+ + CH_3\text{-}CH_2\text{-}CH_2\text{-}CH_3 \longrightarrow C_4H_{10} + CH_3\text{-}\overset{+}{C}H\text{-}CH_2\text{-}CH_3$$

$$CH_3\text{-}\overset{+}{C}H\text{-}CH_2\text{-}CH_3 \longrightarrow \underset{\underset{CH_3}{|}}{CH_3\text{-}CH\text{-}CH_2^+} \longrightarrow \underset{\underset{CH_3}{|}}{CH_3\text{-}\overset{+}{C}\text{-}CH_3}$$

$$\underset{\underset{CH_3}{|}}{CH_3\text{-}\overset{+}{C}\text{-}CH_3} + CH_3\text{-}CH_2\text{-}CH_2\text{-}CH_3 \longrightarrow \underset{\underset{H}{\overset{|}{\underset{|}{C}}}{\overset{CH_3}{|}}}{CH_3\text{-}C\text{-}CH_3} + CH_3\text{-}\overset{+}{C}H\text{-}CH_2\text{-}CH_3$$

Since carbonium ions are continually regenerated, it is only necessary for a small amount of alkene to be present.

It can be seen that the reaction requires the presence of a co-catalyst, hydrogen chloride. This is specifically added in some processes, or could be supposed to be produced by the interaction of adventitious water with aluminium chloride. In fact, the inter-action of water with aluminium chloride is complex, and water itself appears to promote the catalyst, possibly by forming a hydroxyaluminium chloride.

About 60% conversion of n-butane to isobutane can be obtained per pass. The isobutane is separated by distillation, and the residual n-butane is recycled.

Isomerisation of higher alkanes in the presence of aluminium chloride tends to be accompanied by side-reactions leading to dis-proportionation and tar formation. It has been used to a limited extent for octane number improvement of C_5 and C_6 straight-run gasoline fractions. Recently processes have been developed in which disproportionation reactions are reduced or eliminated, and isomerisation of C_5 and C_6 fractions may become more important in the future.

Isomerisation capacity in the USA in 1970 was 110 000 barrels per day.

Catalytic Reforming

Catalytic reforming is used in refining to improve the octane number of feeds in the gasoline boiling range, and in chemical manufacture for the production of aromatic hydrocarbons. The process was first used in refining in 1940; it is second only to catalytic cracking in importance in the USA, and in Europe, where the higher boiling petroleum fractions are relatively more important than in the USA, it is the major refining process. Typically, it increases the octane number of a feed with a boiling range 340 to 460 K from about 40 to about 95, and it thus provides a means of converting naphtha into high-quality gasoline.

Catalytic reforming involves reacting feed vapour over a catalyst at temperatures in the range 720 to 820 K and pressures in the range 10 to 50 atm, in the presence of hydrogen. The catalysts used are *dual-function catalysts*, that is, they have both acidic and hydrogenation–dehydrogenation properties. Most commonly used are platinum-on-alumina catalysts in which the platinum is the hydrogenation–dehydrogenation component, and the alumina, which is generally treated with chlorides or fluorides, the acidic component.

The main types of overall reaction occurring during catalytic reforming are as follows:

(1) Dehydrogenation of cyclohexanes to aromatics. Example:

$+ 3H_2$

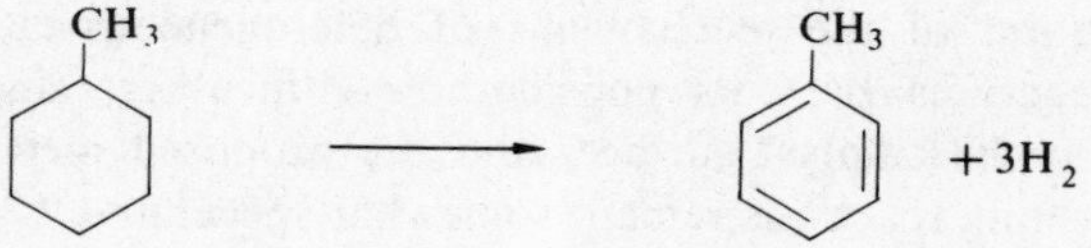

(2) Dehydroisomerisation of cyclopentanes to aromatics. Example:

$+ 3H_2$

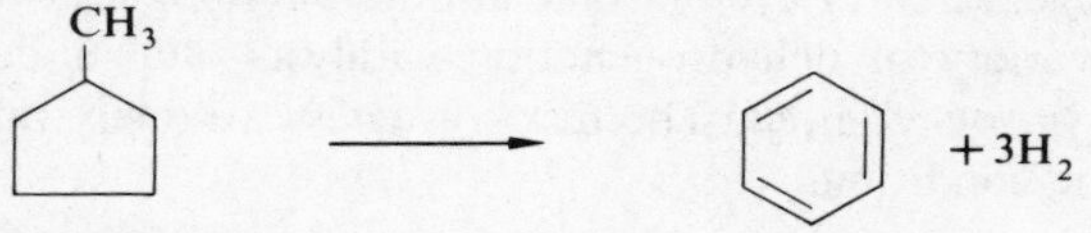

(3) Isomerisation of alkanes. Example:

$$CH_3CH_2CH_2CH_2CH_3 \longrightarrow CH_3\overset{\overset{\displaystyle CH_3}{|}}{CH}CH_2CH_3$$

(4) Dehydrocyclisation of alkanes. Example:

$$n-C_7H_{16} \longrightarrow$$ $+ 4H_2$

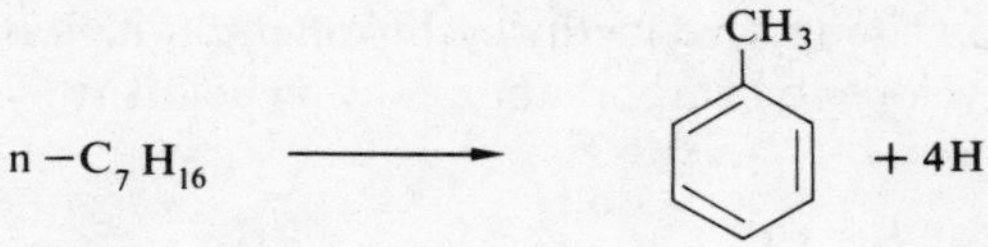

(5) Hydrocracking of alkanes. Example:

$$C_7H_{16} \xrightarrow{H_2} C_3H_8 + C_4H_{10}$$

The most important octane number-improving reactions are those leading to the formation of aromatic hydrocarbons, which have very high octane numbers (see Table 2-1).

The function of the added hydrogen in this process is not immediately obvious, since the reactions in which aromatics are formed are dehydrogenations, and will be favoured by *low* partial pressures of hydrogen. Hydrogen addition is, in fact, necessary to suppress the formation of high molecular weight carbonaceous material, called 'coke', which otherwise deposits on, and inactivates, the catalyst. The formation of coke, which is presumed to involve a combination of polymerisation and dehydrogenation reactions, is a common feature of high-temperature gas-phase reactions.

MECHANISMS OF REFORMING REACTIONS
Investigation of the mechanisms of heterogeneous-catalysed gas-phase reactions presents considerable difficulties, since reaction occurs on the catalyst surface, and the proposed mechanisms for the reforming reactions remain somewhat speculative.

Dehydrogenation of Cyclohexanes
Dehydrogenation of cyclohexane and its homologues occurs readily over a variety of dehydrogenation catalysts, and a dual-function catalyst is not required. The reaction goes extremely rapidly under reforming conditions.

Dehydroisomerisation of Cyclopentanes
Cyclopentane itself cannot undergo this reaction since it contains only five carbon atoms.

Both types of catalyst site are necessary in dehydroisomerisation, and it is suggested that a combination of dehydrogenation and carbonium ion reactions occurs. Thus, using methylcyclopentane as an example, dehydrogenation occurs at a hydrogenation–dehydrogenation site to give methylcyclopentene, which is protonated to a methylcyclopentyl carbonium ion at an acid site:

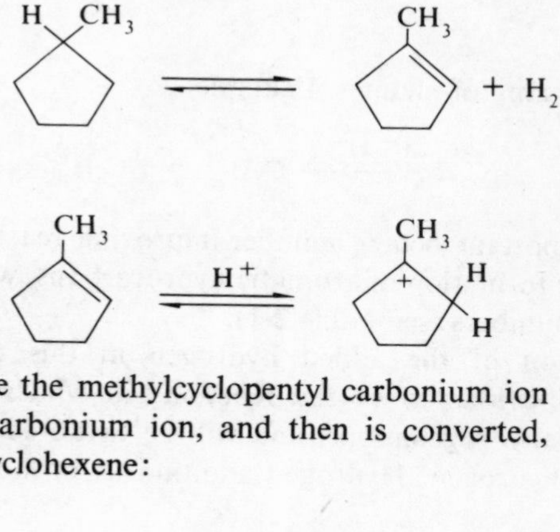

At the acid site the methylcyclopentyl carbonium ion isomerises to a cyclohexyl carbonium ion, and then is converted, by loss of a proton, into cyclohexene:

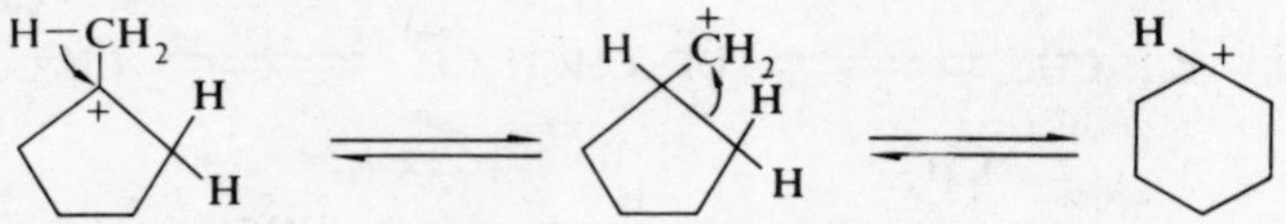

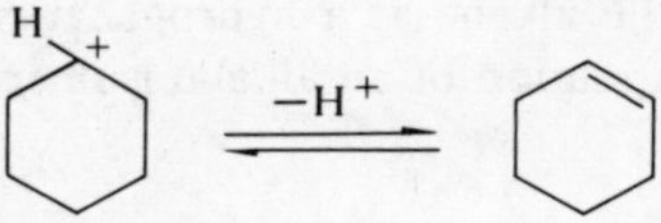

The cyclohexene is then dehydrogenated to benzene at a hydrogenation–dehydrogenation site:

A side reaction which can occur is ring opening of the cyclopentyl carbonium ion:

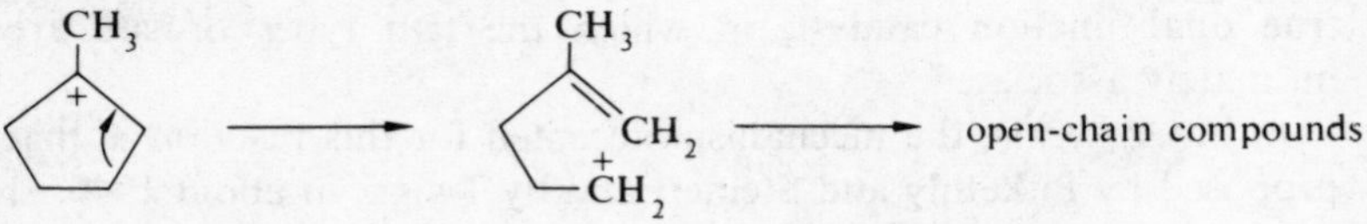

This type of reaction occurs to an appreciable extent, and the yield of benzene from methylcyclopentane is considerably smaller than that from cyclohexane under similar conditions.

Isomerisation of Alkanes
These reactions are thought to involve both types of site. The alkane is dehydrogenated at a hydrogenation–dehydrogenation site, and the resulting alkene is protonated at an acid site:

$$CH_3CH_2CH_2CH_2CH_3 \rightleftharpoons CH_3CH_2CH=CHCH_3 + H_2$$

$$CH_3CH_2CH=CHCH_3 \underset{}{\overset{H^+}{\rightleftharpoons}} CH_3CH_2\overset{+}{C}HCH_2CH_3$$

Rearrangement of the carbonium ion followed by loss of a proton leads to the production of an alkene isomeric with that originally formed by dehydrogenation:

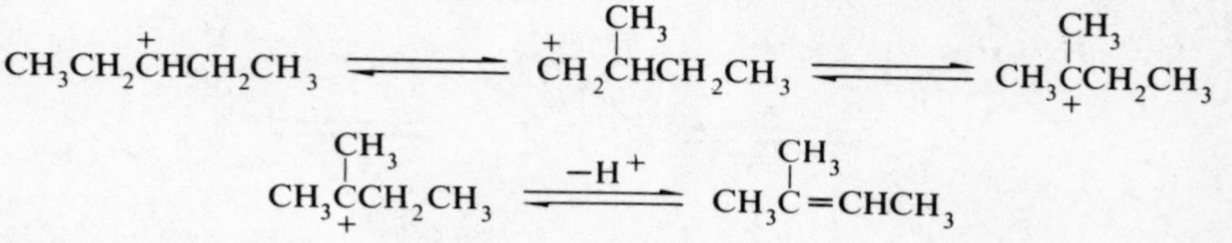

$$CH_3CH_2\overset{+}{C}HCH_2CH_3 \rightleftharpoons \overset{+}{C}H_2\overset{CH_3}{\overset{|}{C}}HCH_2CH_3 \rightleftharpoons CH_3\overset{CH_3}{\overset{|}{\underset{+}{C}}}CH_2CH_3$$

$$CH_3\overset{CH_3}{\overset{|}{\underset{+}{C}}}CH_2CH_3 \underset{-H^+}{\rightleftharpoons} CH_3\overset{CH_3}{\overset{|}{C}}=CHCH_3$$

Hydrogenation of the alkene at a hydrogenation–dehydrogenation site results in the formation of an alkane isomeric with the original alkane:

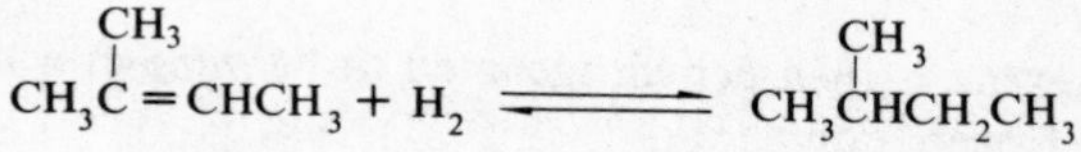

$$CH_3\overset{CH_3}{\overset{|}{C}}=CHCH_3 + H_2 \rightleftharpoons CH_3\overset{CH_3}{\overset{|}{C}}HCH_2CH_3$$

Dehydrocyclisation of Alkanes

This is the most complex of the reactions occurring in reforming, and probably the least well understood. Both types of site are necessary for this reaction to occur, and whereas dehydroisomerisation of substituted cyclopentanes will occur over a mixture of acidic and dehydrogenation catalysts, dehydrocyclisation requires a true dual-function catalyst, in which the two types of site are intimately associated.

In broad terms, the mechanism accepted for this reaction is that proposed by Pitkethly and Steiner, and by Twigg, in about 1940. It is suggested that the alkane is dehydrogenated to an alkene, which is adsorbed on the catalyst at the two carbon atoms involved in the double bond, and that ring closure occurs between one of the adsorbed carbon atoms and a carbon atom in the gas phase, e.g.:

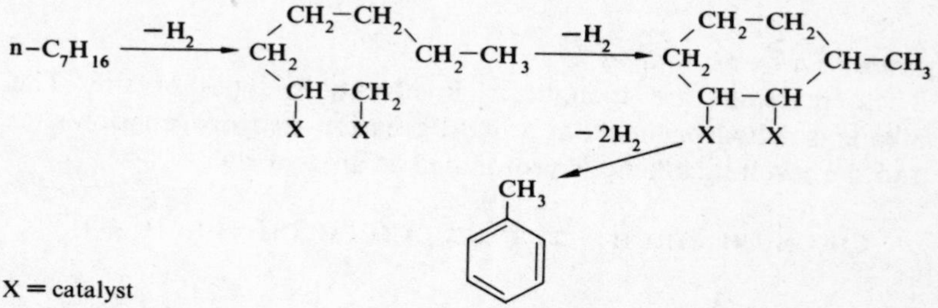

X = catalyst

The details of the ring closure reaction are not at all well understood.

On the basis of this mechanism, one can predict the products to be expected from the dehydrocyclisation of a particular alkane. For instance, dehydrocyclisation of n-octane would be expected to give ethylbenzene and *o*-xylene:

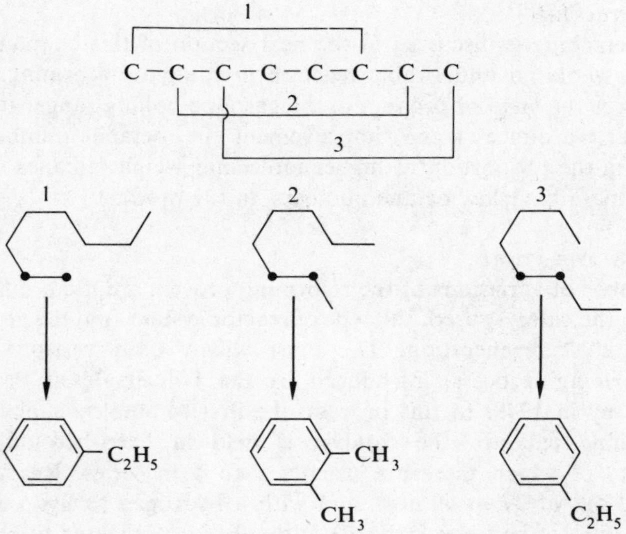

Such results can be achieved in the laboratory, e.g. by operating at low temperatures, or by using a catalyst of low acidity. Thus, reaction of n-octane over a platinum-on-α-alumina catalyst, i.e. a low acidity catalyst, gave mainly ethylbenzene and *o*-xylene (see Table 2-2). However, a much more acidic commercial catalyst gave all possible isomers, and it is assumed that under commercial conditions isomerisation of the alkane precedes dehydrocyclisation.

TABLE 2-2

DEHYDROCYCLISATION OF n-OCTANE
798 K, 1·5 atm H_2

Catalyst	Composition of aromatic fraction (mole %)					
	Benzene	Toluene	Ethyl-benzene	*o*-Xylene	*m*-Xylene	*p*-Xylene
Pt on α-alumina	—	—	39·7	58·3	1·4	0·6
Commercial catalyst	8·8	15·9	12·0	26·3	24·1	13·0

Source of data: Fogelberg, Gore and Ranby, *Acta. Chem. Scand.*, **21**, 2041 (1967). By permission of the Editor, *Acta Chemica Scandinavica*.

The benzene and toluene formed in the reaction over the commercial catalyst are a result of cracking reactions.

Hydrocracking

Hydrocracking is discussed in the next section of this chapter. It is on the whole an undesirable reaction in catalytic reforming, since it reduces the yield of product in the gasoline boiling range. It does, however, produce some improvement in octane number by reducing the proportion of higher molecular weight alkanes, which in the main have low octane numbers, in the product.

PROCESS OPERATION

A number of variations of the reforming process are used, differing, e.g., in the catalyst used, the type of reactor system, and the necessity for catalyst regeneration. The most widely used version is the 'Platforming' process, introduced by the Universal Oil Products Company in 1949. In this process, the first to employ a platinum-containing catalyst, the catalyst is held in fixed-bed adiabatic reactors of which there are usually 3 to 5 in series. Reaction is carried out at 27 to 40 atm, and with a hydrogen to hydrocarbon molar ratio of between 5 and 10:1; the feed is preheated to between 720 and 810 K before being fed to the first reactor. Reforming is, overall, an endothermic process, so that the temperature of the gas stream falls as it passes through the reactors. The heat of reaction is supplied, and the desired reaction temperature maintained, by passing the gas through reheaters between the reactors.

Catalytic reforming capacities in the USA and the UK in 1970 were approximately 2 600 000 and 320 000 barrels per day respectively.

Catalytic Cracking

Cracking of petroleum fractions, that is, subjecting them to treatment which produces material of lower molecular weight than the feed, was first carried out in the USA in 1912, the process used being that of *thermal cracking*. Initially the objective was to increase the yield of gasoline from a given crude oil, but later, as compression ratios of automobile engines began to be increased, another advantage to be obtained by this process became evident, that is, the gasoline produced by thermal cracking is of much higher octane number than straight-run gasoline. A very considerable use of this process built up in the period up to the early

1940s. However, the process of catalytic cracking, introduced in 1936, has major advantages over thermal cracking in both yield and quality of gasoline produced, and has now to a great extent supplanted it in refinery operations. A form of thermal cracking designed to give high yields of low molecular weight products is of great importance in the manufacture of alkenes, and is discussed later in this chapter.

TABLE 2-3

CATALYTIC CRACKING OF VACUUM DISTIL-LATE (b.p. 620–820 K) FROM MIDDLE EAST CRUDE OIL

Product	Yield (Weight %)
H_2S	0·75
H_2	0·25
CH_4	1·4
C_2H_4	0·6
C_2H_6	1·2
C_3H_6	4·2
C_3H_8	1·9
$i-C_4H_8$	1·9
$n-C_4H_8$	3·5
$i-C_4H_{10}$	3·3
$n-C_4H_{10}$	1·0
Gasoline	36·5
Gas oil	15·0
Residue (above 610 K)	22·7
Coke	5·8

Source of data: R.J.H. Gilbert and W.N.N. Knight (British Petroleum Company Ltd.), Preprint No. 67, *Symposium on Quality Criteria for Catalytic Cracking Stocks and Methods of Preparation*, American Institute of Chemical Engineers, December 1959. By permission of the British Petroleum Company Ltd.

The feeds used for catalytic cracking are gas oils (see Fig. 2-1), and fractions obtained by vacuum distillation of the heavy fuel oils (b.p. >620 K) obtained as the residue in primary crude oil distillation. Cracking is carried out by contacting the feed vapour with an acidic catalyst at 730 to 790 K, and approximately atmospheric pressure, for a few seconds. The most commonly used catalysts until very recently were silica–alumina catalysts, and various naturally occurring clays. Since 1964, molecular sieve or zeolite catalysts, which have much higher activities, have become of increasing importance.

The product range obtained, and in particular the ratio of gasoline to heavier liquid fuel fractions, can be varied quite substantially to suit market requirements, by varying the conditions. The product range obtained in one commercial operation is shown in Table 2-3. One of the characteristics of catalytic cracking, which is illustrated by the data in Table 2-3, is that only small proportions of methane, ethane, and ethylene are produced. This contrasts with thermal cracking which produces substantially larger proportions of these compounds.

MECHANISM

The feeds used for catalytic cracking are extremely complex mixtures, and investigation of the reactions involved in cracking a commercially realistic feedstock is not feasible with currently available techniques. Present knowledge of the mechanism of catalytic cracking derives from experiments with pure model compounds. Even with such compounds the reactions occurring are very complicated and far from fully understood, and it is not proposed to discuss them exhaustively here. We shall confine ourselves to considering the major reactions which lead to molecular weight reduction, and to indicating some of the other reactions which occur.

It is generally agreed that catalytic cracking involves carbonium ion intermediates, and that the catalyst functions by generating carbonium ions at its surface. The nature of the active sites is not known with certainty. One suggestion as to their nature in silica–alumina catalysts is that they occur at $Al^{3+}:O:Si^{4+}$ groups at the surface. In such groups, the aluminium atom would be expected to have Lewis-acid activity, or, in the presence of water, Brønsted-acid activity:

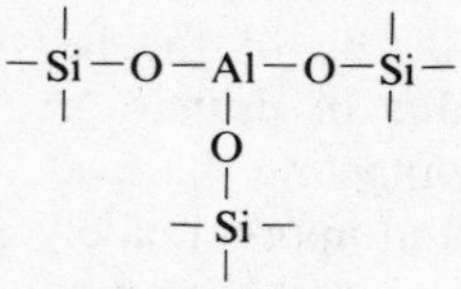

Lewis-acid site 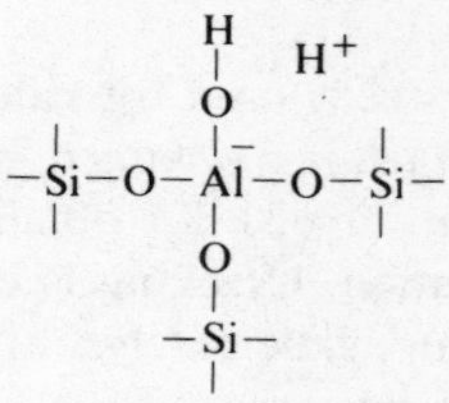   Brønsted-acid site

The sites in molecular sieve catalysts are more complex.

Primary Cracking Reactions

Let us consider firstly the cracking of a straight-chain alkane, using n-hexadecane as an example. The first question that arises is the manner in which carbonium ions are generated in the system. There are a number of suggestions as to how this takes place. One is that a small amount of alkene is produced by thermal cracking of some of the feed (see page 52), and that this is protonated by the catalyst:

$$C_{16}H_{34} \xrightarrow[\text{cracking}]{\text{thermal}} C_n H_{2n} \xrightarrow{\quad H^+ \quad} C_n H_{2n+1}^+$$

Another suggestion is that the catalyst abstracts a hydride ion from the alkane:

$$C_{16}H_{34} + \text{Cat} \longrightarrow C_{16}H_{33}^+ + \text{Cat } H^-$$

Since alkenes are produced in the cracking reaction there will be a plentiful supply of carbonium ions once the reaction is under way.

The first step in the reaction of a particular n-hexadecane molecule is abstraction from it of a hydride ion, by a carbonium ion either generated as discussed above or arising in the reaction chain (see below), or by the catalyst. Removal of a methylene hydrogen, to give a secondary carbonium ion, proceeds substantially more readily than removal of a methyl hydrogen to give a less stable primary carbonium ion. Further, there are 28 methylene hydrogens compared with only 6 methyl hydrogens in n-hexadecane, so that secondary carbonium ions will greatly predominate in the initial products of hydride ion abstraction:

$$C_{16}H_{34} + R^+ \longrightarrow \text{(e.g.) } C_5H_{11} \overset{+}{C}HC_{10}H_{21} + RH$$

Even if a primary carbonium ion is formed in the abstraction, isomerisation to a more stable secondary ion by 1,2-hydride shifts can very readily occur. Overall, it is reasonable to expect a large preponderance of secondary ions. (Skeletal rearrangements will also occur. To simplify the discussion, we will not consider these at this stage, but it will be seen later that they have important effects on the properties of the gasoline produced.)

One of the characteristics of carbonium ions, of basic importance in catalytic cracking, is their tendency to undergo *beta*-cleavage. that is, to cleave across the carbon–carbon bond *beta* to the charged carbon atom to give an alkene and a smaller carbonium ion:

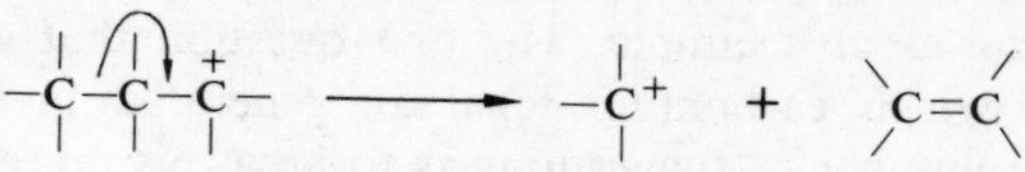

$$-\overset{|}{\underset{|}{C}}-\overset{|}{\underset{|}{C}}-\overset{+}{\underset{|}{C}}- \longrightarrow -\overset{|}{\underset{|}{C}}{}^{+} + \underset{\diagup}{\overset{\diagdown}{C}}=\underset{\diagdown}{\overset{\diagup}{C}}$$

The hexadecyl carbonium ion shown above can undergo *beta*-cleavage in two directions; one mode is shown below:

$$C_5H_{11}-\overset{+}{C}H-CH_2-CH_2-C_8H_{17} \longrightarrow C_5H_{11}-CH{=}CH_2 + \overset{+}{C}H_2-C_8H_{17}$$

The primary carbonium ion formed rapidly isomerises by 1,2-hydride shifts to a secondary ion:

$$C_8H_{17}CH_2^+ \longrightarrow \text{(e.g.) } C_4H_9CH_2CH_2\overset{+}{C}HCH_2CH_3$$

It is postulated that this occurs in conjunction with the cleavage so that the energetic requirements of cleavage correspond to those of the production of a secondary ion and the alkene.

It can readily be seen that the smallest alkene which can be produced by *beta*-cleavage of a secondary carbonium ion is propene, and that the formation of ethylene requires the cleavage of a primary carbonium ion:

$$\sim\sim\sim-CH_2CH_2CH_2^+ \longrightarrow \sim\sim\sim-CH_2^+ + CH_2{=}CH_2$$

Since the concentration of primary carbonium ions in the system is low, little ethylene is formed.

The carbonium ion formed by *beta*-cleavage and rearrangement can undergo further cleavage. Thus, taking the C_9 carbonium ion shown above as an example:

$$C_4H_9CH_2-CH_2-\overset{+}{C}HCH_2CH_3 \longrightarrow C_4H_9CH_2^+ + CH_2{=}CHCH_2CH_3$$

$$C_4H_9CH_2^+ \longrightarrow \text{(e.g.) } CH_3CH_2\overset{+}{C}HCH_2CH_3$$

In this case, the alternative mode of cleavage involves the formation of a methyl carbonium ion:

$$\overset{+}{C_4H_9\,CH_2CH_2CH \!-\! CH_2 \!-\! CH_3} \longrightarrow C_4H_9CH_2\,CH_2\,CH=CH_2 \; + CH_3^+$$

The energetic requirements for this reaction are much higher than for the removal of an ion containing three or more carbon atoms, and this mode of cleavage does not occur to any appreciable extent, nor does cleavage to ethyl carbonium ions. Thus, when successive *beta*-cleavage reactions have resulted in the formation of a carbonium ion which cannot cleave to produce a carbonium ion containing three or more carbon atoms, this ion abstracts a hydride ion from a molecule of substrate to give a new hexadecyl carbonium ion, and a small alkane molecule:

$$C_5H_{11}^+ \; + \; C_{16}H_{34} \longrightarrow C_5H_{12} \; + \; C_{16}H_{33}^+$$

The reaction is thus a chain reaction, and once it has started much of the production of hexadecyl carbonium ions is by hydride ion abstraction by small carbonium ions.

It has already been mentioned that skeletal rearrangement of carbonium ions will occur in the system. As a consequence of this, the small carbonium ions which carry out the hydride ion abstractions will, if they contain four or more carbon atoms, tend to be tertiary ions, and the alkanes produced will, in the main, be branched-chain alkanes:

$$CH_3CH_2\,\overset{+}{C}HCH_2CH_3 \xrightarrow[\text{methide shifts}]{\text{1,2-hydride and}} CH_3CH_2\underset{+}{\overset{\overset{\displaystyle CH_3}{|}}{C}}CH_3$$

$$CH_3CH_2\underset{+}{\overset{\overset{\displaystyle CH_3}{|}}{C}}CH_3 \; + \; C_{16}H_{34} \longrightarrow CH_3CH_2\underset{H}{\overset{\overset{\displaystyle CH_3}{|}}{\underset{|}{C}}}CH_3 \; + \; C_{16}H_{33}^+$$

The production of branched-chain alkanes is highly desirable since these have higher octane numbers than the corresponding n-alkanes. Thermal cracking produces a substantially higher proportion of n-alkanes than catalytic cracking.

The above discussion has referred to cracking of a straight-chain alkane. It can be seen that cracking of branched-chain alkanes will involve analogous reactions. Cracking of cycloalkanes is more complicated, and has not been so well characterised. Cracking occurs both in side chains and in the ring.

Alkyl aromatics with two or more carbon atoms in the side chain readily undergo cracking resulting in loss of the side chain as an alkene. This reaction is in effect the reverse of the Friedel–Crafts alkylation of aromatic hydrocarbons with alkenes (see Chapter 5). It involves protonation of the aromatic ring to an areneonium ion, which then eliminates a carbonium ion:

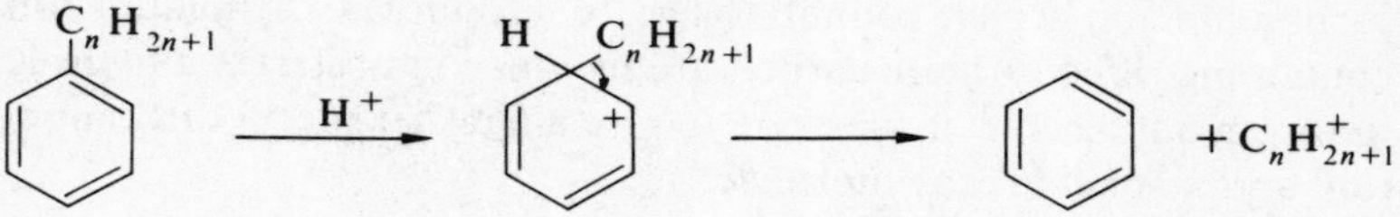

Loss of a proton from the carbonium ion gives the alkene:

$$C_nH_{2n+1}^+ \xrightarrow{-H^+} C_nH_{2n}$$

The aromatic ring itself is essentially unaffected under cracking conditions.

Secondary Reactions

The alkenes formed in cracking can undergo various further reactions. The simplest of these is migration of the double bond towards the centre of the chain, which can readily be envisaged as occurring by protonation followed by deprotonation, e.g.:

$$C_4H_9CH_2CH=CH_2 \xrightarrow{H^+} C_4H_9CH_2\overset{+}{C}HCH_3 \xrightarrow{-H^+} C_4H_9CH=CHCH_3 \xrightarrow{etc.}$$

Similarly, skeletal isomerisation of the carbonium ion prior to deprotonation results in the formation of a branched-chain alkene, e.g.:

$$C_4H_9CH_2\overset{+}{C}HCH_3 \xrightarrow[\text{methide shifts}]{\text{1,2-hydride and}} C_3H_7\overset{CH_3}{\underset{+}{C}}CH_2CH_3 \xrightarrow{-H^+} C_3H_7\overset{CH_3}{C}=CHCH_3$$

Both these reactions are desirable in that they lead to an increase in octane number of the product.

Cracking of alkenes can evidently occur, in that protonation readily converts them into carbonium ions which can then undergo cracking reactions of the type already discussed. Overall, the cracking of an alkene to a smaller alkene and an alkane involves the withdrawal of hydrogen from some other part of the system. Thus:

$$C_8H_{16} \xrightarrow{H^+} C_8H_{17}^+ \longrightarrow C_4H_8 + C_4H_9^+$$

$$C_4H_9^+ + RH \longrightarrow C_4H_{10} + R^+$$

overall:

$$C_8H_{16} \xrightarrow{2H} C_4H_8 + C_4H_{10}$$

This type of reaction must therefore occur in conjunction with the dehydrogenation of other components of the reaction mix. Similar hydrogen transfer reactions can also lead to the conversion of alkenes to alkanes.

A variety of species undergo dehydrogenation in hydrogen transfer reactions of this type. For instance, it is suggested that cyclohexane and its homologues are converted to aromatics by a series of reactions of the type shown below:

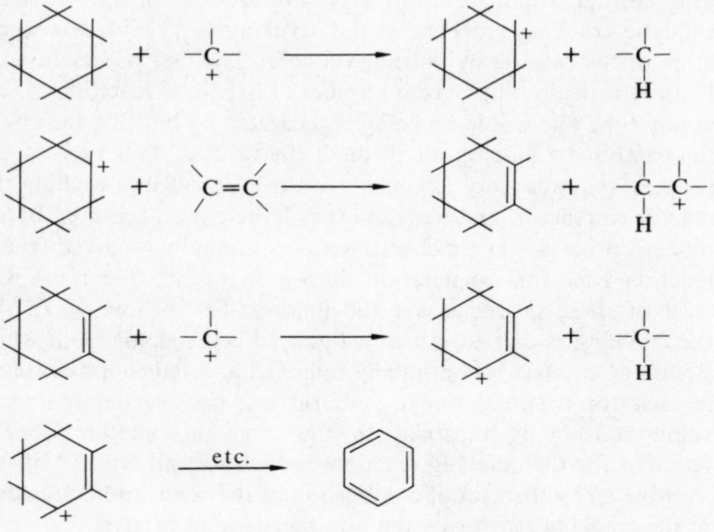

Formation of aromatics from open-chain compounds is observed during cracking. A possible mechanism by which this occurs is cyclisation of an alkenyl carbonium ion, followed by hydrogen transfer reactions of the type just discussed, e.g.:

$$CH_3CH_2CH_2CH_2CH_2CH=CHCH_3 + R^+ \longrightarrow CH_3\overset{+}{C}HCH_2CH_2CH_2CH=CHCH_3 + RH$$

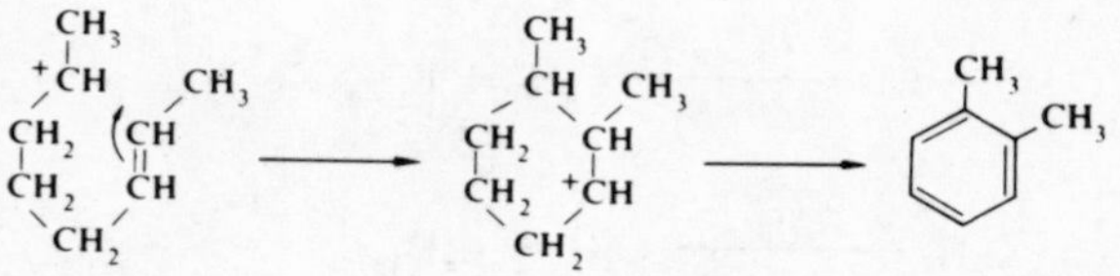

Secondary reactions of very major technological significance are those leading to the formation of high molecular weight material of high carbon content, or 'coke', since this rapidly deposits on, and inactivates, the catalyst. Coke formation is thought to involve a combination of polymerisation and hydrogen transfer reactions.

PROCESS OPERATION

The consideration of overriding importance in the design of catalytic cracking processes is the necessity to provide for regeneration of the catalyst by burning off coke. The first process used, the Houdry process, employed a number of fixed-bed reactors, of which at any time one would be being regenerated by burning the coke off the catalyst by passing air through the reactor. This type of cyclic process requires very complex control gear for switching from reactor to reactor, and has now very largely been replaced by more modern processes in which catalyst is continually removed from the reaction zone for regeneration during operation. The most widely used of these processes are the fluidised-bed processes. In these, the cracking is carried out in a fluidised-bed reactor from which a stream of catalyst is continually removed and fed to a fluidised-bed regenerator where the coke is burnt off, the regenerated catalyst being continually returned to the cracking reactor. The heat required for the cracking reactions, which overall are endothermic, is provided by the heat of combustion of the coke, and is transferred to the cracking reactor by the hot, regenerated catalyst.

Catalytic cracking capacities in the USA and UK in 1970 were approximately 5 900 000 and 230 000 barrels per day respectively.

Hydrocracking

As its name implies, the process of hydrocracking involves cracking in the presence of hydrogen. It has come into prominence in petroleum refining during the past ten years, and its importance is growing rapidly.

Hydrocracking involves treating the feed with hydrogen over a dual-function catalyst with both hydrogenation and cracking activity, at 610 to 690 K and 65 to 135 atm. The cracking function of the catalyst is provided by silica–alumina or zeolites, and the hydrogenation component may be, e.g., nickel, tungsten, platinum or palladium. It is suggested that the primary cracking reactions are essentially the same as those in catalytic cracking, but that the alkenes thus formed are very rapidly hydrogenated to alkanes, so that the secondary reactions are effectively suppressed, one important consequence being that coke formation is eliminated.

The major advantages of hydrocracking over catalytic cracking are that it can be operated on very high boiling feedstocks, and that it allows a considerable degree of control over the product distribution obtained. It can, for instance, be operated to produce high yields of gasoline, or of products in the kerosine–gas oil boiling range. Its disadvantage is that it is a more expensive procedure, because of the high pressure involved, and because of its hydrogen consumption.

The process technology is quite different from that of catalytic cracking. Since coke deposition does not occur there is no necessity to provide for catalyst regeneration, and fixed-bed reactors can be used. Overall, the process is exothermic, so that cooling rather than heating is required in the reactor.

Hydrocracking capacity in the USA in 1970 was 800 000 barrels per day. At the time of writing the process is not much used in Europe, but a number of plants are under construction, including one in the UK.

PETROCHEMICAL PROCESSES

The importance of processes for the manufacture of organic chemicals from petroleum fractions has already been indicated in

Chapter 1. In this chapter some of the major processes used will be discussed in more detail. Other processes involving reactions of alkanes (chlorination, oxidation, and dehydrogenation) are dealt with in later chapters.

Thermal Cracking for Alkenes

This process is of very great importance, and provides the basis for the manufacture of a wide range of products (see Figs. 1-3, 1-4 and 1-5). In the USA the most common feedstocks for thermal cracking are ethane and propane; in Europe and Japan, where these gases are not available in such large quantities, naphtha is the usual feedstock.

In cracking for alkenes, the feedstock vapour, diluted with steam, is passed through an empty tube in a furnace, at temperatures in the region of 1 100 K. Residence time is 1 second or less, and the cracked gases are rapidly quenched after leaving the furnace. When ethane is used as feed the major products are ethylene and hydrogen, with small amounts of methane and higher hydrocarbons. Propane and butane give lower yields of ethylene, and naphtha gives a complex mixture of products, including propene, butenes, butadiene, and substantial amounts of a gasoline fraction of high octane number. At the present time crackers are operated primarily to obtain ethylene.

ETHANE CRACKING REACTIONS

Thermal cracking involves free-radical chain reactions. In ethane cracking it is suggested that initiation of reaction chains is brought about by homolysis of ethane to methyl radicals which then abstract hydrogen atoms from ethane:

$$CH_3-CH_3 \longrightarrow 2CH_3 \cdot \qquad (4)$$

$$CH_3 \cdot + H:CH_2-CH_3 \longrightarrow CH_4 + \cdot CH_2-CH_3 \qquad (5)$$

The ethyl radical formed eliminates a hydrogen atom, forming ethylene, and the hydrogen atom abstracts another hydrogen atom from a molecule of ethane to give a hydrogen molecule and another ethyl radical:

$$CH_3{-}CH_2\cdot \longrightarrow CH_2{=}CH_2 + H\cdot \qquad (6)$$

$$H\cdot + CH_3{-}CH_3 \longrightarrow CH_3{-}CH_2\cdot + H_2 \qquad (7)$$

Reactions (6) and (7) are the propagation reactions of the kinetic chain, and continue until the chain is terminated.

The most important termination reactions are thought to be combination of ethyl radicals, and combination of ethyl radicals and hydrogen atoms:

$$2CH_3{-}CH_2\cdot \longrightarrow CH_3{-}CH_2{-}CH_2{-}CH_3 \qquad (8)$$

$$CH_3{-}CH_2\cdot + H\cdot \longrightarrow CH_3{-}CH_3 \qquad (9)$$

It should be noted that most of the ethyl radicals are formed by reaction (7), and not by reaction (5). It would, of course, be possible to postulate a non-chain reaction scheme leading to ethylene, for example, one involving reactions (4), (5), (6) and (9), but this would not be in accord with the observed kinetics of the reaction and would lead to the formation of much more methane than is in fact produced (see Fig. 2-2).

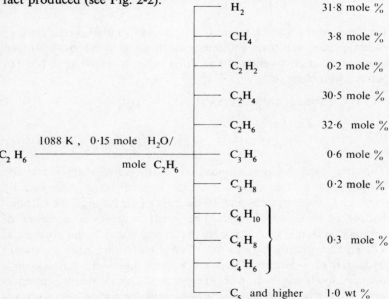

H_2	31·8 mole %
CH_4	3·8 mole %
C_2H_2	0·2 mole %
C_2H_4	30·5 mole %
C_2H_6	32·6 mole %
C_3H_6	0·6 mole %
C_3H_8	0·2 mole %
C_4H_{10} C_4H_8 C_4H_6	0·3 mole %
C_5 and higher	1·0 wt %

C_2H_6 — 1088 K, 0·15 mole H_2O/mole C_2H_6

Fig. 2-2. *Products from ethane cracking.*

Source of data: H.C. Schutt, *Chem. Eng. Progr.*, **55** (1), 68, 1959. By permission of the Editor, *Chemical Engineering Progress*.

Products containing more than two carbon atoms are formed in small amounts, presumably in termination reactions and by polymerisation reactions involving ethylene, e.g.:

$$CH_3-CH_2\cdot + CH_2=CH_2 \longrightarrow CH_3-CH_2-CH_2-CH_2\cdot$$

NAPHTHA CRACKING REACTIONS
Present knowledge of the reactions which occur during naphtha cracking is mainly derived from experiments with model compounds.

Cracking of n-Alkanes
We will consider as an example the cracking of n-decane.
Initiation of reaction chains is brought about by thermal cleavage of carbon–carbon bonds:

$$CH_3CH_2CH_2CH_2CH_2CH_2CH_2CH_2CH_2CH_3 \longrightarrow$$

$$(e.g.)\quad CH_3CH_2CH_2CH_2\cdot + \cdot CH_2CH_2CH_2CH_2CH_2CH_3 \quad (10)$$

However, once the reaction is under way, most production of radicals from substrate molecules is thought to be by hydrogen atom abstraction by methyl and ethyl radicals which arise later in the reaction chain, e.g.:

$$CH_3CH_2CH_2CH_2CH_2CH_2CH_2CH_2CH_2CH_3 + CH_3\cdot \longrightarrow$$

$$CH_3CH_2CH_2CH_2CH_2CH_2CH_2\dot{C}HCH_2CH_3 + CH_4 \quad (11)$$

This attack will occur at all positions on the substrate molecule, the proportions of the various radicals formed being governed by statistical considerations and by the ease of cleavage of the carbon–hydrogen bonds concerned. The relative rates of removal of primary, secondary and tertiary hydrogen atoms from alkanes, at 770 K, have been shown to be $1:3\cdot66:13\cdot4$. Attack on a secondary hydrogen of n-decane is thus much more likely than attack on a primary hydrogen, both from the point of view of relative numbers and of relative ease of abstraction, and formation of secondary radicals predominates.

The radical produced by hydrogen atom abstraction readily undergoes *beta*-cleavage to an alkene and a smaller alkyl radical:

$$CH_3CH_2CH_2CH_2CH_2CH_2 \overset{\curvearrowright}{:} CH_2 - \overset{\curvearrowright}{C}HCH_2CH_3 \longrightarrow$$

$$CH_3CH_2CH_2CH_2CH_2CH_2{\cdot} + CH_2 = CHCH_2CH_3$$

There are two possible modes of cleavage, that shown, and the alternative one to give non-1-ene and a methyl radical. This latter mode, which involves the formation of a relatively unstable radical and presumably proceeds through a higher energy transition state than does cleavage to give a primary radical, occurs to only a minor extent.

The primary alkyl radical formed by this initial *beta*-cleavage, or by the homolysis of an n-decane molecule [reaction (10)], can now undergo a series of cleavages each giving rise to a molecule of ethylene, until finally a methyl or ethyl radical is formed:

$$CH_3CH_2CH_2CH_2CH_2CH_2{\cdot} \longrightarrow CH_3CH_2CH_2CH_2{\cdot} + CH_2 = CH_2$$

$$CH_3CH_2CH_2CH_2{\cdot} \longrightarrow CH_3CH_2{\cdot} + CH_2 = CH_2$$

Methyl radicals thus formed abstract hydrogen atoms from substrate molecules [reaction (11)] or, less often, enter into termination reactions by coupling with other radicals. In the case of ethyl radicals, loss of a hydrogen atom to give ethylene, i.e. reaction (6) discussed under 'Ethane Cracking', can also occur.

A number of reactions which lead to a reduction in the amount of ethylene formed can occur. Thus, intermolecular chain transfer, in which an intermediate primary radical abstracts a hydrogen atom from a substrate molecule will most often result in the formation of a secondary radical:

$$RCH_2{\cdot} + C_{10}H_{22} \longrightarrow RCH_3 + (e.g.) \ C_7H_{15}CH_2\overset{\cdot}{C}HCH_3 \qquad (12)$$

The first *beta*-cleavage of this newly formed radical will give an alkene other than ethylene:

$$C_7H_{15}CH_2\overset{\cdot}{C}HCH_3 \longrightarrow C_7H_{15}{\cdot} + CH_2 = CHCH_3$$

It is suggested that intramolecular chain transfer can occur by a 'coiling' mechanism, in radicals of six or more carbon atoms, e.g.:

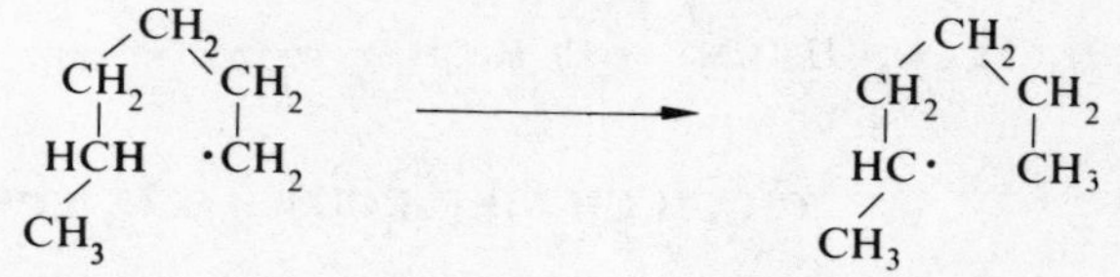

$$
\begin{array}{c}
\quad\quad \text{CH}_2 \\
\text{CH}_2 \quad\; \text{CH}_2 \\
\mid \qquad \mid \\
\text{HCH} \quad \cdot\text{CH}_2 \\
\quad \text{CH}_3
\end{array}
\qquad\longrightarrow\qquad
\begin{array}{c}
\quad\quad \text{CH}_2 \\
\text{CH}_2 \quad\; \text{CH}_2 \\
\mid \qquad \mid \\
\text{HC}\cdot \quad \text{CH}_3 \\
\quad \text{CH}_3
\end{array}
$$

This results in the conversion of a primary radical to a secondary radical, and the production of an alkene other than ethylene in the subsequent *beta*-cleavage:

$$CH_3 \, \dot{C}HCH_2 \, CH_2CH_2CH_3 \longrightarrow CH_3CH{=}CH_2 + \cdot CH_2CH_2CH_3$$

The yield is also reduced by the occurrence of coupling of alkyl radicals (other than methyl):

$$RCH_2\cdot + \cdot CH_2\, R' \longrightarrow RCH_2CH_2\, R' \qquad (13)$$

Although the alkanes formed in such reactions can undergo further cracking, hydrogen atom abstraction is likely to give a secondary rather than a primary radical, so that the first *beta*-cleavage of the resulting radical will give an alkene other than ethylene.

Since reactions (12) and (13) are bimolecular, while the desired *beta*-cleavage reactions are unimolecular, the relative extent to

TABLE 2-4

EFFECT OF PRESSURE ON CRACKING OF n-HEXADECANE
AT 773 K

Pressure (atm)	1	21
Conversion (%)	42·4	47·5
Products	Moles/100 moles hexadecane reacted	
H_2	16·5	17·4
CH_4	50·9	23·3
C_2H_4	84·0	14·9
C_2H_6	56·6	38·0
C_3H_6	59·0	27·1
C_3H_8	16·5	36·5
C_4H_6	1·7	0·9
C_4H_8	20·1	19·1
C_4H_{10}	3·3	13·8
C_5H_{10}	14·8	14·3
C_5H_{12}	1·0	8·1
C_6 and above	92·9	111·9

Source of data: H.H. Voge and G.M. Good, *J. Amer. Chem. Soc.*, **71**, 593 (1949). By permission of the American Chemical Society.

which they occur may be reduced by reducing the partial pressure of hydrocarbon. Table 2-4 illustrates the effect of pressure on product distribution in thermal cracking of n-hexadecane. It can be seen that reduction in pressure leads to a very marked increase in yield of ethylene. In commercial operations, low partial pressures of hydrocarbon are conveniently obtained by using steam as a diluent. The more obvious method of operating the plant under reduced pressure is potentially hazardous, in that a leak would result in air being drawn into the plant, and the possible formation of an explosive mixture.

Cracking of Branched-Chain Alkanes

It can readily be seen that branched-chain alkanes will give a lower yield of ethylene than straight-chain alkanes in thermal cracking. If we consider as an example the cracking of 4-ethyloctane, then one reaction path is as follows:

$$
\underset{\text{CH}_2\text{CH}_3}{\text{CH}_3\text{CH}_2\text{CH}_2\text{CHCH}_2\text{CH}_2\text{CH}_3} + \text{CH}_3^{\cdot} \longrightarrow \underset{\text{CH}_2\text{CH}_3}{\text{CH}_3\text{CH}_2\text{CH}_2\dot{\text{C}}\text{HCH}_2\dot{\text{C}}\text{HCH}_2\text{CH}_3} + \text{CH}_4
$$

$$
\underset{\text{CH}_2\text{CH}_3}{\text{CH}_3\text{CH}_2\text{CH}_2\dot{\text{C}}\text{HCH}_2\dot{\text{C}}\text{HCH}_2\text{CH}_3} \longrightarrow \underset{\text{CH}_2\text{CH}_3}{\text{CH}_3\text{CH}_2\text{CH}_2\dot{\text{C}}\text{H}} + \text{CH}_2 = \text{CHCH}_2\text{CH}_3
$$

$$
\text{CH}_3\text{CH}_2\text{CH}_2\dot{\text{C}}\text{HCH}_2\text{CH}_3 \longrightarrow \text{CH}_3\text{CH}_2^{\cdot} + \text{CH}_2 = \text{CHCH}_2\text{CH}_3
$$

$$
\text{CH}_3\text{CH}_2^{\cdot} + \text{C}_{10}\text{H}_{22} \begin{cases} \longrightarrow \text{CH}_3\text{CH}_3 + \text{C}_{10}\text{H}_{21}^{\cdot} \\ \longrightarrow \text{CH}_2 = \text{CH}_2 + \text{H}^{\cdot} \end{cases}
$$

The reader can readily satisfy himself, by considering initial hydrogen atom abstraction from other carbon atoms, that cracking of 4-ethyloctane will give a lower yield of ethylene than cracking of n-decane.

Cracking of Cycloalkanes

The cracking of cycloalkanes has not been as extensively investigated as that of alkanes, and is not so well understood. However, it is well established that feedstocks containing a high proportion of cycloalkanes (naphthenic feeds) give lower yields of ethylene and higher yields of butadiene than highly paraffinic feeds. This fits in quite reasonably with the postulated mechanism for thermal cracking. For example, *beta*-cleavage of a cyclohexyl radical will produce a hexenyl radical, which can be easily envisaged as giving rise to a molecule each of ethylene and butadiene:

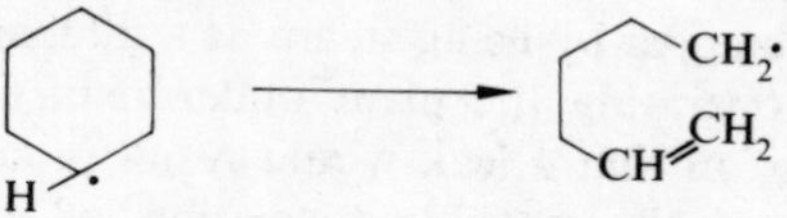

$$\cdot CH_2CH_2CH_2CH_2CH=CH_2 \longrightarrow CH_2=CH_2 + \cdot CH_2CH_2CH=CH_2$$

$$\cdot CH_2CH_2CH=CH_2 \xrightarrow{\ -H\cdot\ } CH_2=CHCH=CH_2$$

Cracking of Aromatic Hydrocarbons

The benzene ring is highly resistant to cracking, and aromatic hydrocarbons undergo cracking only in side-chains. Again, much lower yields of ethylene are obtained than from n-alkanes.

Secondary Reactions

The alkenes produced in the reactions described above can undergo various further reactions under cracking conditions, e.g., further cracking, dehydrogenation, polymerisation, and cyclisation to mono- and polynuclear aromatic compounds. Some extra ethylene is formed by alkene cracking, but in the main these secondary reactions cause a reduction in the yield of lower alkenes, and are undesirable. Particularly undesirable are reactions leading to the formation of 'coke', which tends to cause blockages in the equipment.

In general, the secondary reactions have lower activation energies, and are of higher orders, than the primary cracking reactions. The latter are therefore favoured in relation to the secondary reactions by using high temperatures and low partial pressures.

Process Operation

A cracking reactor consists essentially of a tube, e.g., of about 5 cm diameter and of total length about 60 metres, in a gas- or oil-fired furnace, the whole being termed a 'cracking furnace'. (The tube is coiled or folded so that the furnace may be made reasonably compact.) A modern naphtha cracking plant of capacity 500 000 tons per annum ethylene might have 15 to 20 such furnaces, each containing a number of tubes. On emerging from the furnace, the cracked gases are quenched to about 670 K in a 'quench boiler', the heat being used to generate steam for use in the process. After passing through the quench boiler, the gases are further cooled by an oil spray.

TABLE 2-5

TYPICAL PRODUCT DISTRIBUTION FROM
NAPHTHA CRACKING

Products	Yield (wt% of feed)
hydrogen	1·0
methane	15·5
ethylene	29·8
ethane	4·0
propene	14·6
propane	0·8
butadiene	4·1
butenes and butanes	4·7
gasoline	20·9
fuel oil	4·6

It was indicated above that high yields of ethylene are favoured by the use of high temperatures. In practice, the temperatures which can be used are limited by the maximum working temperatures of the alloys from which the furnace tubes are made, and the maximum temperature reached by the gas is usually in the range 1 020 to 1 100 K. Steam/naphtha ratios (weight) of about 0·5:1 are used. A typical product distribution from the cracking of a full range naphtha under conditions designed for high ethylene yields is shown in Table 2-5.

The separation section of a naphtha cracking plant is, as might be expected, complex. Separation of the products is mainly achieved by distillation, the lower molecular weight products requiring distillation at low temperature and under pressure.

The economics of a naphtha cracking operation are very much affected by the value which can be attached to the co-products, and every effort is made to find high-value outlets for these products.

Naphtha cracking plants normally form part of an integrated complex in which not only the ethylene, but also the propene and often the butenes and butadiene are used as feedstocks for chemical processes. The gasoline fraction has a high octane number, and is used for motor spirit. Some operators separate benzene, toluene and xylenes from it.

The scale of manufacture of ethylene has increased very greatly since cracking operations began. The first manufacture of ethylene by cracking in the UK was by British Celanese in 1942, at a scale of 20 000 tons per annum. In 1969 ICI brought on-stream a plant with a capacity of 450 000 tons per annum of ethylene, and Shell are currently planning one with a capacity of 500 000 tons per annum.* The cost of making ethylene in a modern plant in the UK has been estimated to be about £20–25 per ton (1970).

WAX CRACKING

Another thermal cracking process operated commercially, though on a much smaller scale, is paraffin wax cracking for the production of higher 1-alkenes.

Paraffin wax, obtained from the high-boiling residues from the primary fractionation of crude oil, consists essentially of n-alkanes from about C_{18} to C_{56}. Cracking at fairly low temperatures, and with relatively long residence times, gives quite high yields of 1-alkenes with five carbon atoms and above. Thus, cracking at 810 to 840 K for 7 to 15 seconds gives a 60% (weight) yield of C_5–C_{20} alkenes, of which 95% are 1-alkenes.

Wax cracking is operated mainly to provide alkenes for detergent manufacture (see page 168).

Acetylene Processes

As was indicated in Chapter 1, the manufacture of acetylene from coke *via* calcium carbide is a relatively expensive procedure, and for many years work concerned with the development of methods of making acetylene from hydrocarbons has been in progress. There are now a number of such methods available.

The preparation of acetylene from lower alkanes is thermodynamically feasible at temperatures above about 1 200 K (1 500 K for methane), and acetylene can in fact be made by high-temperature

* It is reported that the Shell plant will be able to use either naphtha or gas oil as feed. This reflects the fact that naphtha has shown some tendency to go into short supply in recent years.

pyrolysis of alkanes. However, acetylene is unstable with respect to its elements at temperatures up to about 4 200 K, and consequently if the reaction is allowed to proceed to equilibrium the products are carbon and hydrogen. The essential features of a process for the manufacture of acetylene from alkanes are that the feed should be heated to a very high temperature and then quenched, within a very short time (0·1 second or less depending on the temperature being used). The formation of acetylene is favoured by low partial pressures of hydrocarbon. The mechanism by which acetylene is formed is not known; one fact for which any mechanism must account is that acetylene can be made in this way from methane.

There are very considerable engineering difficulties in attaining the conditions necessary for acetylene production. Use of a tubular reactor of the type used in cracking for alkenes is not feasible, since it is not possible to construct a tube that will withstand the temperatures involved, and transmit heat at the very high rates necessary. Acetylene processes that have been operated to date fall into three classes.

ELECTRIC ARC PROCESSES

An electric arc provides a convenient method of rapidly heating a gas stream, and a number of arc processes have been developed. The most well-established of these is that operated by I.G. Farben at Hüls, in West Germany, since 1940. This plant was reported to have a capacity of 100 000 tons per annum in 1965.

A variety of feedstocks from methane to naphtha are used in the I.G. process. The arc, which is about 1 metre long, is struck between a water-cooled cathode and a water-cooled tubular anode. Feed vapour passes longitudinally through part of the arc at a velocity of about 1 kilometre per second, and the residence time is about 0·002 second. The peak temperature reached by the gas is about 1 900 K. The gases are quenched to about 1 300 K by injecting into them a C_2–C_4 hydrocarbon stream which undergoes cracking to ethylene and other alkenes, and finally to 470 K by injecting water. There are 17 arc units in the Hüls plant.

The yield of acetylene is about 45%. About 13 kWh electricity per kg acetylene produced is consumed by the process. Higher yields and lower electricity consumptions have been claimed for a number of other arc processes in various stages of development. However, all these processes suffer from the disadvantage that electricity is generally an expensive way of supplying energy.

REGENERATIVE PYROLYSIS PROCESSES

In these processes the feed is heated by contact with a refractory material which has itself been previously heated by burning fuel.

The best known such process is the Wulff process, which has been under development since about 1930. Cracking is carried out in pairs of furnaces filled with refractory brickwork. These are operated in a cyclic manner, one being heated up by burning fuel in air in it while the other is on the cracking cycle. The feeds are then reversed, and the reheated furnace goes on to the 'make' cycle, while the other furnace, which has been cooled down by the endothermic reaction goes onto the 'reheat' cycle. The cracked gases are quenched with water; residence time is about 0·1 second, of which about 0·03 second is at the peak temperature (about 1 500 K). The furnaces are operated at about 0·5 atm and steam dilution is used to further reduce the hydrocarbon partial pressure. Propane is the preferred feedstock for this process, and gives an acetylene yield of about 30% together with substantial quantities of ethylene. Methane is not a very suitable feedstock because of the limitations on operating temperatures imposed by the maximum service temperatures of available refractories. The use of naphtha results in the production of more tar and carbon than when lighter hydrocarbons are used, and this can give rise to operational difficulties.

Two Wulff plants, using naphtha as feedstock, were constructed in the UK in the mid 1960s, for British Oxygen Company Ltd., and for British Geon Ltd., both with design capacities of about 30 000 tons per annum. It appears that very considerable difficulties were encountered in starting these plants up, and that their performance subsequently has been disappointing. One major source of difficulties has been blockages caused by tar and carbon. The British Geon plant, which by this time belonged to BP, was shut down at the end of 1970.

FLAME PROCESSES

In these processes, which fall into two classes, the heat of a flame is transferred direct to the feedstock. In one-stage processes, the hydrocarbon feed is burnt in an insufficient supply of oxygen, when the fraction of the feed which is not burnt is cracked by the heat of combustion of the rest. In the two-stage processes, feed is injected into a flame formed by burning fuel and oxygen. In both cases the gases are quenched by water or oil sprays after a few thousandths of a second at the reaction temperature of about 1 750 K. It is necessary to use oxygen rather than air in order to avoid excessive dilution of the product; even using oxygen, the acetylene content of the cracked gas is only about 7–10%, since large amounts of carbon monoxide are formed in the combustion. Yields vary depending on the process and the feedstock. One of the most widely used one-stage processes gives a yield of about 30% when using methane.

Flame processes, mainly using methane as feedstock, account for most hydrocarbon-based acetylene production. It has been estimated that in 1962, 350 000 tons of acetylene were made by this type of process.

In the UK, ICI constructed a plant to use a flame process with naphtha as feedstock in the mid 1960s, but it was never brought into satisfactory operation, and has now been abandoned. As in the case of the Wulff plants already discussed, the operating difficulties encountered appear to have been at least partly associated with the formation of carbon and tars, which occur to a much greater extent when naphtha rather than methane is used as feed.

PRODUCT SEPARATION

Acetylene is thermodynamically unstable at normal temperatures, and given the right conditions it will decompose to its elements with the evolution of a great deal of heat:

$$C_2H_2 = 2C + H_2 \qquad \Delta H = -226 \ \text{kJ/mol}$$

The decomposition can occur relatively slowly, or can be a violent detonation, depending on circumstances, and handling acetylene on the large scale requires great care. Increase in pressure very much increases the tendency of acetylene to undergo explosive decomposition, and this limits the pressure under which it can safely be used. Liquid acetylene is a very hazardous material, and cannot be safely handled on the large scale.

The simplest and cheapest method of separating acetylene from the cracked gases, fractional distillation, is thus ruled out, since this requires both operation under pressure and the presence of acetylene in the liquid phase. The method of separation normally used is selective absorption in solvents.

ECONOMICS OF ACETYLENE MANUFACTURE

In many of its applications acetylene is in direct competition with ethylene, so it is highly relevant to consider the relative costs involved in the manufacture of these two intermediates. It is not difficult to see that under present-day conditions acetylene is intrinsically more expensive to manufacture than ethylene. There are a number of factors which contribute to the difference.

The production of acetylene consumes very much more energy than the production of ethylene. Not only is the heat of reaction higher, but because of the higher temperatures required, more heat has to be supplied to heat up the reactants. Further, most of the heat content of the cracked gas stream is lost in the quench liquid, whereas in ethylene manufacture, where such a high rate of cooling is not necessary, and where the cracked gases do not contain so much carbon and tar, a substantial amount of heat is recovered by generating steam in the quench boiler.

As will have been gathered from the discussion of reactor systems used in acetylene manufacture, the engineering required for the construction of such systems is much more complex, and therefore expensive, than that involved in the construction of a cracking furnace for ethylene manufacture. Also, because acetylene cannot be isolated by distillation, the capital and operating costs associated with the isolation of acetylene from the cracked gases are substantially higher than those in ethylene manufacture. In addition, the hazardous nature of acetylene adds to the cost of constructing a plant.

The overall effect is that both capital and production costs for acetylene are much higher than those for ethylene. An idea of the relative costs of acetylene and ethylene plants can be obtained by comparing the ICI acetylene plant mentioned above (projected capacity 60 000 tons per annum) with an ethylene plant of capacity 250 000 tons per annum built for BP at about the same time. The capital costs were reported to be £7·5 million and £8 million respectively. The cost of making acetylene in a modern hydrocarbon-based plant is suggested to be about £70 per ton compared with about £20–25 per ton for ethylene. As will be seen elsewhere in this book, the relatively high cost of acetylene has led to widespread replacement of acetylene-based routes by routes based on ethylene and other hydrocarbons. An indication of the current relative importance of ethylene and acetylene is given by the fact that in 1969 there were under construction in Europe 3 acetylene plants with a total projected capacity of 85 000 tons acetylene per annum, and 28 ethylene plants with a total projected capacity of nearly 7 million tons ethylene per annum.

Catalytic Reforming for Aromatics

The processes used for manufacture of aromatic hydrocarbons by catalytic reforming of naphtha are essentially the same as those used in petroleum refining, except that they are specifically designed to produce high yields of benzene, toluene, and xylenes. The preferred feed is a naphtha containing a high proportion of cycloalkanes, and consisting mainly of C_6 to C_8 compounds. The aromatic products are generally separated from the reformate by extraction with a solvent which dissolves aromatic hydrocarbons but not alkanes or cycloalkanes, diethylene glycol containing about 8 to 10% of water being often used. Distillation of the extract gives the mixed aromatic product, from which benzene, toluene, and a C_8 fraction may be isolated by fractional distillation. Separation of individual compounds from the C_8 fraction is discussed in Chapter 5 (see page 173).

Benzene, toluene, and C_8 aromatics are produced by catalytic reforming in the approximate ratio 17:33:50. This product distribution does not correspond well with the demand for these materials for chemical applications, in that benzene, which is produced in the smallest quantities by reforming, is required in by far the largest quantities by the chemical industry. The process of *hydrodealkylation* provides a means of redressing this imbalance by converting toluene or xylenes to benzene (see Chapter 6).

Steam Reforming

Steam reforming of alkanes is of very great importance as a method of preparing synthesis gas (carbon monoxide/hydrogen mixtures) for use, (i) in ammonia and methanol manufacture, (ii) in the Oxo process (see page 219), and (iii) as a source of hydrogen for hydrogenations and other processes. It has been used in the USA with C_1 to C_4 alkanes, particularly methane, as feedstock since the 1930s, but in countries with no substantial supplies of these materials reforming did not become an important method of making synthesis gas until the late 1950s, when ICI developed a process for the steam reforming of light naphtha.

REFORMING REACTIONS

Steam reforming is carried out by passing a mixture of the hydrocarbon and steam over a promoted nickel catalyst; it involves a fairly complex combination of reactions, of which the more important are indicated below.

In reforming of methane, the basic reforming reaction is:

$$CH_4 + H_2O \rightleftharpoons CO + 3H_2 \qquad \Delta H = +205 \ kJ/mol$$

The so-called 'shift reaction' between carbon monoxide and steam proceeds readily under the conditions used:

$$CO + H_2O \rightleftharpoons CO_2 + H_2 \qquad \Delta H = -42 \ kJ/mol \qquad (14)$$

Carbon can be formed by the following reactions:

$$2CO \rightleftharpoons C + CO_2$$

$$CO + H_2 \rightleftharpoons C + H_2O \qquad (15)$$

$$CH_4 \rightleftharpoons C + 2H_2$$

It is important to avoid carbon formation since it deposits on, and inactivates, the catalyst.

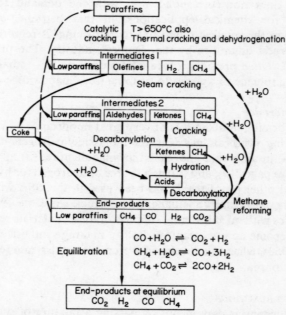

Fig. 2-3. *Steam reforming of higher alkanes. Reproduced from* Catalyst Handbook, *Imperial Chemical Industries Ltd., Wolfe Scientific Books, London,* 1970. *By permission of Imperial Chemical Industries Ltd. and Wolfe Publishing Ltd.*

The reforming of naphtha is much more complicated than that of methane. A suggested scheme for the major reactions involved is shown in Fig. 2-3. The tendency to carbon formation is much greater with naphtha than with methane.

SYNTHESIS GAS FOR AMMONIA MANUFACTURE

The mode of operation of a reforming process depends on the purpose for which the synthesis gas is required. Since the most important chemical application of steam reforming is in the production of synthesis gas for ammonia manufacture, we shall discuss this procedure in most detail.

Reforming catalysts are readily poisoned by sulphur, and consequently the first stage in a steam reforming process is normally treatment of the feed to remove sulphur compounds. Natural gas most often contains hydrogen sulphide and low boiling mercaptans and sulphides, and these can be removed by passing the gas through beds of active charcoal, or through zinc oxide at 620 to 670 K. The sulphur compounds in naphtha are less readily removed, and naphtha is most commonly subjected to *hydrodesulphurisation* which consists of treatment with hydrogen in the presence of cobalt molybdate and zinc oxide catalysts at about 640 K. The sulphur compounds are hydrogenolysed, and the resulting hydrogen sulphide is absorbed by the zinc oxide.

The desulphurised feed and steam are passed to the *primary reformer*, in which the catalyst is held in tubes, typically of about 10 cm diameter, in a furnace heated by burning oil or gas. The ratio of steam to hydrocarbon used is determined by the necessity to avoid the formation of carbon. A high steam ratio discourages carbon deposition by reducing carbon monoxide concentration by the shift reaction [reaction (14)], and by removing deposited carbon by the water gas reaction, i.e. the reverse of reaction (15). However, there is an economic penalty in using high steam ratios in the increased fuel consumption involved. In practice, molar steam/carbon ratios of about 3:1 are used.

The major factor in developing a reforming process that could use naphtha as raw material was the development of a catalyst which minimised carbon deposition at steam ratios about the same as those used for methane. The essential feature of the catalyst developed by ICI is that it contains potassium carbonate, which it is suggested acts by neutralising the acidity of the catalyst support and so reducing the extent of catalytic cracking of the feed, and by catalysing the reaction of deposited carbon with steam. Analyses of methane and naphtha reforming catalysts are shown in Table 2-6.

The operating temperature of the primary reformer is mainly determined by the maximum working temperature of the alloy of which the tubes are made, and is normally in the range 1 020 to 1 120 K.

TABLE 2-6

ICI REFORMING CATALYSTS — TYPICAL ANALYSES

Component	Weight %	
	Methane reforming catalyst	Naphtha reforming catalyst
NiO	32	21
CaO	14	11
SiO$_2$	0·1	16
Al$_2$O$_3$	54	32
MgO	—	13
K$_2$O	—	7

Source of data: *Catalyst Handbook*, Imperial Chemical Industries Ltd., Wolfe Scientific Books, London, 1970. By permission of Imperial Chemical Industries Ltd. and Wolfe Publishing Ltd.

The reaction is usually carried out at about 30 atm. This adversely affects the position of the equilibrium of the reforming reaction since there is an increase in the number of molecules in the reaction, and also increases the tendency to carbon formation. However, it leads to a saving in compression costs, since the synthesis gas is finally required at high pressure, and compression of the feed is cheaper than compression of the higher volume of reformed gas.

The product from the primary reformer consists of steam, carbon monoxide, hydrogen, some residual methane, and carbon dioxide formed by the shift reaction. This mixture is passed to the *secondary reformer*, an adiabatic reactor containing a single catalyst bed, where it is mixed with a quantity of air controlled so as to give a hydrogen/nitrogen molar ratio of 3:1 in the final product. The air oxidises some of the gas and the heat of combustion raises the temperature in the secondary reformer to about 1 270 K. At this temperature, reforming goes practically to completion and the gas leaving the secondary reformer contains steam, hydrogen, nitrogen, carbon monoxide, carbon dioxide, and only traces of methane.

The catalyst in the secondary reformer is nickel-based, but designed specifically to withstand the high temperatures encountered.

The gas stream from the secondary reformer is cooled to about 620 K and is passed to the *shift converters* in which most of the carbon monoxide is converted to carbon dioxide and further hydrogen by the shift reaction [reaction (14)]. Since this reaction is exothermic, the lower the temperature, the higher the conversion to carbon dioxide at equilibrium. At the same time, however, the lower the temperature, the slower is the reaction. The shift reaction is generally carried out in two stages, the so-called high-temperature shift at about 620 K, and the low-temperature shift at about 470 K. The high-temperature shift is carried out over an iron oxide catalyst, but this is not active enough to bring about the reaction at the lower temperature, and a copper catalyst is used for the low-temperature shift reaction. The product from the shift reactors consists of steam, hydrogen, nitrogen, carbon dioxide and very small amounts of carbon monoxide and methane.

The final stages of gas treatment are concerned with the removal of carbon dioxide and traces of carbon monoxide from the gas stream. Carbon dioxide is normally removed by scrubbing out with a solvent from which the carbon dioxide is subsequently recovered. A number of scrubbing liquids are used; one which is widely employed is aqueous potassium carbonate solution. After scrubbing, the gas still contains small amounts of carbon monoxide and carbon dioxide, and since these are both poisons for the ammonia synthesis catalyst, the gas requires a final treatment to reduce their concentration to a very low level. In many modern plants this involves passing the gas through a *methanation* stage, in which the carbon oxides are catalytically converted to methane:

$$CO + 3H_2 \rightleftharpoons CH_4 + H_2O$$

$$CO_2 + 4H_2 \rightleftharpoons CH_4 + 2H_2O$$

Methanation is carried out at about 620 K over a nickel-containing catalyst. Practically complete conversion of carbon oxides to methane is obtained. The small amount of methane formed is left in the gas; it does not interfere with the synthesis reaction.

Ammonia synthesis is carried out over an iron catalyst, at about 770 K and 300 atm:

$$N_2 + 3H_2 \rightleftharpoons 2NH_3$$

Operating and capital costs are somewhat higher for a naphtha-based than for a methane-based process. There are three main reasons for this:

(i) Naphtha requires a more elaborate desulphurisation treatment than methane, since it generally contains more sulphur, and this is in less easily removed forms than in methane.

(ii) Naphtha has to be vaporised before being fed to the process, and this requires the provision of equipment and the utilisation of energy.

(iii) A substantial amount more carbon dioxide is produced when naphtha is used:

$$\text{naphtha} + 2H_2O \longrightarrow CO_2 + 3{\cdot}05\,H_2$$
$$(ca.\,CH_{2{\cdot}1})$$

$$CH_4 + 2H_2O \longrightarrow CO_2 + 4H_2$$

The costs involved in carbon dioxide removal are consequently higher.

Thus, when methane is available at a cost comparable to that of naphtha, it is the preferred raw material for steam reforming. Following the discovery of natural gas in the North Sea during the 1960s, ICI converted its steam reforming plants in the UK from naphtha-based to methane-based operation.

Steam reforming provides the main route to ammonia, and the process is carried out for this purpose on a very large scale. Modern plants commonly have capacities of 1 000 tons ammonia per day. All synthetic ammonia in the UK is made by steam reforming, by ICI and Shell, with capacities of about 1 370 000 tons per annum and 350 000 tons per annum respectively. A small amount is obtained as a by-product of coal carbonisation.

SYNTHESIS GAS FOR OTHER APPLICATIONS

When steam reforming is employed for purposes other than ammonia manufacture, secondary reforming, which would introduce unwanted nitrogen into the gas-stream, is omitted. In hydrogen manufacture, the reformed gases are subjected to the shift reaction, followed by carbon dioxide and carbon monoxide removal. In making synthesis gas for methanol synthesis, and for the Oxo process, carbon monoxide/hydrogen molar ratios of $2:1$ and *ca.* $1:1$ respectively are required, whereas the gas produced by methane reforming contains a ratio of $3:1$. The ratio can be adjusted by a number of methods, a favoured one being to add carbon dioxide to the feed to the reformer. The carbon dioxide undergoes the reverse shift reaction during reforming:

$$CO_2 + H_2 \rightleftharpoons CO + H_2O$$

Evidently the gas is not subjected to the shift reaction after reforming, but is used for synthesis direct from the primary reformer.

Methanol Manufacture
The catalytic conversion of synthesis gas provides by far the most important route to methanol:

$$CO + 2H_2 \rightleftharpoons CH_3OH \qquad \Delta H = -92 \quad kJ/mol$$

Since the reaction is exothermic, conversion to methanol is favoured by low temperatures. It can be seen from Table 2-7 that

TABLE 2-7

EQUILIBRIUM CONSTANTS (CALCULATED)
FOR THE REACTION $CO + 2H_2 \rightleftharpoons CH_3OH$

Temperature ($^{\circ}K$)	K
273	527 450
373	10·84
473	$1·695 \times 10^{-2}$
573	$2·316 \times 10^{-4}$
673	$1·091 \times 10^{-5}$

Source of data: R.H. Ewell, *Industrial and Engineering Chemistry*, **32**, 147 (1940). By permission of the American Chemical Society.

at 273 K the equilibrium is highly favourable to the formation of methanol, but becomes progressively less favourable with increase in temperature. At above 570 K at atmospheric pressure, the conversion to methanol at equilibrium will be negligible. It is, therefore, desirable to run the reaction at a low temperature. On the other hand, the rate of reaction rises with increase in temperature. Until very recently, the only practicable catalysts available for this process were zinc oxide–chromium oxide catalysts, first developed in about

1920. These require a reaction temperature in the range 570 to 670 K to give an adequate rate of reaction, and in order to achieve an acceptable degree of conversion the reaction has to be carried out at high pressure, to displace the equilibrium to the right. The effect of pressure on the position of the equilibrium is shown in Table 2-8. Typically, a pressure of about 300 atm is used. As was

TABLE 2-8

EFFECT OF PRESSURE ON EQUILIBRIUM CONVERSION TO METHANOL AT 573 K

Pressure (atm)	Conversion to methanol at equilibrium (%)
10	0
25	1·7
50	8·0
100	24·2
200	48·7
300	62·3

Source of data: R.H. Ewell, Industrial and Engineering Chemistry, 32, 147 (1940). By permission of the American Chemical Society.

indicated in Chapter 1, the use of high pressures carries penalties in capital and operating costs.

In the late 1950s a programme of research and development work was started at ICI with the aim of producing a catalyst which would bring about the synthesis reaction at lower temperatures, and thus allow lower pressures to be used. A copper-based catalyst which gives an acceptable rate of reaction at about 520 K was developed, and on this was based the ICI 'low-pressure' methanol process which operates at 50 to 100 atm and shows considerable savings in capital and operating costs over the conventional process. It was first brought into operation in 1966, and is now used in a number of locations around the world. ICI operate the process in the UK at a scale of about 120 000 tons per annum and are currently planning a 320 000 tons per annum plant; they also operate the high-pressure process at the time of writing. BP operate the high-pressure process at a scale of 60 000 tons per annum.

The major use of methanol is in the manufacture of formaldehyde (see page 216). It has a variety of other chemical uses, e.g. in the manufacture of methyl esters, and finds some use as a solvent.

Chapter 3

Oxidation

OXIDATION reactions are extremely important in industrial organic chemistry, and are used at all scales from a few kilogrammes to many thousands of tons a year.

As was pointed out in Chapter 1, the larger the scale on which a product is made, the more money it is worth spending on research and development into methods of manufacture, and consequently methods used for carrying out oxidations vary very much depending on the scale of operation. In the case of dyestuffs, pharmaceuticals, and similar products, relatively complicated compounds made at rates of up to a few hundred tons per annum, the methods used tend to be adaptations of laboratory procedures, using reagents such as potassium permanganate, sodium dichromate, chlorine, sodium hypochlorite, etc. If the product is made at the rate of many thousands of tons a year, there is a very considerable incentive to use cheaper oxidising agents than these. The cheapest possible agent is air, which is free, apart from the small costs involved in handling it, and for many large-tonnage processes this is what is used. It is worth pointing out that although air is free, oxygen is not. If oxygen is required it has to be separated from air by distillation, and there are capital and production costs associated with this operation. Oxygen is therefore only used when air is for some reason not suitable.

Because of their importance, and because small-tonnage oxidations are often carried out by methods very similar to those used in classical organic chemistry, we shall consider only oxidations using air or oxygen. Such oxidations fall into four classes:

  (i) free-radical liquid-phase oxidations
  (ii) free-radical gas-phase oxidations
 (iii) liquid-phase non-free-radical oxidations
 (iv) heterogeneous-catalysed gas-phase oxidations

FREE-RADICAL LIQUID-PHASE OXIDATION

This type of oxidation is of major importance in chemical manufacture and is used, for example, in the manufacture of phenol, acetic
acid, acetone, nylon intermediates, and terephthalic acid. Free-
radical oxidation of this type is also important as a cause of in-service
deterioration of many materials, e.g. rubber and other polymers, and
in the drying of oil-based paints. Reaction is brought about by
contacting air or oxygen with the substrate at moderate temperatures,
often in the presence of a soluble catalyst. An example which the
reader may have observed in the laboratory is the oxidation of
benzaldehyde to crystals of benzoic acid around the stopper of a
storage bottle.

Although this type of reaction has been used commercially for a
considerable number of years, an understanding of the mechanisms
involved has only been achieved relatively recently.

Generalised Mechanism

Discussion of these reactions is simplified by the fact that, up to
a certain stage, the free-radical liquid-phase oxidation of a large
number of compounds appears to involve the same steps, and so a
generalised mechanism can be postulated.

Let us consider a generalised substrate, containing a carbon–
hydrogen bond:

$$\diagdown C - H$$

The first step in the oxidation of a particular molecule of substrate
involves the cleavage of the carbon–hydrogen bond. This cleavage is
brought about by abstraction of the hydrogen atom by some species.
At this stage, let us not worry about what the species is:

$$\diagdown C - H + X\cdot \longrightarrow \diagdown C\cdot + XH \qquad (1)$$

Evidently, for any particular abstracting species under a particular
set of conditions, the ease with which this reaction occurs will depend
on the strength of the carbon–hydrogen bond. Strengths of carbon–
hydrogen bonds vary considerably depending on the overall structure
of the substrate.

The free radical produced in reaction (1) can undergo combination
with oxygen, itself a bi-radical:

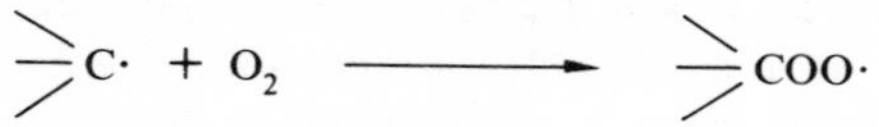

$$\ce{>C. + O2 -> >COO.} \tag{2}$$

The product of this reaction, a peroxy radical, can abstract hydrogen from another molecule of substrate to give a *hydroperoxide* and another substrate radical:

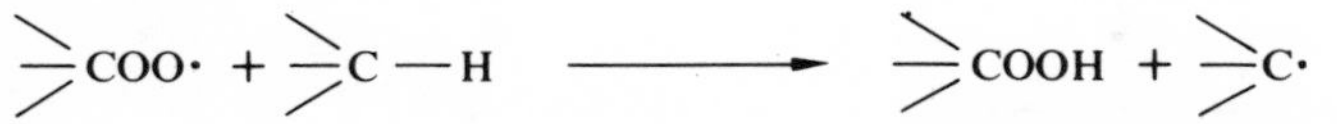

$$\ce{>COO. + >C-H -> >COOH + >C.} \tag{3}$$
hydroperoxide

The substrate radical produced in reaction (3) can react with more oxygen to give another peroxy radical. Thus, we have a chain reaction in which (2) and (3) are the propagation steps, and the primary product is $\ce{>COOH}$, a hydroperoxide. As in all free-radical chain reactions, the number of times reactions (2) and (3) occur in any particular chain, i.e. the chain length, is limited by the occurrence of termination reactions in which free radicals are eliminated:

$$\ce{>C. + .C< -> >C-C<} \tag{4}$$

$$\ce{>C. + .OOC< -> >COOC<} \tag{5}$$

$$\ce{>COO. + .OOC< -> >COOC< + O2} \tag{6}$$

If the reaction is being carried out with a plentiful supply of oxygen, reaction (2), the reaction of substrate radicals with oxygen, is very rapid and consequently the concentration of substrate radicals is very low. Under such conditions reactions (4) and (5) are considered to be of negligible importance as termination reactions compared with reaction (6).

If reactions (1), (2), (3), and (6) were the only reactions occurring in the system, the products obtained would be hydroperoxide from reaction (3) and peroxide from reaction (6). The relative amounts of these products would depend on the chain length of the reaction, i.e. the number of cycles of reactions (2) and (3) which occur before termination of the chain, and as most important oxidations have high chain lengths, the main product would be expected to be hydroperoxide. In fact, in most, but not all, cases a hydroperoxide is *not* the product obtained at the end of the reaction. Oxidation of cyclohexane, for instance, gives a mixture of cyclohexanol and cyclohexanone, together with other products. A further point to consider is that we have not yet said what X· is in reaction (1).

The oxygen–oxygen bond in hydroperoxides is weak and easily undergoes homolysis. Thus,

$$\text{>COOH} \longrightarrow \text{>CO·} + \text{·OH} \tag{7}$$

The radicals formed in this reaction can both abstract hydrogen atoms from substrate molecules:

$$\text{>C—H} + \text{>CO·} \longrightarrow \text{>C·} + \text{>COH} \tag{8}$$

$$\text{>C—H} + \text{·OH} \longrightarrow \text{>C·} + H_2O \tag{9}$$

The main products in most free-radical liquid-phase oxidations are products of further reaction of the radical produced in reaction (7), or of ionic or molecular reactions of the hydroperoxide.

Reactions (8) and (9) correspond to reaction (1). Thus, once the oxidation has started, initiation of the chains is by radicals formed by the cleavage of hydroperoxide. Note that the reaction chains 'breed'. If all the hydroperoxide formed undergoes cleavage across the O–O bond, then in going through reactions (2), (3), and (7), one free radical gives rise to three free radicals; this is called *chain branching*. Because it occurs these oxidations are strongly autocatalytic.

We are still left with the question what causes reaction (1) right at the start of the oxidation, before any hydroperoxide has been formed, and this is a question that has not yet been answered in a completely satisfactory manner. One suggestion is that abstraction of hydrogen by molecular oxygen occurs:

$$\text{>C—H} + O_2 \longrightarrow \text{>C·} + HO_2·$$

Once the oxidation is under way, this reaction is swamped by reactions (8) and (9).

In industrial processes the reactor system is normally arranged so that there is always some hydroperoxide present, so that the reaction proceeds at a satisfactory rate throughout. Stirred flow reactors are often used for these reactions.

Although many free-radical liquid-phase oxidations will proceed in the absence of catalysts, processes generally involve the use of catalysts, usually salts of cobalt or manganese which are soluble in the reaction medium, e.g. cobalt naphthenate.* These catalysts act by increasing the rate of cleavage of hydroperoxide:

$$\geqslant COOH + Co^{2+} \longrightarrow \geqslant CO\cdot + OH^- + Co^{3+}$$

$$\geqslant COOH + Co^{3+} \longrightarrow \geqslant COO\cdot + H^+ + Co^{2+}$$

They may also, in some cases, take part in initiation reactions:

$$\geqslant C{-}H + Co^{3+} \longrightarrow \geqslant C\cdot + Co^{2+} + H^+$$

Let us now consider a number of important processes involving this type of oxidation.

Oxidation of Cumene

This is the simplest possible case, the desired product being the hydroperoxide, which is required as an intermediate in the cumene process for the manufacture of phenol. This process follows the overall route:

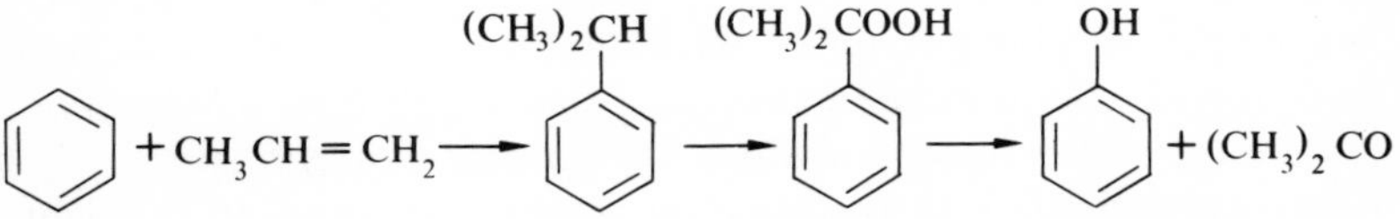

| cumene | cumene hydroperoxide | phenol | acetone |

Manufacture of cumene, by a Friedel–Crafts reaction between propene and benzene, is discussed in Chapter 5.

Of the twelve hydrogen atoms in cumene, that on the carbon atom attached to the benzene ring is by far the most readily removed. Its

---

* Napthenic acids are carboxylic acids of somewhat complex structures, which are isolated from petroleum. Their heavy metal salts have high solubilities in organic materials.

abstraction gives rise to a tertiary, resonance-stabilised radical, and can be assumed to involve a lower-energy transition state than removal of any of the other hydrogens:

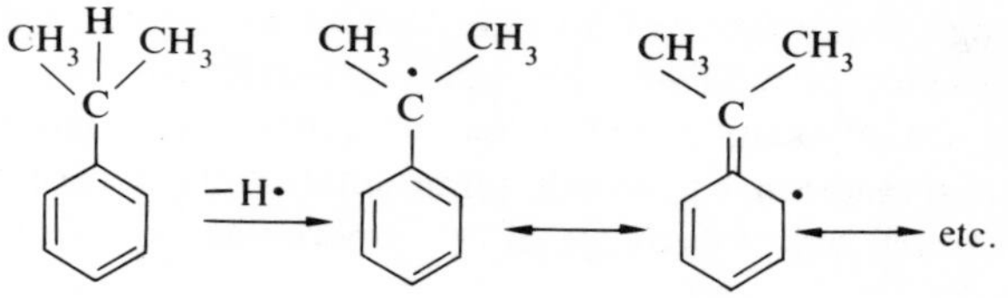

Under the conditions normally used, only this hydrogen is attacked. Addition of oxygen to the cumyl radical thus formed, followed by abstraction of a hydrogen atom from a further molecule of cumene leads to the formation of cumene hydroperoxide. Thus, the first three steps of the reaction are the analogues of reactions (1), (2), and (3) in the generalised scheme:

$$\underset{CH_3}{\overset{CH_3}{PhCH}} + X\cdot \longrightarrow \underset{CH_3}{\overset{CH_3}{PhC\cdot}} + XH$$

$$\underset{CH_3}{\overset{CH_3}{PhC\cdot}} + O_2 \longrightarrow \underset{CH_3}{\overset{CH_3}{PhCOO\cdot}}$$

$$\underset{CH_3}{\overset{CH_3}{PhCOO\cdot}} + \underset{CH_3}{\overset{CH_3}{PhCH}} \longrightarrow \underset{CH_3}{\overset{CH_3}{PhCOOH}} + \underset{CH_3}{\overset{CH_3}{PhC\cdot}}$$

cumene hydroperoxide

In so far as the hydroperoxide is the desired product, it is desirable to avoid its cleavage. However, as we have seen, the autocatalytic nature of free-radical liquid-phase oxidations is due to chain branching produced by cleavage of hydroperoxide, and in order that an acceptable rate of reaction be achieved, it is necessary that some cleavage of hydroperoxide should occur. Thus:

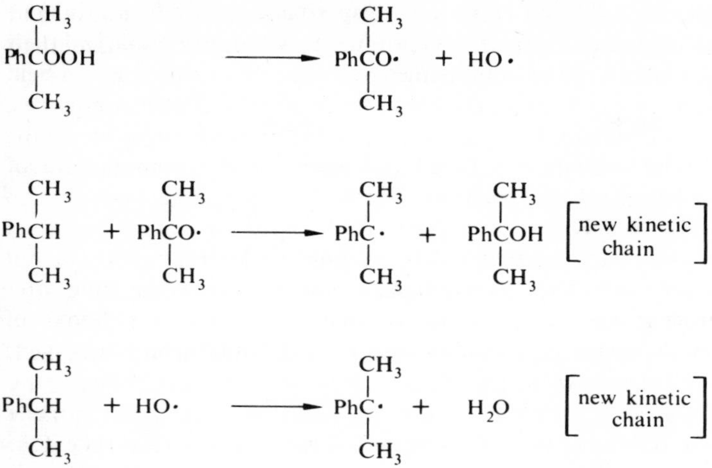

$$\underset{\overset{|}{CH_3}}{\overset{CH_3}{\overset{|}{PhCOOH}}} \longrightarrow \underset{\overset{|}{CH_3}}{\overset{CH_3}{\overset{|}{PhCO\cdot}}} + HO\cdot$$

$$\underset{\overset{|}{CH_3}}{\overset{CH_3}{\overset{|}{PhCH}}} + \underset{\overset{|}{CH_3}}{\overset{CH_3}{\overset{|}{PhCO\cdot}}} \longrightarrow \underset{\overset{|}{CH_3}}{\overset{CH_3}{\overset{|}{PhC\cdot}}} + \underset{\overset{|}{CH_3}}{\overset{CH_3}{\overset{|}{PhCOH}}} \quad \left[\begin{array}{c}\text{new kinetic}\\\text{chain}\end{array}\right] \quad (10)$$

$$\underset{\overset{|}{CH_3}}{\overset{CH_3}{\overset{|}{PhCH}}} + HO\cdot \longrightarrow \underset{\overset{|}{CH_3}}{\overset{CH_3}{\overset{|}{PhC\cdot}}} + H_2O \quad \left[\begin{array}{c}\text{new kinetic}\\\text{chain}\end{array}\right]$$

Very careful adjustment of reaction conditions is necessary to achieve adequate reaction rates without excessive loss of yield. In fact, yields of over 90% at up to about 30% conversion are obtainable.

As we would expect from reaction (10), 2-phenylpropan-2-ol, $PhC(CH_3)_2OH$, is one of the by-products of this stage of the process. Acetophenone, another by-product, is formed by *beta*-cleavage of cumoxy radicals:

$$\underset{\overset{|}{CH_3}}{\overset{CH_3}{\overset{|}{PhC\!-\!O\cdot}}} \longrightarrow \underset{}{\overset{CH_3}{\overset{|}{PhC}}}=O + \cdot CH_3 \quad (11)$$

The methyl radicals formed in the above reaction undergo oxidation leading to the production of methyl hydroperoxide, methanol, formaldehyde, and formic acid. (The types of reaction involved in the formation of these products are discussed later in this chapter.)

Most of the unconverted cumene is removed from the oxidation product by vacuum distillation, and is recycled to the oxidation reactor.

The next stage in the process, the cleavage of the hydroperoxide to phenol and acetone, is not an oxidation, but we will consider it for the sake of completeness, and because it was the discovery of this reaction which led to the development of the cumene process. The reaction was discovered by two German chemists, Hock and Lang, in 1944. They treated cumene hydroperoxide with dilute sulphuric acid and found that phenol was produced. When they published their results, chemists in two companies recognised that this reaction held the promise of a very attractive route to phenol. These companies, Distillers Company Ltd., in the UK, and Hercules Powder Inc. in the USA independently developed processes for the manufacture of phenol based on this reaction. The first plant was in operation in 1952.

The cleavage is carried out by adding the hydroperoxide to hot dilute sulphuric acid. Its mechanism, investigated some time after the process was in operation, is widely quoted in text-books of organic chemistry as an example of a 1,2 shift from carbon to oxygen:

$$PhC(CH_3)_2-O-OH \underset{-H^+}{\overset{+H^+}{\rightleftharpoons}} PhC(CH_3)_2-O-OH_2^+ \overset{-H_2O}{\longrightarrow} PhC(CH_3)_2-O^+ \overset{1,2\ shift}{\longrightarrow} PhO-C^+(CH_3)_2$$

$$\overset{H_2O}{\longrightarrow} PhO-C(CH_3)_2-OH_2^+ \underset{+H^+}{\overset{-H^+}{\rightleftharpoons}} PhO-C(CH_3)_2-OH \overset{H^+ catalysed}{\longrightarrow} PhOH + (CH_3)_2C=O$$

Note that it is highly undesirable that this type of cleavage should occur during the oxidation stage because phenols are potent inhibitors of free-radical oxidations (see page 294). It is therefore essential that the presence of acidic materials be avoided at the oxidation stage. In the process developed by the Distillers Company, the cumene is oxidised as an emulsion in dilute aqueous sodium carbonate, which neutralises any acidic by-products, e.g. formic acid, from further oxidation of methyl radicals formed in reaction (11).

The product from the cleavage stage contains phenol, acetone, cumene, acetophenone, 2-phenylpropan-2-ol, α-methylstyrene, formed by acid-catalysed dehydration of the 2-phenylpropan-2-ol, and various other by-products. Phenol and acetone are separated by distillation.

The process gives a yield of phenol of about 93 % based on cumene, and about 84 % based on benzene. 0·6 ton of acetone is produced for every ton of phenol made, so that the economics of the process are heavily dependent on the price obtainable for the acetone. At the present time, demand for acetone appears to be growing faster than demand for phenol, and there are no major problems in its disposal. The process is the favoured process for the manufacture of phenol. In the UK it is operated by ICI and BP in plants of capacities of 80 000 and 63 000 tons per annum respectively, and accounts for all manufacture of synthetic phenol.

Phenol has a variety of uses, and their relative importance varies from country to country. In the UK and the USA the major outlet is in the manufacture of phenol-formaldehyde resins (see Chapter 7), which accounts for over 50% of consumption in the USA. Some of the more important other uses are in the manufacture of bisphenol A (see page 174), alkylphenols (see page 169), and as a source of cyclohexanol for use in the manufacture of caprolactam and adipic acid. In this latter use it is being displaced by processes involving the oxidation of cyclohexane (see later in this chapter).

Oxidation of Acetaldehyde

Aldehydes are very readily oxidised, attack being generally exclusively on the aldehydic hydrogen under conditions normally used. The oxidation of acetaldehyde is by far the most important such reaction, and provides one of the major routes to acetic acid and acetic anhydride. It has been used for acetic acid production since 1911.

The reaction follows the generalised mechanism up to the formation of the hydroperoxide, peracetic acid:

$$\underset{}{CH_3}\overset{O}{\overset{\|}{C}}H + X\cdot \longrightarrow \underset{}{CH_3}\overset{O}{\overset{\|}{C}}\cdot + XH$$

$$CH_3\overset{O}{\overset{\|}{C}}\cdot + O_2 \longrightarrow CH_3\overset{O}{\overset{\|}{C}}OO\cdot$$

$$CH_3\overset{O}{\overset{\|}{C}}H + CH_3\overset{O}{\overset{\|}{C}}OO\cdot \longrightarrow CH_3\overset{O}{\overset{\|}{C}}O\cdot + CH_3\overset{O}{\overset{\|}{C}}OOH$$
$$\phantom{CH_3\overset{O}{\overset{\|}{C}}O\cdot + CH_3\overset{O}{\overset{\|}{C}}OOH aaaaaaaaaaaaaa} \textit{peracetic}$$
$$\phantom{CH_3\overset{O}{\overset{\|}{C}}O\cdot + CH_3\overset{O}{\overset{\|}{C}}OOH aaaaaaaaaaaaaaaaa} \textit{acid}$$

Like other hydroperoxides, peracetic acid cleaves readily across the oxygen–oxygen bond to give fragments that can participate in initiation reactions:

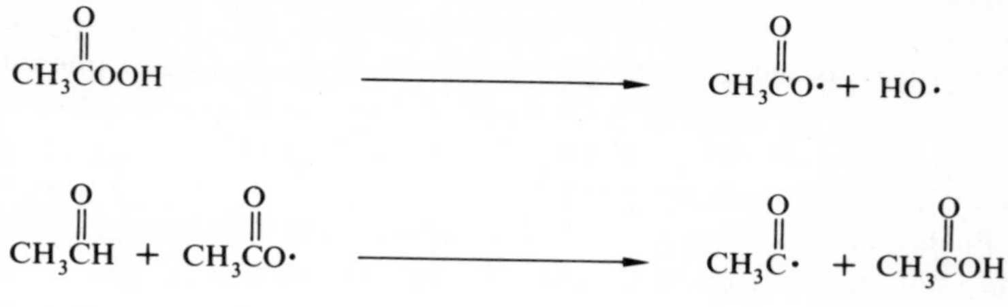

$$CH_3\overset{O}{\overset{\|}{C}}OOH \longrightarrow CH_3\overset{O}{\overset{\|}{C}}O\cdot + HO\cdot$$

$$CH_3\overset{O}{\overset{\|}{C}}H + CH_3\overset{O}{\overset{\|}{C}}O\cdot \longrightarrow CH_3\overset{O}{\overset{\|}{C}}\cdot + CH_3\overset{O}{\overset{\|}{C}}OH$$

However, it is thought that this is not the main route by which acetic acid is formed, but rather that the peracetic acid reacts with acetaldehyde to give α-hydroxyethyl peracetate, which then decomposes by a non-free-radical mechanism to acetic acid. The mechanisms involved are presumably as shown below:

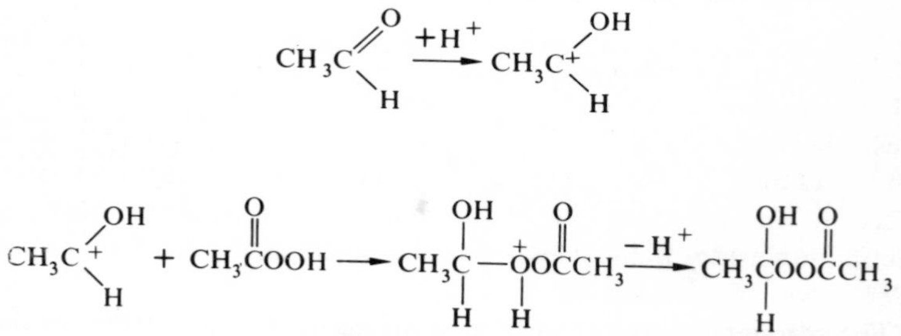

$$CH_3C\overset{O}{\underset{H}{\diagdown}} \xrightarrow{+H^+} CH_3C\overset{OH}{\underset{H}{\diagup}}^+$$

$$CH_3C\overset{OH}{\underset{H}{\diagup}}^+ + CH_3\overset{O}{\overset{\|}{C}}OOH \longrightarrow CH_3\underset{\underset{H}{|}}{\overset{OH}{\overset{|}{C}}}-\overset{+}{O}O\overset{O}{\overset{\|}{C}}CH_3 \xrightarrow{-H^+} CH_3\underset{\underset{H}{|}}{\overset{OH}{\overset{|}{C}}}OO\overset{O}{\overset{\|}{C}}CH_3$$

α−hydroxyethyl
peracetate

$$\underset{HO}{\overset{CH_3}{\diagdown}}\overset{H}{\underset{\diagup}{C}} \quad \overset{O}{\underset{O-O}{\diagdown}}C-CH_3 \longrightarrow 2CH_3CO_2H$$

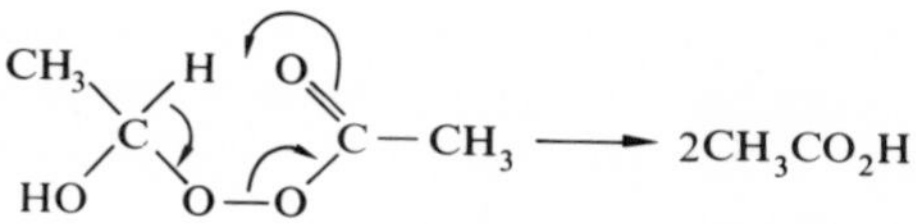

The oxidation is normally carried out with air, at temperatures in the range 330 to 355 K, manganese or cobalt acetate being used as the catalyst. Yields of about 95% are obtained, the main loss being through *beta*-cleavage of acetoxy radicals:

$$CH_3\overset{O}{\overset{\|}{C}}O\cdot \longrightarrow CO_2 + CH_3\cdot \longrightarrow CO_2 + H_2O + \text{other products}$$

If the oxidation is carried out in the presence of a mixture of cobalt and copper acetates, a mixture of acetic acid and acetic aynhdride is obtained:

$$CH_3\overset{\overset{\displaystyle O}{\|}}{C}\cdot\ +\ Cu^{2+}\ \longrightarrow\ CH_3\overset{\overset{\displaystyle O}{\|}}{C}^{+}\ +\ Cu^{+}$$

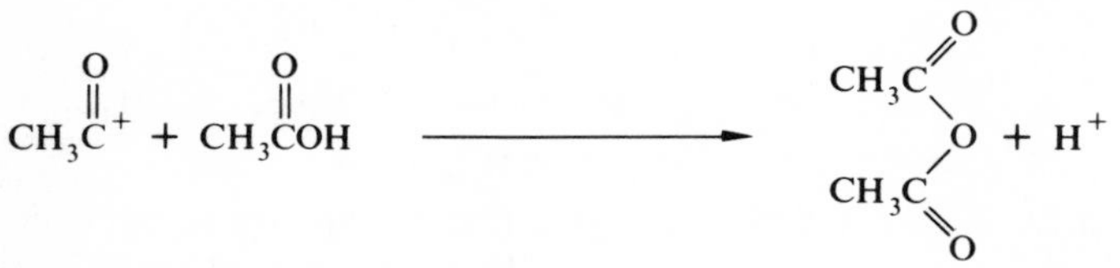

$$CH_3\overset{\overset{\displaystyle O}{\|}}{C}^{+}\ +\ CH_3\overset{\overset{\displaystyle O}{\|}}{C}OH\ \longrightarrow\ \begin{array}{c} CH_3C\overset{\displaystyle O}{\diagup}\\ \diagdown O\\ CH_3C\diagup \\ \overset{\diagdown}{O} \end{array}\ +\ H^{+}$$

$$CH_3\overset{\overset{\displaystyle O}{\|}}{C}OOH\ +\ Cu^{+}\ \longrightarrow\ CH_3\overset{\overset{\displaystyle O}{\|}}{C}O\cdot\ +\ Cu^{2+}\ +\ OH^{-}$$

This is the basis for one of the commercial processes for the manufacture of acetic anhydride.

As will be seen later in this chapter, these acetaldehyde-based routes to acetic acid and acetic anhydride are now under competition from direct routes from petroleum fractions.

Oxidation of Cyclohexane

The hydrogens of cyclohexane are much more difficult to remove than the aldehydic hydrogen of acetaldehyde, or the tertiary hydrogen of cumene, and more severe conditions are required for its oxidation than in the cases of these compounds. Partly as a consequence of this and the fact that the products are more readily oxidised than the cyclohexane itself, the chemistry involved is much more complex than in the two previous cases.

The oxidation of cyclohexane can give rise to a large number of products. It is carried out commercially on a very large scale for the manufacture of cyclohexanol/cyclohexanone mixtures ('mixed oils'), which are intermediates in nylon manufacture. The reaction follows the typical steps up to the formation of the hydroperoxide. (Note: for the sake of simplicity, only those hydrogen atoms which are involved in the reaction are shown in the following reaction schemes.)

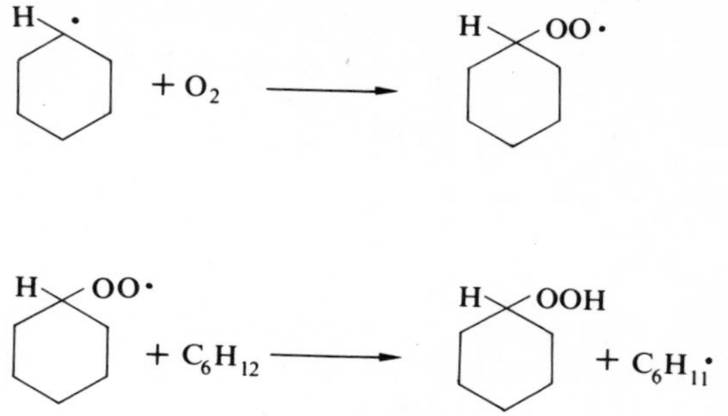

It is possible to account for the formation of cyclohexanol and cyclohexanone by reactions similar to those we have considered up to now:

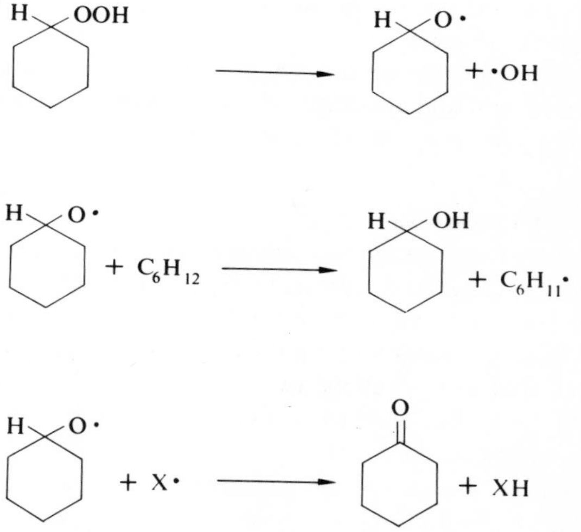

In reality, there are probably also a number of other routes by which the hydroperoxide gives cyclohexanol and cyclohexanone.

A certain amount of *beta*-cleavage of cyclohexyloxy radicals occurs:

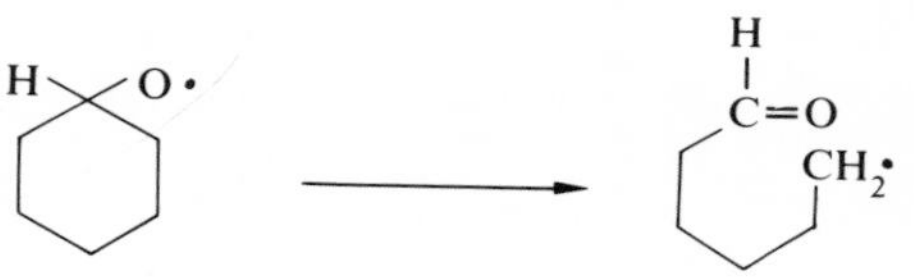

This leads to the formation of open-chain products, e.g. caproic acid. Thus:

$$\underset{\text{(CH}_2)_4}{\overset{O}{\overset{\|}{HC}}}\text{(CH}_2)_4\text{CH}_2\text{·} \xrightarrow{\text{RH}} \overset{O}{\overset{\|}{HC}}\text{(CH}_2)_4\text{CH}_3 \xrightarrow{\text{oxidation}} \underset{\textit{caproic acid}}{HO_2C\,(CH_2)_4CH_3}$$

FURTHER REACTIONS OF CYCLOHEXANOL AND CYCLOHEXANONE
Both cylohexanol and cyclohexanone are attacked more readily than cyclohexane and undergo further reaction in the system. Cyclohexanol probably gives initially hydroxycyclohexyl hydroperoxide:

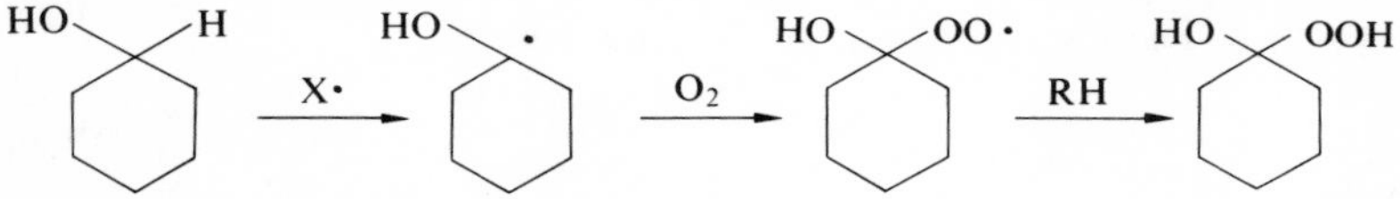

This undergoes further reaction to give cyclohexanone, and, to a minor extent, open-chain compounds, including e.g. adipic acid and caproic acid:

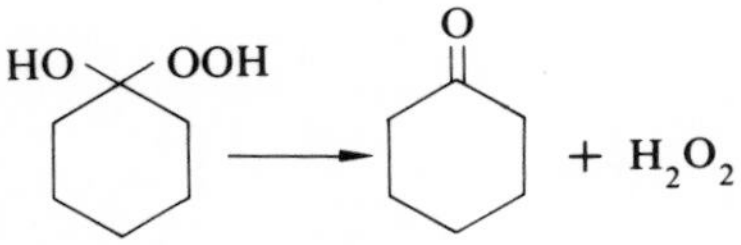

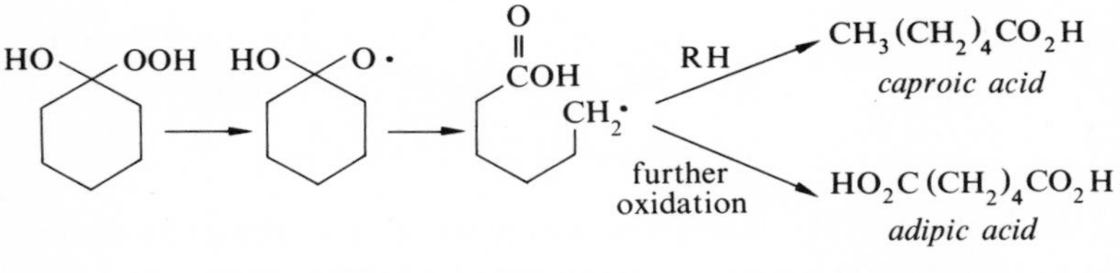

Heterolytic reaction of the hydroperoxide to give open-chain structures can also occur:

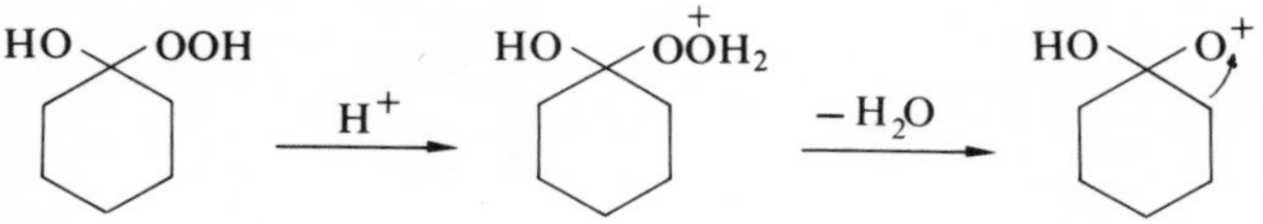

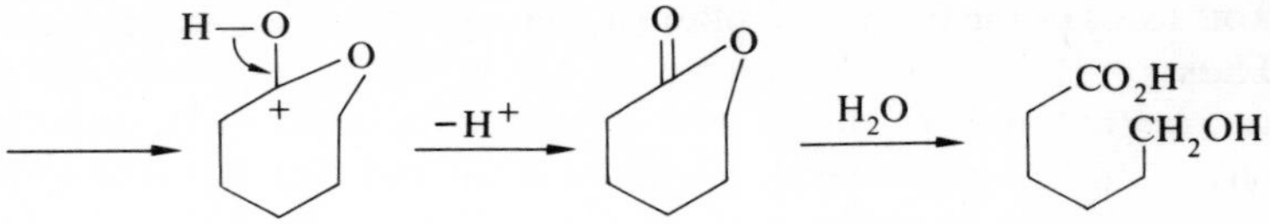

Further oxidation of cyclohexanone also gives a hydroperoxide:

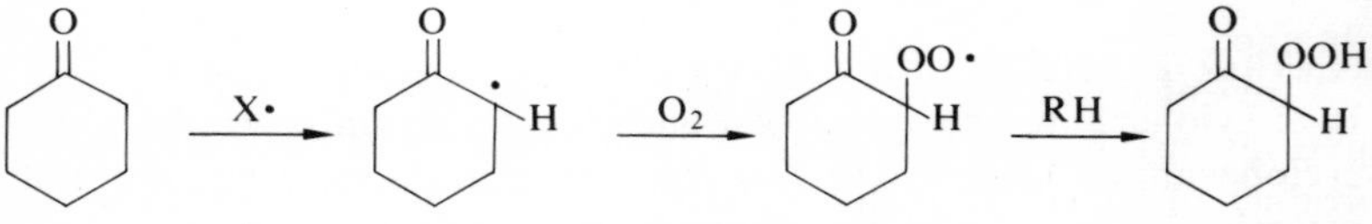

This undergoes a number of ionic, free-radical, and molecular reactions leading to opening of the ring, e.g.:

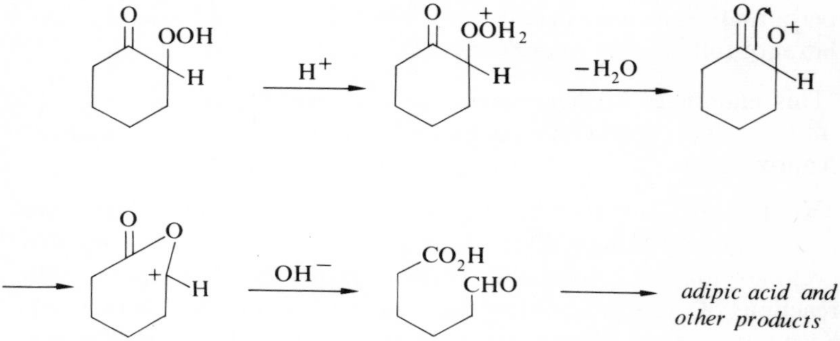

The overall reactions occurring in cyclohexane oxidation are summarised below:

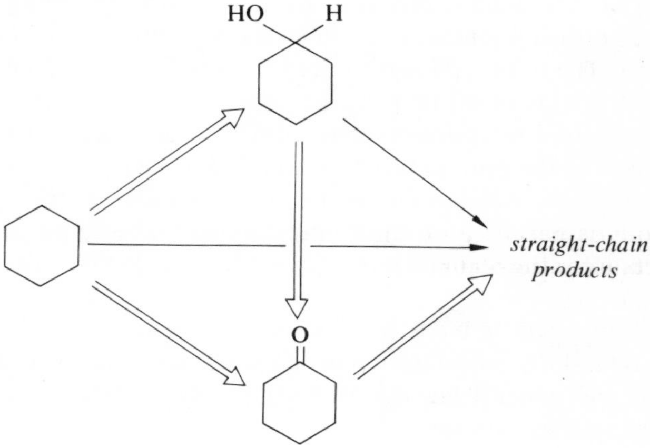

PROCESS OPERATION

Oxidation is carried out with air, at 420 to 430 K and under 8 to 9 atm, in the presence of a cobalt naphthenate catalyst. In order to achieve an acceptable yield of cyclohexanol and cyclohexanone, the reaction has to be operated at low conversions (*ca.* 10%), so that further oxidation of these products is minimised. Yields of about 75% cyclohexanol+cyclohexanone in a ratio of 1 to 2:1 are obtained.

A recent development which leads to substantially higher yields involves carrying out the oxidation in the presence of boric acid. The product obtained from the oxidation in this case is a mixture of cyclohexyl borate with some cyclohexanone and free cyclohexanol. The cyclohexyl borate is thought to be produced by reaction between cyclohexyl hydroperoxide and boric acid, possibly through cyclohexyl perborate as an intermediate, and not by the formation of cyclo-hexanol followed by esterification:

$$C_6H_{11}OOH \xrightarrow{\quad H_3BO_3 \quad} (C_6H_{11}O)_3B$$

cyclohexyl borate

Two factors contribute to the improvement in yield. Firstly, the main reaction path is diverted to a route which does not involve cyclohexyloxy radicals as intermediates so that the occurrence of side reactions of these radicals leading to open-chain products is reduced. Secondly, the cyclohexyl borate is stable to oxidation and is not further attacked in the system.

The reaction product is treated with water to hydrolyse the cyclo-hexyl borate, the organic products are separated from the aqueous material and the cyclohexanol/cyclohexanone mixture is isolated by distillation. The boric acid is recovered from the aqueous phase and is recycled. Yields of 90 to 95% are obtained at 10% conversion. The cyclohexanol to cyclohexanone ratio is in the range 5 to 10:1. This version of the process has been widely adopted.

Production of cyclohexanol/cyclohexanone mixtures is usually carried out as part of an overall operation to make adipic acid or caprolactam for the manufacture of nylon 66 or nylon 6 respectively (see Chapter 7).

Most adipic acid is made by oxidising the mixed oil with nitric acid of about 40% concentration at temperatures ranging from 320 to 420 K, and under a few atmospheres pressure, with copper and vanadium salts as catalysts:

$$C_6H_{11}OH + C_6H_{10}O \xrightarrow{\;HNO_3\;} HO_2C(CH_2)_4CO_2H$$

Yields of 90% or more are obtained.

This process consumes substantial amounts of nitric acid, and it would obviously be desirable to carry out the whole oxidation with air. As we have seen in the preceding discussion, adipic acid or its precursors can arise in various reaction chains that occur during the oxidation of cyclohexane, cyclohexanol, and cyclohexanone, and there are processes for the one- and two-stage air oxidation of cyclohexane to adipic acid. However, because of the large number of reactions which can occur in the system, and because adipic acid is subject to further oxidation, a large number of by-products are formed and it is difficult to obtain good yields. Such processes appear to be of relatively minor importance at present.

Caprolactam is made by a variety of routes, one of which is outlined below:

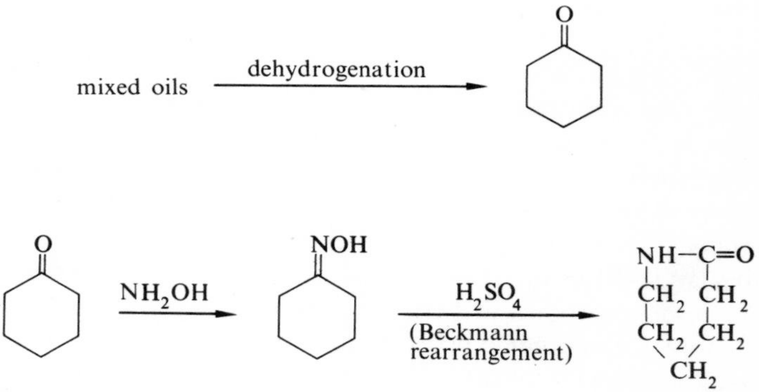

mixed oils $\xrightarrow{\text{dehydrogenation}}$

$\xrightarrow{NH_2OH}$ $\xrightarrow[\substack{(Beckmann \\ rearrangement)}]{H_2SO_4}$

$$\begin{array}{c} NH-C=O \\ | \quad\;\; | \\ CH_2 \; CH_2 \\ | \quad\;\; | \\ CH_2 \; CH_2 \\ \backslash \;/ \\ CH_2 \end{array}$$

caprolactam

Adipic acid and caprolactam manufacture are sometimes based on phenol rather than on cyclohexane, the phenol being hydrogenated to give cyclohexanone and/or cyclohexanol. This type of route, which is not used in the UK, has probably now been rendered unattractive for new manufacture by the developments in cyclohexane oxidation technology.

Oxidation of Alkanes

Currently, the most important commercial application of the free-radical liquid-phase oxidation of alkanes is in the manufacture of acetic acid. In the USA substantial quantities of acetic acid are made by liquid-phase oxidation of n-butane:

$$C_4H_{10} \xrightarrow{\text{air, Co}^{2+}, 420\text{-}520 \text{ K, 50 atm}} CH_3CO_2H \text{ (ca. 50\% theory)}$$
$$+ \text{other organic products}$$
$$+ CO_2 + H_2O + CO$$

There is a very large number of reactions which can occur in this system and it would not be appropriate to consider them all here. Let us just look at one of the ways in which the generalised liquid-phase oxidation mechanism can account for the formation of acetic acid:

$$CH_3CH_2CH_2CH_3 + X\cdot \longrightarrow CH_3CH_2\overset{\cdot}{C}HCH_3 + XH$$

$$CH_3CH_2\overset{\cdot}{C}HCH_3 + O_2 \longrightarrow CH_3CH_2\underset{\underset{OO\cdot}{|}}{C}HCH_3$$

$$CH_3CH_2\underset{\underset{OO\cdot}{|}}{C}HCH_3 + C_4H_{10} \longrightarrow CH_3CH_2\underset{\underset{OOH}{|}}{C}HCH_3 + C_4H_9\cdot$$

$$CH_3CH_2\underset{\underset{OOH}{|}}{C}HCH_3 + Co^{2+} \longrightarrow CH_3CH_2\underset{\underset{O\cdot}{|}}{C}HCH_3 + OH^- + Co^{3+}$$

$$CH_3CH_2\underset{\underset{O\cdot}{|}}{C}HCH_3 \longrightarrow CH_3CH_2\cdot + CH_3CHO$$

As we have already seen, acetaldehyde is readily oxidised to acetic acid:

$$CH_3CHO \xrightarrow{\text{further oxidation}} CH_3CO_2H$$

Ethyl radicals can undergo a number of further reactions, of which a series which leads to the formation of acetic acid is shown below:

$$CH_3CH_2\cdot + O_2 \longrightarrow CH_3CH_2OO\cdot$$

$$CH_3CH_2OO\cdot + RH \longrightarrow CH_3CH_2OOH + R\cdot$$

$$CH_3CH_2OOH \ + \ Co^{2+} \ \longrightarrow \ CH_3CH_2O\cdot \ + \ OH^- \ + \ Co^{3+}$$

$$CH_3CH_2O\cdot \ + \ X\cdot \ \longrightarrow \ CH_3CHO \ + \ XH$$

$$CH_3CHO \ \xrightarrow{\text{further oxidation}} \ CH_3CO_2H$$

We have considered only one particular set of reactions which can follow the initial abstraction of a methylene hydrogen to give a secondary radical. Such a radical can participate in a number of other reaction chains. Further, initial attack involving abstraction of a methyl hydrogen will occur. In view of the complexity of the situation, it seems somewhat surprising that fairly good yields of acetic acid can be obtained. This is made possible by the relative stability of acetic acid to oxidation, the process being run under conditions such that intermediate products such as methyl ethyl ketone and propionic acid are mainly oxidised further.

It has already been pointed out that in Europe supplies of butane are not so plentiful as in the USA, and that in general naphtha is a preferred feedstock for chemical manufacture. In the UK, acetic acid is made by the liquid-phase air-oxidation of light naphtha. In this case, added to the complexities we have seen above is the added one that the feedstock is itself a complex mixture. Lower yields of acetic acid are obtained than from butane, and large amounts of by-products are formed, the main ones being acetone, formic acid, and propionic acid (about 0·33, 0·30, and 0·12 tons per ton of acetic acid respectively). The process is operated by BP, at Hull, at a scale of 90 000 tons per annum, and provides the major source of acetic acid in the UK.

The main use of acetic acid is in the manufacture of acetates of which cellulose acetate (see page 200) and vinyl acetate (see page 101) are the most important. For cellulose acetate manufacture, the acid has to be converted into acetic anhydride. This is brought about by coverting it to ketene and reacting this with further acetic acid:

$$CH_3CO_2H \ \xrightarrow[\substack{\text{triethyl phosphate} \\ \text{(catalyst)}}]{970-1070 \text{ K}} \ \underset{\textit{ketene}}{CH_2{=}C{=}O} \ + \ H_2O$$

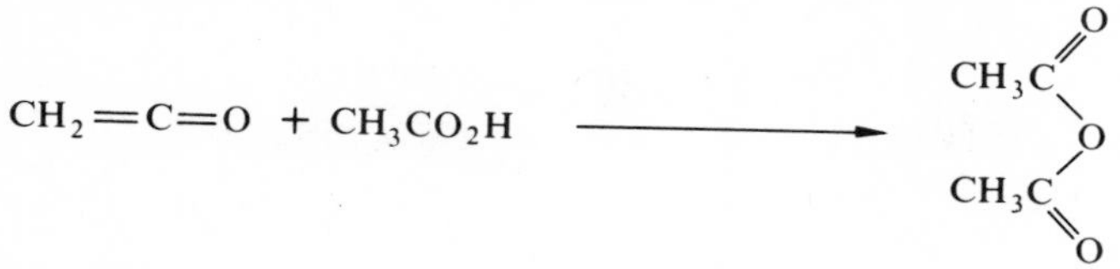

$$CH_2 = C = O \ + \ CH_3CO_2H \longrightarrow$$

$$\begin{array}{c} CH_3C \overset{\displaystyle O}{\underset{\displaystyle O}{\diagdown}} \\ CH_3C \overset{\displaystyle O}{\underset{\displaystyle O}{\diagup}} \end{array}$$

Oxidation of long straight chain alkanes, e.g. from paraffin wax, can be made to give moderate yields of long chain fatty acids, but this method of manufacture does not appear to be competitive with manufacture from fats and oils at present.

Oxidation of isobutane to t-butyl hydroperoxide may become of considerable importance as a stage in a recently developed process for propylene oxide manufacture (see 'Propylene Oxide').

Oxidation of Toluene and Xylene
Toluene can be oxidised by air, in the liquid phase, to benzoic acid:

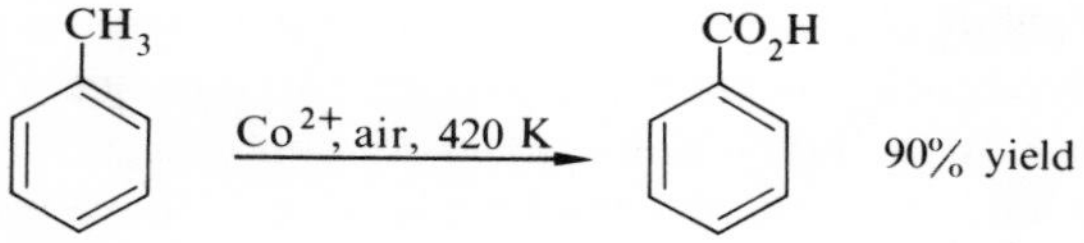

The probable mechanism can be readily worked out from what we have seen previously.

This is now the preferred method of manufacture of benzoic acid.

Oxidation of p-xylene provides an important route to terephthalic acid, which is required in large quantities for the manufacture of poly(ethylene terephthalate), an important fibre-forming polymer (see Chapters 7 and 8). When commercial manufacture of this acid first started, it was found difficult to carry out the oxidation with air, in that although one methyl group was readily oxidised, the p-toluic acid formed as an intermediate was very resistant to oxidation. This has been attributed to the electron-withdrawing effect of the carboxyl group, it being suggested that this reduces the ease of abstraction of a hydrogen atom from the methyl group. As a consequence, in the early years of terephthalic acid manufacture, various routes other than direct air oxidation had to be used, e.g. oxidation with nitric acid:

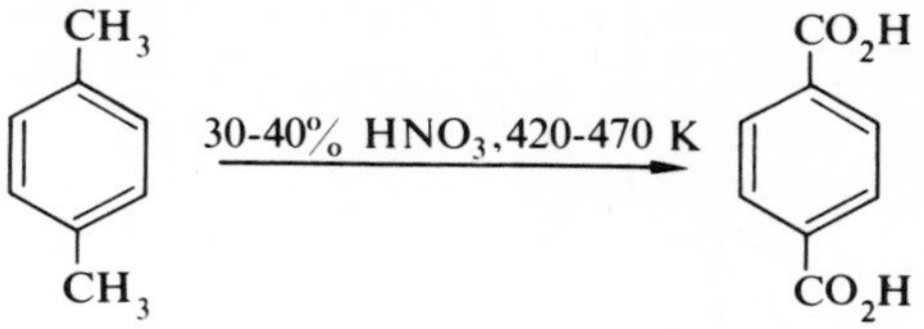

$$\text{CH}_3\text{-C}_6\text{H}_4\text{-CH}_3 \xrightarrow{\text{30-40\% HNO}_3, \text{420-470 K}} \text{HO}_2\text{C-C}_6\text{H}_4\text{-CO}_2\text{H}$$

More recently a number of processes have been developed by which direct air oxidation can be carried out. Of these, the most important at present is a process in which the oxidation is carried out in acetic acid solution at about 470 K and 15 atm, in the presence of manganese and cobalt bromides. High yields (*ca.* 95% theory) are obtained.

It is thought that bromine atoms are generated in the system and act as efficient hydrogen abstraction agents, and that propagation steps of the type indicated below are involved (the HBr arises from the interaction of bromide ions with acetic acid):

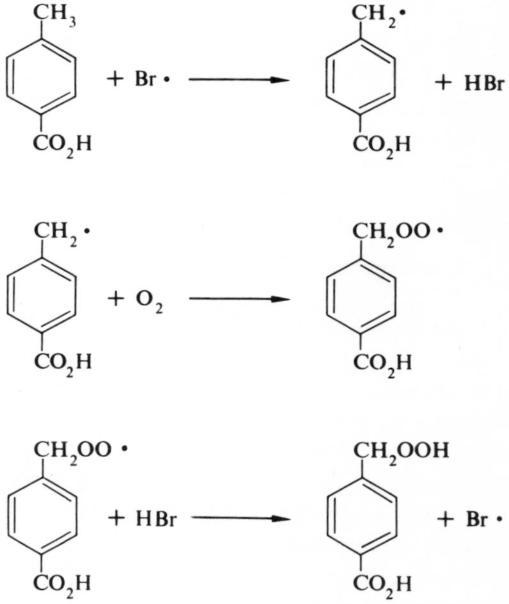

$$\text{CH}_3\text{-C}_6\text{H}_4\text{-CO}_2\text{H} + \text{Br}\cdot \longrightarrow \text{CH}_2\cdot\text{-C}_6\text{H}_4\text{-CO}_2\text{H} + \text{HBr}$$

$$\text{CH}_2\cdot\text{-C}_6\text{H}_4\text{-CO}_2\text{H} + \text{O}_2 \longrightarrow \text{CH}_2\text{OO}\cdot\text{-C}_6\text{H}_4\text{-CO}_2\text{H}$$

$$\text{CH}_2\text{OO}\cdot\text{-C}_6\text{H}_4\text{-CO}_2\text{H} + \text{HBr} \longrightarrow \text{CH}_2\text{OOH-C}_6\text{H}_4\text{-CO}_2\text{H} + \text{Br}\cdot$$

The hydroperoxide reacts further to give terephthalaldehyde, which is then oxidised to terephthalic acid.

This method can be used for the oxidation of other aromatic hydrocarbons to acids.

Oxidation of Isopropanol

This reaction does not follow the generalised mechanism, but proceeds as shown below:

$$\underset{CH_3}{\overset{CH_3}{>}}CHOH \;+\; X\cdot \quad\longrightarrow\quad \underset{CH_3}{\overset{CH_3}{>}}\overset{\displaystyle\cdot}{C}OH \;+\; XH$$

$$\underset{CH_3}{\overset{CH_3}{>}}\overset{\displaystyle\cdot}{C}OH \;+\; O_2 \quad\longrightarrow\quad \underset{CH_3}{\overset{CH_3}{>}}C{=}O \;+\; HOO\cdot$$

$$\underset{CH_3}{\overset{CH_3}{>}}CHOH \;+\; HOO\cdot \quad\longrightarrow\quad \underset{CH_3}{\overset{CH_3}{>}}\overset{\displaystyle\cdot}{C}OH \;+\; H_2O_2$$

In this case, oxygen abstracts a hydrogen atom from the substrate radical rather than adds to it.

This forms the basis of a process for the manufacture of hydrogen peroxide and acetone:

$$(CH_3)_2CHOH \quad\xrightarrow[\text{13-20 atm}]{\text{air, 260-400 K}}\quad CH_3COCH_3 \;+\; H_2O_2$$

The yield of acetone is about 93% of theory, and that of hydrogen peroxide is about 87% of theory. The process is operated by Shell in the USA.

Propylene Oxide

At the present time most propylene oxide is made by the chlorohydrin route (see page 135). This is very wasteful of chlorine, and efforts have been going on for a number of years to find a 'direct oxidation' route. Whereas ethylene oxide can be made by oxidation of ethylene in the gas phase over a silver catalyst (see page 106), this method is not effective for the production of propylene oxide.

In the late 1960s Halcon International announced the development of a number of processes in which propylene oxide is made by the reaction of a hydroperoxide, produced by liquid-phase air oxidation of a hydrocarbon, with propene. The two main variants of this class of process involve the use of t-butyl hydroperoxide and ethylbenzene hydroperoxide:

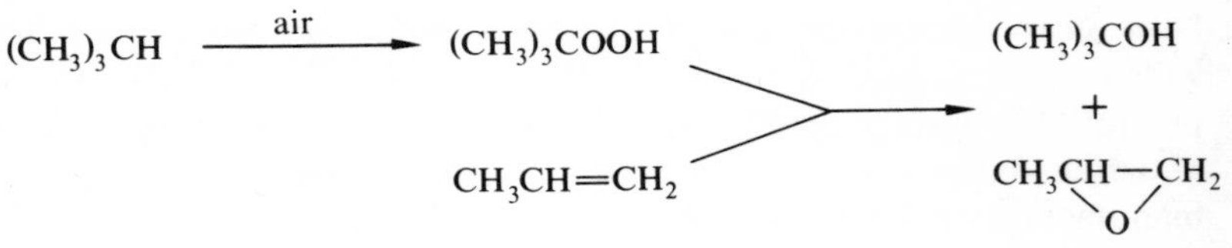

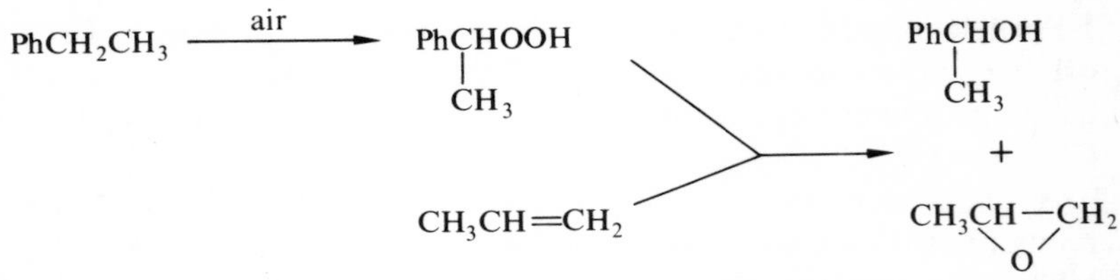

$(CH_3)_3CH \xrightarrow{\text{air}} (CH_3)_3COOH$ $\qquad (CH_3)_3COH$
$+$
$CH_3CH{=}CH_2$ $\qquad CH_3\overset{\displaystyle CH}{\underset{O}{\diagdown\diagup}}CH_2$

$PhCH_2CH_3 \xrightarrow{\text{air}} PhCHOOH$ $\qquad PhCHOH$
$\qquad\qquad\qquad\qquad\quad CH_3$ $\qquad\qquad\qquad CH_3$
$+$
$CH_3CH{=}CH_2$ $\qquad CH_3\overset{\displaystyle CH}{\underset{O}{\diagdown\diagup}}CH_2$

The epoxidation is carried out at temperatures in the range 350 to 450 K and pressures from 17 to 67 atm, in the presence of a soluble molybdenum catalyst. Present indications are that the catalyst forms a polarised complex with the hydroperoxide, and that electron-deficient oxygen in the complex attacks the π-bond of the propene:

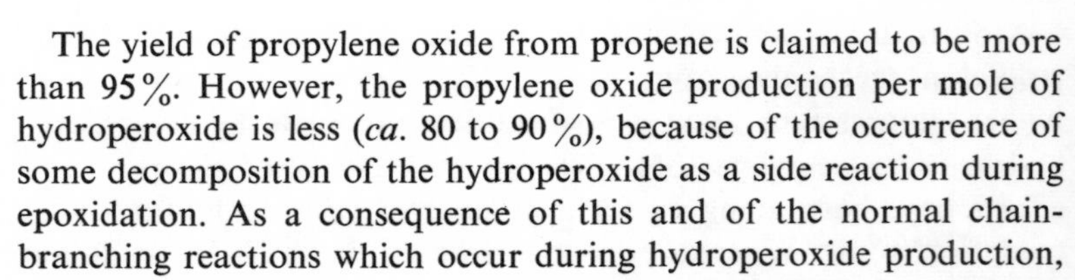

$$ROOH + CH_2{=}CHCH_3 \longrightarrow \left[\begin{array}{c} H_2C-CHCH_3 + RO^- \\ \diagdown O^+ \diagup \\ | \\ H \end{array} \;\; Mo^{n+} cat. \right] \longrightarrow \begin{array}{c} H_2C-CHCH_3 + ROH \\ \diagdown O \diagup \end{array} \;\; Mo^{n+} cat.$$

Mo^{n+} cat.

$$ROH + ROOH \longrightarrow ROH + ROOH$$
Mo^{n+} cat. $\qquad\qquad Mo^{n+}$ cat.

The yield of propylene oxide from propene is claimed to be more than 95%. However, the propylene oxide production per mole of hydroperoxide is less (*ca.* 80 to 90%), because of the occurrence of some decomposition of the hydroperoxide as a side reaction during epoxidation. As a consequence of this and of the normal chain-branching reactions which occur during hydroperoxide production, considerably more than one mole of co-product per mole of propylene oxide is obtained. The proportion in terms of weight is greater, since the co-products have higher molecular weights than propylene oxide. It has been estimated that about 2·2 kg t-butanol or 2·5 kg styrene (produced by dehydration of the 1-phenylethanol, see below) are produced for every kg propylene oxide made.

The economics of these processes are therefore dominated by the necessity to dispose of large amounts of co-product for an adequate return, and although a variety of hydrocarbons can be used as the basis for hydroperoxide generation, only isobutane and ethylbenzene have thus far formed the basis of viable processes. t-Butanol has not previously been a large-tonnage chemical, but it has a number of potential major uses, e.g. as a source of isobutene for gasoline manufacture or for butyl rubber manufacture (see page 252), as an anti-icing agent for gasoline, or for the manufacture of t-butyl acetate for addition to gasoline as a lead tetraethyl synergist. 1-Phenylethanol can readily be converted to styrene by dehydration, e.g. over a titanium oxide catalyst at 470 to 520 K. Styrene, of course, has large tonnage applications in the manufacture of polymers (see Chapter 7).

A plant to make 155 000 tons per annum of propylene oxide, reputedly with t-butanol as co-product, has been constructed in the USA and a similar one is under construction in Holland. A plant to made 32 000 tons per annum of the oxide with styrene as co-product is under construction in Spain. It will be appreciated that the necessity to provide for the disposal of a very large amount of co-product reduces the flexibility of these processes, and restricts the number of companies which can consider their use.

FREE-RADICAL GAS-PHASE OXIDATION

This type of oxidation is much less important as a method of chemical manufacture than liquid-phase oxidation, mainly because it is much less selective. Higher temperatures are required than for liquid-phase oxidation.

Most hydrocarbons oxidise at temperatures below about 670 K by a different mechanism from that by which they undergo oxidation at higher temperatures; the types of product obtained are different:

low-temperature oxidation	$\longrightarrow$	oxygen-containing products (alcohols, aldehydes, ketones)
high-temperature oxidation	$\longrightarrow$	hydrocarbons (products of cracking and dehydrogenation)

As one would expect, the change in mechanism does not occur

sharply. There is an intermediate stage in which both types of reaction are occurring.

Low-Temperature Mechanism

These reactions are not so well understood as those involved in liquid-phase oxidation.

There is fairly general agreement that reactions similar to those in liquid-phase oxidation lead to the formation of peroxy radicals, e.g.:

$$CH_3CH_2CH_2CH_3 \longrightarrow CH_3CH_2\underset{\underset{OO\cdot}{|}}{C}HCH_3$$

There are two main schools of thought as to the fate of these radicals. One considers that they give a hydroperoxide which then undergoes cleavage, as in liquid-phase oxidation. The other suggests that cleavage of the alkylperoxy radical itself occurs:

$$CH_3CH_2\underset{\underset{OO\cdot}{|}}{C}HCH_3 \longrightarrow CH_3CHO + CH_3CH_2O\cdot$$

High yields of single product are not obtained in this type of oxidation, but rather a mixture of alcohols and carbonyl compounds. This marked, and commercially very important difference from liquid-phase oxidation is due to the relatively high temperatures used (which lead to indiscriminate attack on molecules), and to the fact that in the gas phase, intramolecular reactions of radicals are favoured over intermolecular reactions. Further, solvation of intermediates and products can play no part.

The only major commercial use of free-radical gas-phase oxidation is in the production of acetaldehyde, methanol and formaldehyde, by the oxidation of propane and butane. This is carried out in the USA:

$$CH_3CHO \text{ (31 kg/100 kg butane)}$$

$$C_4H_{10} \xrightarrow{\text{air, 640-720 K}} HCHO \text{ (33 kg/100 kg butane)} + \text{other products}$$

$$CH_3OH \text{ (20 kg/100 kg butane)}$$

Although cheap raw materials and a simple reaction system are used, separation of pure products requires a complex and expensive plant.

High-Temperature Mechanism

At higher temperatures, and with low concentrations of oxygen, the products are those of dehydrogenation and cracking reactions. It appears that, at these temperatures, combination of oxygen molecules with alkyl radicals is much slower than abstraction of hydrogen from the radicals by oxygen. It is suggested that important reactions in the high-temperature oxidation of hydrocarbons, using butane as an example, are:

$$CH_3CH_2CH_2CH_3 + O_2 \longrightarrow CH_3\overset{\bullet}{C}HCH_2CH_3 + HO_2\cdot$$

$$CH_3CH_2CH_2CH_3 + O_2 \longrightarrow CH_3CH_2CH_2CH_2\cdot + HO_2\cdot$$

$$C_4H_9\cdot + O_2 \longrightarrow C_4H_8 + HO_2\cdot$$

$$CH_3\overset{\bullet}{C}HCH_2CH_3 \longrightarrow CH_3CH=CH_2 + CH_3\cdot$$

$$CH_3CH_2CH_2CH_2\cdot \longrightarrow CH_3CH_2\cdot + CH_2=CH_2$$

$$CH_3CH_2\cdot + O_2 \longrightarrow CH_2=CH_2 + HO_2\cdot$$

$$CH_3\cdot + C_4H_{10} \longrightarrow CH_4 + C_4H_9\cdot$$

The $HO_2\cdot$ radicals are supposed to be destroyed at the surface of the reactor.

LIQUID-PHASE NON-FREE-RADICAL OXIDATION

There are at the present time three important processes involving this type of oxidation.

Wacker Chemie Process for Acetaldehyde

The development of this process affords an excellent example of what one might call 'creative recognition', that is, the appreciation of the potential technological importance of a newly observed reaction or phenomenon.

In the mid 1950s, chemists at Wacker Chemie in Germany were attempting to make ethylene oxide by passing ethylene and oxygen over palladium-on-charcoal catalysts. No ethylene oxide was formed, but with some catalysts acetaldehyde was produced. Investigation showed that these were catalysts in which reduction of palladium chloride to palladium was not complete. Further, it was shown that palladium chloride solution oxidised ethylene to acetaldehyde in high yield, a reaction which had in fact previously been reported by an American chemist in 1894:

$$C_2H_4 + PdCl_2 + H_2O \longrightarrow CH_3CHO + Pd + 2HCl$$

A programme of work aimed at developing a process for the manufacture of acetaldehyde based on this reaction was started.

The reaction as it stands could not form the basis of a commercial process, in that large quantities of palladium, which has a price of the same order as that of platinum, would be required. Evidently this would be recoverable, but appreciable physical losses would be inevitable; also a very large amount of capital would be tied up.

However, cupric chloride rapidly oxidises palladium to palladium chloride:

$$Pd + 2CuCl_2 \longrightarrow PdCl_2 + (CuCl)_2$$

and oxygen rapidly oxidises cuprous chloride to cupric chloride:

$$(CuCl)_2 + 2HCl + \tfrac{1}{2}O_2 \longrightarrow 2CuCl_2 + H_2O$$

Consequently, if ethylene and oxygen are passed into a solution of palladium chloride and cupric chloride in dilute hydrochloric acid, the ethylene is oxidised to acetaldehyde. This was very rapidly developed into a full scale process, which first came into operation in 1960. During the development work, the mechanism of the reaction was extensively investigated, and investigation by both industrial and academic chemists has continued since.

The essential features of the mechanism are formation of an ethylene–palladium chloride complex, and attack of this by water. Current ideas on the steps involved in the reaction are shown in Fig. 3-1.

The process can be operated in two forms, the so-called one-stage and two-stage processes. In the one-stage process (see Fig. 3-2), ethylene and oxygen, with the ethylene in large excess, so that the mixture remains outside the explosive limits, is passed into the catalyst solution at 330 to 370 K and about 3 atm. The acetaldehyde passes out of the reactor in the stream of excess ethylene, from which

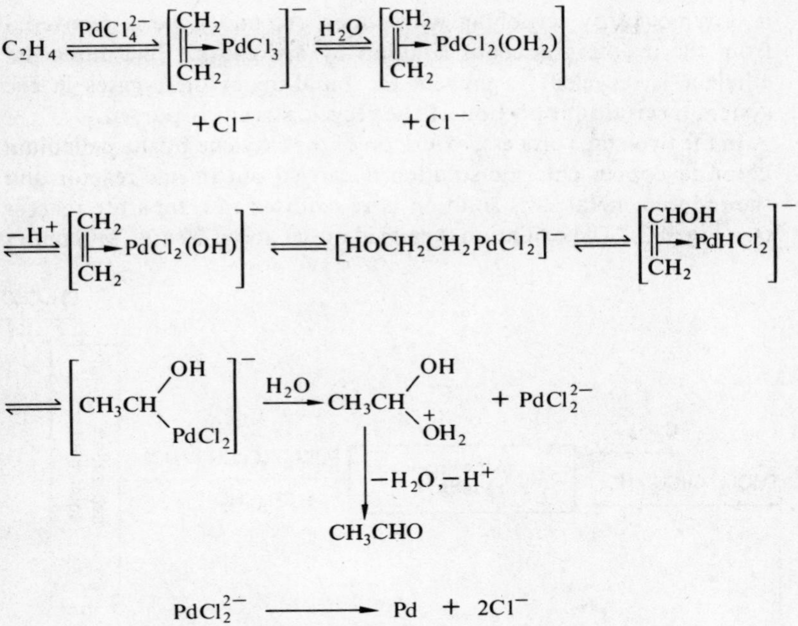

$$C_2H_4 \xrightleftharpoons{\text{PdCl}_4^{2-}} \begin{bmatrix} \text{CH}_2 \\ \| \rightarrow \text{PdCl}_3 \\ \text{CH}_2 \end{bmatrix}^- \xrightleftharpoons{\text{H}_2\text{O}} \begin{bmatrix} \text{CH}_2 \\ \| \rightarrow \text{PdCl}_2(\text{OH}_2) \\ \text{CH}_2 \end{bmatrix}$$

$$+ \text{Cl}^- \qquad\qquad + \text{Cl}^-$$

$$\xrightleftharpoons{-\text{H}^+} \begin{bmatrix} \text{CH}_2 \\ \| \rightarrow \text{PdCl}_2(\text{OH}) \\ \text{CH}_2 \end{bmatrix}^- \rightleftharpoons [\text{HOCH}_2\text{CH}_2\text{PdCl}_2]^- \rightleftharpoons \begin{bmatrix} \text{CHOH} \\ \| \rightarrow \text{PdHCl}_2 \\ \text{CH}_2 \end{bmatrix}^-$$

$$\rightleftharpoons \begin{bmatrix} \text{CH}_3\text{CH} \overset{\text{OH}}{\underset{\text{PdCl}_2}{}} \end{bmatrix}^- \xrightarrow{\text{H}_2\text{O}} \text{CH}_3\text{CH} \overset{\text{OH}}{\underset{\overset{+}{\text{OH}_2}}{}} + \text{PdCl}_2^{2-}$$

$$\downarrow -\text{H}_2\text{O}, -\text{H}^+$$

$$\text{CH}_3\text{CHO}$$

$$\text{PdCl}_2^{2-} \longrightarrow \text{Pd} + 2\text{Cl}^-$$

Fig. 3-1. *Proposed mechanism of Wacker Chemie Process.*

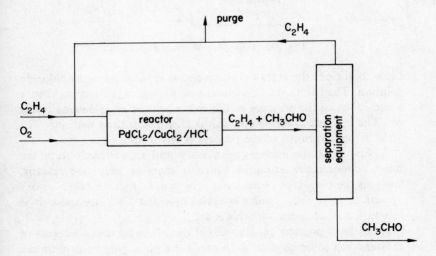

Fig. 3-2. *One-stage Wacker Process.*

it is removed by scrubbing with water. Acetaldehyde is recovered from the resulting aqueous solution by distillation. The unreacted ethylene is recycled: to prevent the build-up of inert gases in the system a certain proportion of the recycle stream is purged.

In the two-stage process, oxidation of the ethylene by the palladium chloride–copper chloride solution is carried out in one reactor and the reduced metal salts solution is re-oxidised in a separate reactor (see Fig. 3-3). The oxidation is carried out at about 10 atm, and under

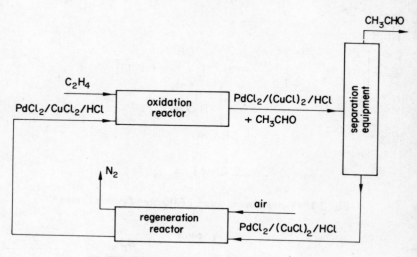

Fig. 3-3. *Two-stage Wacker Process.*

these conditions the acetaldehyde remains in solution in the chloride solution. The solution of chlorides and acetaldehyde is passed to a vessel in which the pressure is reduced, when the acetaldehyde distils off. The reduced chloride solution is then re-oxidised with air. The yield in both versions of the process is about 95%.

At first sight, the one-stage process would appear likely to be the more economically attractive since it requires only one reactor, whereas the two-stage process requires two. In fact the two versions appear to have very similar overall costs and both are used. It is interesting to consider why this is so.

It has been pointed out that in the one-stage process an excess of ethylene has to be used so as to avoid the possibility of explosions, and that ethylene consequently has to be recycled. If air were used as the oxidising gas in this process, the ethylene remaining after separation of acetaldehyde would have to be separated from nitrogen before recycling, e.g. by low-temperature distillation or solvent extraction, and this would be quite an expensive procedure. Oxygen is used to avoid the necessity for such a separation. In the two-stage process, since oxidation of the reduced chloride solution is carried out separately, air can be used. Thus, although the one-stage process only requires one reactor compared with two in the two-stage process, it requires the provision and operation of an oxygen plant. Overall, if the cost of the oxygen plant is included for the one-stage process, the capital costs for the two processes are very similar. Production costs are also similar, but the exact levels of these depend on the particular circumstances at the site involved. The two-stage process has the advantage, in some circumstances, that since there is no recycle of ethylene, high purity ethylene is not required, and ethylene–ethane mixtures can thus be used, with an attendant saving in separation costs.

The catalyst solution is highly corrosive, so that careful choice of materials of construction has to be made. The reactors are brick- and rubber-lined, and titanium is used for much of the pipe-work, pumps, etc.

The Wacker Chemie process shows clear advantage over the other route to acetaldehyde from ethylene, *via* ethanol, and has been very successful: in 1969 there were a total of 19 plants with a total capacity of about 900 000 tons using the process. However, it should be noted that a major outlet for acetaldehyde is in the production of acetic acid and acetic anhydride, so that acetaldehyde processes are subject to competition from acetic acid processes based on other routes. In the UK most acetic acid is made by direct oxidation of light naphtha, and the amount of acetaldehyde made in this country is relatively small. It has not proved worthwhile to construct a new plant to use the Wacker Chemie process in this country.

OTHER ALKENES

Propene and higher 1-alkenes give mainly methyl ketones, with smaller amounts of aldehyde, when oxidised in this type of system. This is unfortunate, in that the production of aldehydes would have been commercially more interesting. The reaction does, however, provide alternative routes to acetone and methyl ethyl ketone.

Vinyl Acetate Process

This is a logical development of the Wacker Chemie acetaldehyde process, and involves the use of a similar catalyst system, but in this case in acetic acid rather than in aqueous solution. The presence of sodium acetate is required, and it is supposed that nucleophilic attack of the complex by acetate ions occurs:

$$\left[\begin{array}{c} CH_2 \\ \| \\ CH_2 \end{array} \!\!\!\!\!\! - PdCl_3 \right]^- \xrightarrow{\;AcO^-\;} \left[\begin{array}{c} CH_2 \\ \| \\ CH_2 \end{array} \!\!\!\!\!\! - PdCl_2\,(OAc) \right]^- \rightleftharpoons \left[AcOCH_2CH_2PdCl_2 \right]^-$$
$$+\; Cl^-$$

$$\rightleftharpoons \left[\begin{array}{c} CHOAc \\ \| \\ CH_2 \end{array} \!\!\!\!\!\! - PdHCl_2 \right]^- \longrightarrow CH_2{=}CHOAc \;+\; Pd \;+\; 2Cl^- \;+\; H^+$$

This reaction was first observed by Moiseev, in Russia; a number of companies have now developed processes based on it:

$$CH_2{=}CH_2 + CH_3CO_2H \xrightarrow[\;370\,K,\;20\,atm\;]{PdCl_2/CuCl_2/CH_3CO_2Na,\; O_2} CH_2{=}CHO_2CCH_3 + H_2O$$

The presence of water in the reaction mixture leads to the formation of acetaldehyde in addition to vinyl acetate, either through attack of the ethylene complex by water, or by hydrolysis of vinyl acetate. Since water is formed during the reaction, this is difficult to avoid. However, advantage can be taken of this side reaction by using the acetaldehyde as feedstock for acetic acid manufacture, thus giving a vinyl acetate process requiring only ethylene as raw material. ICI brought a plant using this process into operation in the UK in 1966. However, it suffered from continuous, very severe corrosion problems, and these led to it being shut down in 1969. The process is being operated successfully in the USA.

The reaction can also be carried out in the vapour phase over a palladium-containing catalyst at 370 to 470 K. In this form the process can be made to produce no acetaldehyde by-product, and this can be an advantage, e.g. if low-cost acetic acid is available from another source. Also, corrosion problems are less than for the liquid-phase process. Yields of over 90% at conversions of ethylene of 10 to 15% are obtained. BP will operate such a process in a 50 000 ton per annum plant at present under construction in Baglan Bay in Wales.

At present most vinyl acetate is made from acetylene and acetic acid:

$$CH{\equiv}CH + CH_3CO_2H \xrightarrow[\text{440-480 K}]{\text{Zn(OAc)}_2\text{on charcoal}} CH_2{=}CHO_2CCH_3$$

This process has the disadvantage of being based on an expensive hydrocarbon, and ethylene-based processes are likely to become increasingly important in the next few years.

Practically all the vinyl acetate made is used in the manufacture of polymers (see Chapter 7).

Dow Phenol from Toluene Process

The reaction on which this process is based was first observed in 1845 when German chemists showed that when cupric benzoate is heated, phenyl benzoate is formed. Ninety to a hundred years later, other workers showed that, in addition, phenol and salicylic acid are formed. Thus:

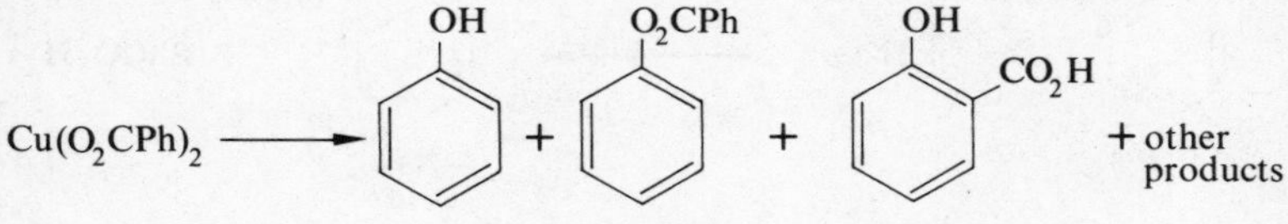

phenyl benzoate   *salicylic acid*

The reaction leads to attachment of oxygen to the aromatic ring, giving either phenol, or materials which could readily be converted into phenol.

The Dow Chemical Company, who were already major manufacturers of phenol (by the monochlorobenzene process), set out, in the 1950s, to investigate the feasibility of a process for the manufacture of phenol based on this reaction. They developed the process outlined below:

$$PhCH_3 \longrightarrow PhCO_2H \xrightarrow[\text{520 K}]{\text{air, H}_2\text{O, Cu(O}_2\text{CPh)}_2} PhOH + CO_2$$

Conversion of toluene to benzoic acid is, as we have already seen, readily brought about by liquid-phase free-radical oxidation. In the second stage, air and steam are passed into molten benzoic acid containing a catalytic amount of cupric benzoate, with magnesium benzoate present as a promoter. At the temperature of reaction, phenol distils out as it is formed, and its destruction by further oxidation is avoided.

The chemistry of the second stage has been extensively investigated, both during the process development work and subsequently. The key step is an oxidation–reduction reaction of cupric benzoate to give *o*-benzoyloxybenzoic acid and cuprous benzoate:

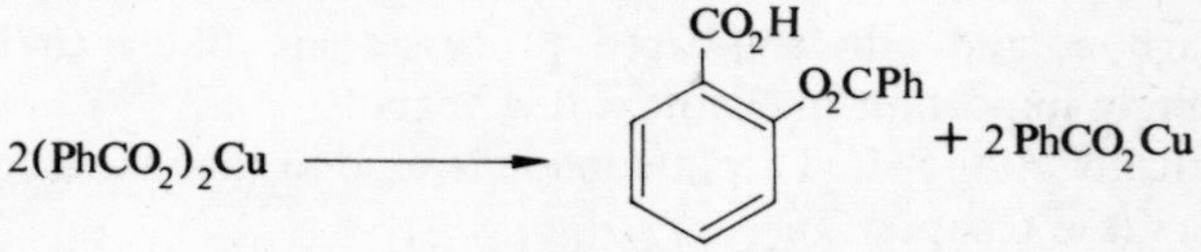

$$2(PhCO_2)_2Cu \longrightarrow$$

o-benzoyloxy-benzoic acid

$$+ \; 2\,PhCO_2Cu$$

The *o*-benzoyloxybenzoic acid hydrolyses to give benzoic acid and salicylic acid, and under the process conditions the salicylic acid decarboxylates to phenol:

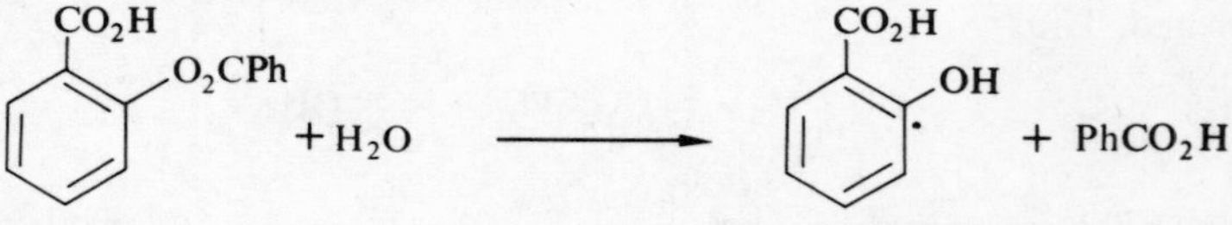

$$+ \, H_2O \longrightarrow$$

salicylic acid

$$+ \; PhCO_2H$$

$$\longrightarrow$$

$$+ \, CO_2$$

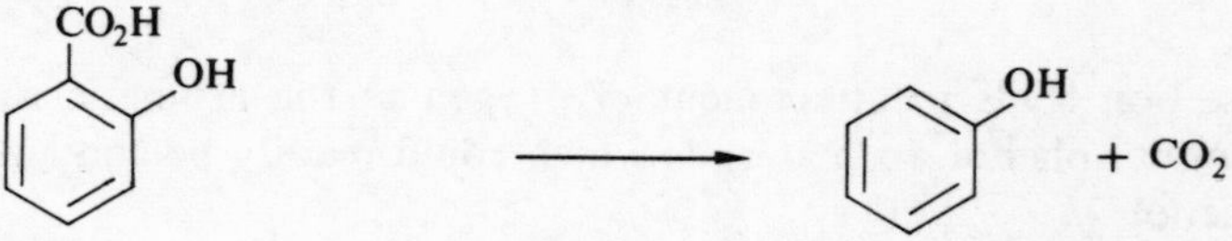

The cuprous benzoate is oxidised back to cupric benzoate by air:

$$2PhCO_2Cu \; + \; 2PhCO_2H \; + \; ½O_2 \longrightarrow 2(PhCO_2)_2Cu \; + \; H_2O$$

A suggested mechanism for the first step is:

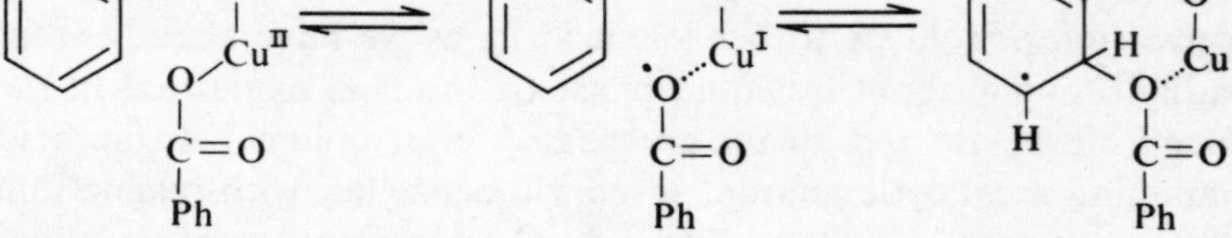

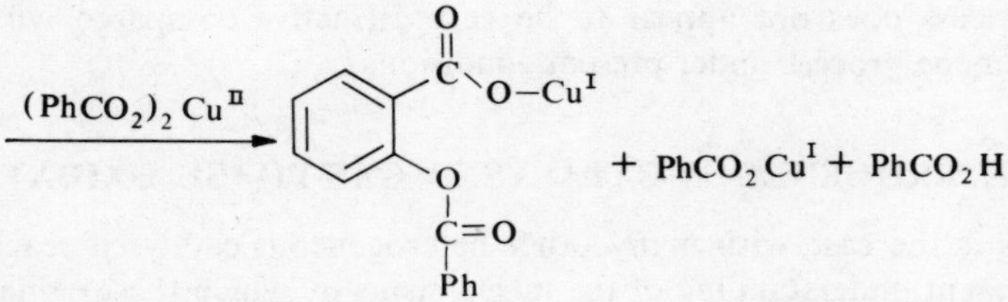

$$\xrightarrow{\text{(PhCO}_2)_2\text{, Cu}^{II}}$$

(structure: benzene ring with $\overset{O}{\underset{}{C}}\text{O}-\text{Cu}^{I}$ ortho to $O-\overset{}{\underset{O}{C}}=O$, Ph) $+ \text{PhCO}_2\text{Cu}^{I} + \text{PhCO}_2\text{H}$

It has been established, by experiments with isotopically labelled benzoic acid, and with toluic acids, that the oxygen becomes attached to the ring *ortho*- to the carboxyl group. Thus:

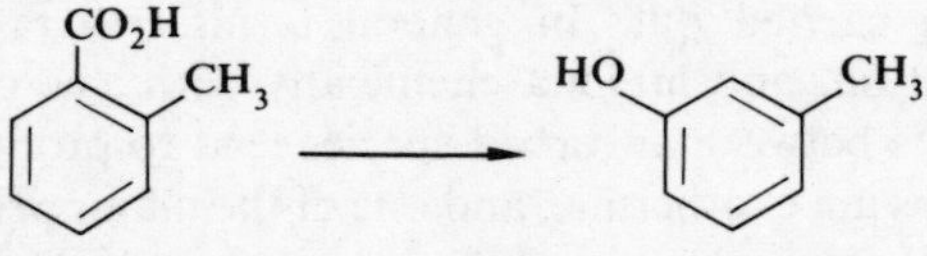

o-toluic acid $\longrightarrow$ *m*-cresol

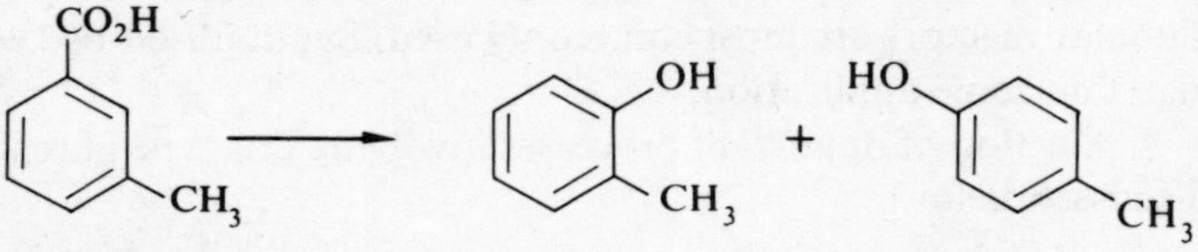

m-toluic acid $\longrightarrow$ *o*-cresol $+$ *p*-cresol

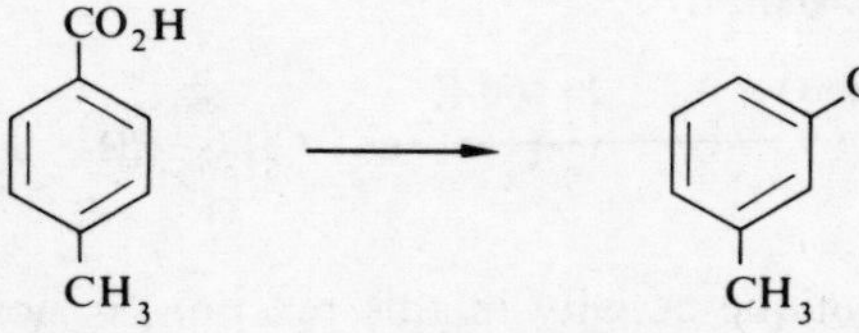

p-toluic acid $\longrightarrow$ *m*-cresol

The process would appear to be wasteful of raw materials in that 15% of the hydrocarbon used is lost as carbon dioxide. However, since toluene is produced in excess of requirements by catalytic reforming of naphtha it is in fact a somewhat cheaper source of the benzene ring than benzene itself.

At the time of writing, there are three plants operating this process, in Canada, the USA, and in the Netherlands, and a further plant is under construction in the Netherlands. However, in general the process does not appear to be very attractive compared with the cumene process under present conditions.

HETEROGENEOUS-CATALYSED GAS-PHASE OXIDATION

As is the case with many other heterogeneous-catalysed reactions, present understanding of the mechanisms by which these oxidations occur is far from complete, and there is not a great deal that can be said about the detailed chemistry of such reactions here. The catalysts used are most often metals or metal oxides. As would be expected, the mode of catalytic action depends on the catalyst and on the oxidation being carried out. In general terms, reactant and oxygen are adsorbed on, and interact chemically with, the catalyst surface, and reactions between adsorbed species lead to products.

Oxidation reactions are exothermic, and one of the major problems in the engineering of heterogeneous-catalysed gas-phase oxidations is the removal of the heat of reaction from the catalyst bed. Failure to do this would lead to overheating, with consequent loss of yield and possibly destruction of the catalyst and, in some cases, the equipment. Tubular reactors are most commonly used, but fluidised-bed reactors also find some application.

A selection of important processes involving this type of reaction is discussed below.

Ethylene Oxide Manufacture
Oxidation of ethylene over a silver catalyst is the favoured method of manufacture of ethylene oxide:

$$CH_2 = CH_2 \xrightarrow{\text{O}_2,\ \text{Ag},\ 520\text{-}600\ \text{K}} \underset{\diagdown \text{O} \diagup}{CH_2 - CH_2}$$

Silver, which has a unique activity in this reaction, is normally used on an inert support, e.g. aluminium oxide. The oxidising gas may be either air or oxygen depending on the particular form of the process being used. Contact times are in the range 1 to 4 seconds. Yields of around 60% are obtained, the other main products being carbon dioxide and water.

The reaction has been extensively investigated, but there is still much uncertainty as to its mechanism. One suggested scheme is given below ([Ag] represents catalyst):

$$[Ag] \qquad\qquad + O_2 \qquad\longrightarrow\quad [Ag].O_2\text{ads.}$$

$$[Ag].O_2\text{ads.} \quad + C_2H_4\text{ads.} \longrightarrow \underset{O}{CH_2{-}CH_2} + [Ag].O\text{ads.}$$

$$4[Ag].O\text{ads.} \quad + C_2H_4\text{ads.} \longrightarrow 2CO + 2H_2O + 4[Ag]$$

$$[Ag].O_2\text{ads.} \quad + 2CO \qquad\longrightarrow 2CO_2 + [Ag]$$

According to this scheme the maximum possible yield of ethylene oxide obtainable is 80%. Actual yields have never reached this.

The main overall competing reaction in this process is complete oxidation of ethylene to carbon dioxide and water. This reaction produces about fourteen times as much heat as the desired reaction:

$$C_2H_4 + \tfrac{1}{2}O_2 \longrightarrow \underset{O}{CH_2{-}CH_2} \qquad \Delta H = -104 \ \text{kJ/mol}$$

$$C_2H_4 + 3O_2 \longrightarrow 2CO_2 + 2H_2O \qquad \Delta H = -1420 \ \text{kJ/mol}$$

The reaction leading to complete combustion has a higher activation energy than that leading to ethylene oxide, and selectivity therefore falls with increasing temperature. The reaction thus has a pronounced tendency to 'run away'. An increase in temperature leads to an increase in rate of reaction, particularly of complete oxidation, with a consequent large increase in heat output, which may cause the temperature to rise further, and so on. Very careful attention to temperature control is required in the design of ethylene oxide plants.

A small amount (*ca.* 1 part per million) of ethylene dichloride is normally added to the feed. This improves selectivity by suppressing the complete oxidation to carbon dioxide and water. The mode of action is not known.

In most countries direct oxidation has displaced the older method of manufacture of ethylene oxide, the chlorohydrin process (see page 135). Although the latter gives a higher yield of ethylene oxide (*ca.* 80%), it consumes large amounts of chlorine and consequently has substantially higher raw material costs. All UK manufacture of ethylene oxide (175 000 tons in 1970) is by direct oxidation.

The main use of ethylene oxide is hydration to ethylene glycol (see page 194), which is required for use as motor car engine antifreeze and for the manufacture of poly(ethylene terephthalate). It has a number of other uses as an intermediate, e.g. in the manufacture of surface-active agents (see page 258), and brake fluids.

Phthalic Anhydride Manufacture

Phthalic anhydride is manufactured by oxidising naphthalene or o-xylene with air over a supported vanadium pentoxide catalyst:

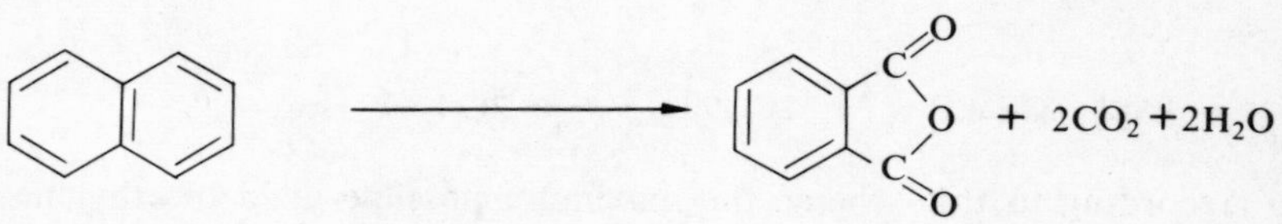

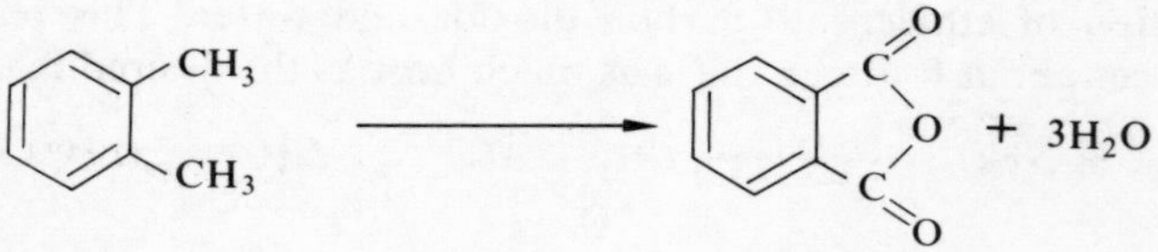

$$+ \; 2CO_2 + 2H_2O$$

$$+ \; 3H_2O$$

In naphthalene-based manufacture both fixed-bed tubular reactors and fluidised-bed reactors are used. In one fixed-bed process the reaction temperature is 620 to 720 K, and the contact time is 0·1 to 0·6 second. In other processes lower temperatures and longer contact times are used. Yields of about 80% are obtained, with complete conversion of the naphthalene.

o-Xylene is becoming of increasing importance as a raw material for phthalic anhydride manufacture, partly because supplies of naphthalene are failing to keep pace with demand. The processes used are fixed-bed processes similar to those used for naphthalene, but rather higher temperatures and shorter contact times are used. The yield obtained is currently somewhat less than obtained from naphthalene (about 75%), but it should be noted that whereas there is an inevitable wastage of two out of ten carbon atoms when naphthalene is used as raw material, potentially all the carbon atoms of o-xylene appear in the product. Thus, at 100% yield, 1 kg of phthalic anhydride would require 0·87 kg naphthalene and only 0·72 kg o-xylene. At current prices and yields, o-xylene appears to be the more attractive raw material for new plants, although a substantial amount of phthalic anhydride is made from naphthalene in existing plants.

Until recently there were seven producers of phthalic anhydride in the UK, but a process of rationalisation is under way, and by 1971 there will be only four producers, with a total capacity of 168 000 tons per annum. Of this, 136 000 tons per annum will be based on *o*-xylene. The major use of phthalic anhydride is in the manufacture of phthalate esters for use as plasticisers (see page 202). The manufacture of unsaturated polyester resins and alkyd resins (see page 267) are also important outlets.

Maleic Anhydride Manufacture

Maleic anhydride is made by the oxidation of benzene over a vanadium pentoxide catalyst under conditions similar to those used in phthalic anhydride manufacture:

$$\text{benzene} \xrightarrow{\text{air}} \underset{\text{CHC}}{\overset{\text{CHC}}{\parallel}} \begin{matrix} \diagup O \\ O \\ \diagdown O \end{matrix} + 2CO_2 + 2H_2O$$

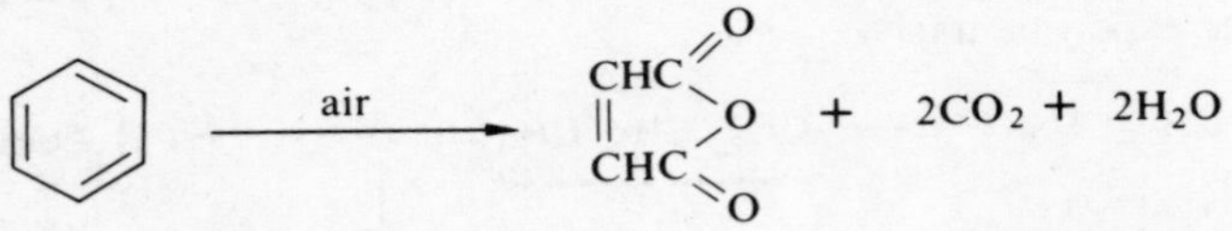

Yields of about 50 to 60% are obtained. Oxidation of n-butenes under similar conditions is also used.

In the UK maleic anhydride is made by Monsanto and ICI, at scales of about 20 000 and 10 000 tons per annum respectively. Its main use is in making unsaturated polyester resins and alkyd resins (see Chapter 7).

Ammoxidation of Propene

This process, which has been spectacularly successful, and is now the preferred method of making acrylonitrile, was developed independently by Sohio in the USA, and the Distillers Company in the UK. It was first brought into operation by Sohio in 1960.

The process involves oxidation of a mixture of propene and ammonia with air over a catalyst at 670 to 770 K:

$$CH_2{=}CHCH_3 + NH_3 \xrightarrow{\text{air}} CH_2{=}CHCN + 3H_2O$$

In the originally developed versions of the process, using e.g. a bismuth molybdate catalyst, yields of about 65% are obtained, and acetonitrile and hydrogen cyanide are formed as by-products in substantial quantities. In a more recent version of the process, using a uranium-containing catalyst, substantially higher yields are obtained.

The ammoxidation process shows very marked advantages in both capital and operating costs over the main alternative route to acrylonitrile, the reaction of acetylene with hydrogen cyanide. A major factor is that it is based on a cheap hydrocarbon, and a cheap nitrogen source, whereas both acetylene and hydrogen cyanide are relatively expensive to make.

The mechanism of the reaction is not known at the present time. Oxidation of propene to acrolein under similar conditions to those used in ammoxidation had been known for some time prior to the development of the ammoxidation reaction, and has been extensively investigated. It is thought that at the catalyst surface one of the methyl hydrogens is removed to give a symmetrically bound allyl group, which can then be attacked by adsorbed oxygen, at either end, to give acrolein. Attack at either end of the propene is supported by tracer experiments:

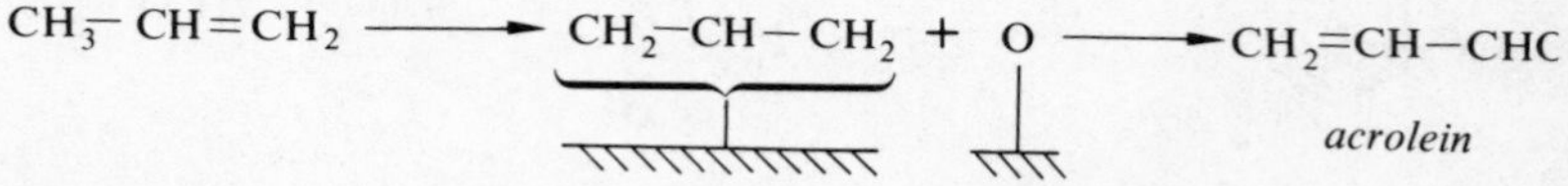

$$CH_3-CH=CH_2 \longrightarrow \underbrace{CH_2-CH-CH_2}_{} + \underset{|}{O} \longrightarrow CH_2=CH-CHC$$

$$\textit{acrolein}$$

It is easy to envisage mechanisms for the ammoxidation reaction in which acrolein is formed and then reacts with ammonia. However, it is not known whether acrolein is in fact an intermediate in this reaction.

When this process came into operation it rendered uneconomic a number of plants operating older processes, including one brought into operation in the UK in 1959. All UK manufacture of acrylonitrile (UK capacity 1970—119 000 tons per annum) is by ammoxidation.

The major use of acrylonitrile (about 85% of total consumption in the UK) is in the manufacture of acrylic fibres (see page 252). It is also used in the manufacture of a number of other polymers, e.g. nitrile rubbers.

Chapter 4

Halogen Compounds

IN the main, halogen atoms are introduced into organic compounds for two reasons in commercial operations, for use in subsequent synthetic operations, and in order to obtain products with desirable technological properties.

The versatility of halogen compounds in synthetic applications is of course well known to all students of organic chemistry, and it is not surprising that they find many similar uses in commercial manufacture. However, whereas cost is normally a minor factor in the choice between available methods of synthesis in the laboratory, it is normally the major factor in choosing a commercial route, and this imposes its pattern on the range of uses of halogen compounds as intermediates in manufacturing operations. In particular, since synthetic routes involving the substitution or elimination of a halogen atom often lead to the downgrading in value of a halogen source, and also often require the use of other ancillary raw materials, they tend to be avoided where possible on the large scale. On the small scale, however, e.g. in dyestuffs and pharmaceuticals manufacture, where the cost of the halogen source is a smaller proportion of the total production costs, where potential savings from the development of alternative routes are smaller, and where more complicated molecules are often involved, they are widely used, often under conditions very similar to those used in the laboratory. Chlorine, being by a substantial margin the cheapest halogen (see Table 4-1), is used where possible, but bromine, and much less frequently, iodine, are sometimes used.

The range of end-use products in which the presence of halogen atoms confers desirable technological properties is very wide, and covers, for instance, polymers, solvents, dyestuffs, drugs, anaesthetics, insecticides and herbicides. Again, chlorine-containing compounds are considerably more important than those containing other halogens.

111

It would be impossible in a book of this type to consider all halogen compounds with commercial applications, so that, in the main, only the more important compounds are discussed. Most of

TABLE 4-1
HALOGEN PRICES, UK 1970 (approx.)

Halogen source	Price (£/ton)
Hydrogen fluoride (anhydrous)	220
Chlorine	32
Bromine	140
Iodine	1 900

the discussion concerns chlorine compounds, but the more important applications of the other halogens in the organic chemical industry are briefly considered.

CHLORINE COMPOUNDS

Chlorine-containing organic compounds are of very considerable commercial importance: their manufacture provides the main outlet for chlorine.

The question of the economics of manufacture of such compounds on the large scale is a complex one. There are often a number of possible organic starting materials of different costs available, and these may require different chlorine sources. Evidently one low-cost chlorinating agent is chlorine. However, an important factor is that the use of chlorine often leads to the generation of hydrogen chloride as a by-product. The value of this then becomes an important factor in the process economics. In a situation where there is no outlet for the hydrogen chloride, so that its value is zero (or negative, if costs have to be incurred in its disposal), its production represents a wastage of part of the chlorine fed; on the other hand, if there is an outlet then its value may be equivalent to that of its contained chlorine. As hydrogen chloride cannot economically be transported in large quantities, the outlet must be near the producing plant. One obvious outlet is to dissolve the hydrogen chloride in water to produce hydrochloric acid. However, the market for hydrochloric acid is limited, and this does not provide a sink for unlimited quantities of hydrogen chloride.

The economics of a particular process for a chlorine-containing organic compound therefore depend markedly on the circumstances at the site involved; manufacture of chlorine compounds is often carried out as an integrated operation.

Chloromethanes

These four compounds, methyl chloride CH_3Cl, methylene chloride CH_2Cl_2, chloroform $CHCl_3$, and carbon tetrachloride CCl_4, all find fairly substantial use, in a variety of applications. It is logical to consider these products as a group from the point of view of manufacturing operations since a major method of production for them all is the chlorination of methane.

CHLORINATION OF METHANE

Chlorine reacts with alkanes in general by free-radical chain reactions. The mechanism is summarised by the following scheme:

initiation:
$$Cl_2 \longrightarrow 2Cl\cdot$$

propagation:
$$RH + Cl\cdot \longrightarrow R\cdot + HCl$$
$$R\cdot + Cl_2 \longrightarrow RCl + Cl\cdot$$

termination:
$$R\cdot + R\cdot \longrightarrow R{-}R$$
$$R\cdot + Cl\cdot \longrightarrow R{-}Cl$$

The necessary homolysis of chlorine can be brought about photolytically, or thermally at temperatures of 520 K or higher: free radical producing agents will also promote the reaction. The reaction will proceed in either the gas phase or the liquid phase. Overall the reaction is exothermic.

Chlorination of methane is carried out in the gas phase at temperatures ranging from 620 to 1 020 K. The reaction is a classic example of the situation where initial products can undergo further reaction, and where the product distribution obtained depends on the conversion to which the reaction is taken:

$$CH_4 \xrightarrow{Cl_2} \underset{+\ HCl}{CH_3Cl} \xrightarrow{Cl_2} \underset{+\ HCl}{CH_2Cl_2} \xrightarrow{Cl_2} \underset{+\ HCl}{CHCl_3} \xrightarrow{Cl_2} \underset{+\ HCl}{CCl_4}$$

The conversion of methane is generally controlled by the ratio of chlorine to methane used, chlorine being completely consumed under the process conditions. Operation at low conversion favours the production of the less highly chlorinated products, and high conversion, the more highly chlorinated products (see Fig. 4-1).

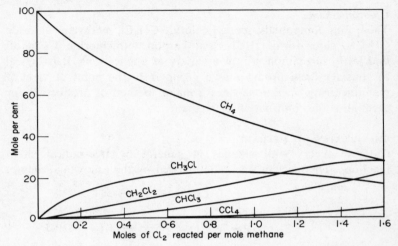

Fig. 4-1. *Chlorination of methane. W. Hirschkind*, Ind. Eng. Chem., **41**, 2749 (1949). *Copyright (1949) by the American Chemical Society. Reprinted by permission of the copyright owner.*

As can be seen from Fig. 4-1, it is not possible to operate the process so as to obtain only methyl chloride, since this is attacked by chlorine much more readily than methane itself. This is attributed to resonance stabilisation of the incipient chloromethyl radical in the hydrogen abstraction stage, which is presumed to lead to an increased ease of abstraction of the hydrogen atom compared with abstraction from methane:

$$H - \overset{\displaystyle\cdot}{\underset{\displaystyle H}{C}} - \overset{\displaystyle\cdot\cdot}{\underset{\displaystyle\cdot\cdot}{C}}l: \quad\longleftrightarrow\quad H - \overset{\displaystyle-}{\underset{\displaystyle H}{C}} - \overset{\displaystyle+}{\underset{\displaystyle\cdot\cdot}{C}}l:$$

However, within this limitation, there is considerable flexibility, and the process can be operated to give mixtures of methyl chloride, methylene chloride, and chloroform, all four products, or carbon tetrachloride alone.

A major factor in the economics of this process is that half the chlorine fed is converted to hydrogen chloride; as we have seen, the value of this depends on individual circumstances. Another factor is that the price of methane varies considerably from country to country, depending on whether there is access to natural gas. In the UK, for example, methane was relatively expensive before the discovery of North Sea gas. Where methane is cheap, and there is an outlet for the hydrogen chloride produced, e.g. in vinyl chloride manufacture (see below), the process is attractive. There are, however, other processes in use for individual chloromethanes. These are briefly discussed below, together with the major uses for these compounds.

METHYL CHLORIDE

This is also made by the reaction of methanol and hydrogen chloride:

$$CH_3OH + HCl \longrightarrow CH_3Cl + H_2O$$

The reaction may be carried out in the vapour phase over a catalyst, e.g. alumina gel or zinc oxide on pumice, at 620 to 720 K, or in the liquid phase at about 370 to 420 K in the presence of a catalyst, e.g. zinc chloride. The catalyst is presumed to act by facilitating heterolysis of the carbon–oxygen bond. Yields of about 95% are obtained.

This process does not involve any problems of hydrogen chloride disposal; it can in fact act as a hydrogen chloride sink for other processes. On the other hand, methanol is in general a more expensive raw material than methane.

The main use of methyl chloride is in the preparation of chloromethylsilanes, which are important intermediates in silicone manufacture:

$$CH_3Cl + Si \xrightarrow[\text{Cu catalyst}]{570\ K} \left. \begin{array}{c} (CH_3)_2SiCl_2 \\ \text{dichlorodimethyl-} \\ \text{silane} \\ \text{(main product)} \\ + \\ CH_3SiCl_3 \\ \text{methyltrichloro-} \\ \text{silane} \\ + \\ (CH_3)_3SiCl \\ \text{trimethylchloro-} \\ \text{silane} \end{array} \right\} \longrightarrow \textit{silicones}$$

It finds a number of other uses as a methylating agent, and a relatively small amount (possibly 10% of total production in the UK) is used as a refrigerant fluid, though in the main it has been replaced by fluorocarbons in this application. Production of methyl chloride in the USA in 1967 was 123 000 tons.

METHYLENE CHLORIDE

Some methylene chloride is made by the chlorination of methyl chloride, which as we have seen can be derived from methanol, but most is made by the direct chlorination of methane.

Methylene chloride is a very powerful solvent, and this is the basis for practically all its commercial use. It is non-flammable, an important advantage in many solvent applications. Major uses are in paint strippers, solvent extraction, as a solvent in the spinning of cellulose triacetate fibres, and in various metal cleaning applications. USA production in 1967 was 116 800 tons.

CHLOROFORM

A small amount of chloroform is still made by the method by which this compound was originally prepared, the reaction of acetone and calcium hypochlorite. However, this process is very wasteful of raw materials, and most manufacture is now by the chlorination of methane.

Chloroform is now used as an anaesthetic on only a very limited scale. Its major use is as an intermediate in the manufacture of chlorofluoromethanes (see 'Fluorine Compounds'). It has some other uses as an intermediate, and finds some application as a solvent. Production in the USA in 1967 was 85 200 tons.

CARBON TETRACHLORIDE

Substantial quantities of carbon tetrachloride are made by the chlorination of carbon disulphide, this process having the advantage (in some circumstances) of producing no by-product hydrogen chloride:

$$CS_2 + 3Cl_2 \xrightarrow{\text{Fe, 300 K}} CCl_4 + S_2Cl_2$$

The sulphur chloride formed can be reacted with more carbon disulphide to give carbon tetrachloride and sulphur:

$$CS_2 + 2S_2Cl_2 \longrightarrow CCl_4 + 6S$$

Carbon tetrachloride can also be made by chlorinolysis of a variety of materials, e.g. propene (see 'Perchloroethylene'). Chlorinolysis of propene will be used in a new carbon tetrachloride plant at present being constructed by ICI at Runcorn.

The most important use of carbon tetrachloride is as starting material in the production of chlorofluoromethanes for use as refrigerants and aerosol propellants. It is also used in fire extinguishers, and in some solvent applications. Production in the USA in 1967 was 318 700 tons.

Ethyl Chloride

Chlorination of ethane can be made to give high yields of ethyl chloride, since this is much less readily chlorinated than ethane, and the reaction system can be readily arranged so that it does not undergo further attack:

$$C_2H_6 + Cl_2 \longrightarrow C_2H_5Cl + HCl$$

At first sight this seems a little strange when compared with the chlorination of methane, for, as we have seen, methyl chloride is more readily attacked than methane. It appears that rate of abstraction of hydrogen atoms by chlorine atoms is not only influenced by the strengths of the carbon–hydrogen bonds involved, but that polar factors are also important. Thus, the presence of chlorine in the ethyl chloride molecule generally discourages attack by chlorine atoms. Attack, when it does occur, is mainly on the most easily removed hydrogen atoms, on the carbon bearing the chlorine, where the resonance effects discussed above come into play, so that 1,1-dichloroethane is the major dichlorination product. Methane is much less readily attacked than ethane, and it is presumed that the resonance effect in methyl chloride overrides the polar effect.

Chlorination of ethane is carried out at temperatures in the range 570 to 670 K. As is the case with chloromethane manufacture, the economics of this operation depend markedly on the value of the hydrogen chloride obtained as co-product.

Ethyl chloride is also made by the addition of hydrogen chloride to ethylene. The reaction can be carried out in either the gas phase or the liquid phase, e.g. in solution in ethyl chloride, in the presence of catalysts such as aluminium chloride:

$$C_2H_4 + HCl \longrightarrow C_2H_5Cl$$

In the gas phase temperatures in the range 400 to 520 K are used: in the liquid phase much lower temperatures (265 to 325 K) are required. The catalyst functions by polarising the hydrogen chloride and increasing its protonating activity. (In the presence of aluminium chloride, hydrogen chloride is sometimes visualised as acting as the hypothetical very strong acid $HAlCl_4$.)

The choice between ethylene and ethane as feedstock depends on the relative costs of the two hydrocarbons and on the value of hydrogen chloride in the particular circumstances involved. In the USA, most manufacture of ethyl chloride is based on ethylene. In the UK, the so-called integrated process is used. This takes advantage of the fact that reaction of chlorine with ethylene in the gas phase is much slower than reaction with ethane. Thus, a mixture of ethane and ethylene is treated with chlorine at about 670 K. The chlorine reacts with ethane to give ethyl chloride, which is separated, and the remaining gases consisting of ethylene, unconverted ethane, and hydrogen chloride are passed to a second reactor where reaction between ethylene and hydrogen chloride takes place under conditions of the type described above.

Manufacture of ethyl chloride from ethanol, although formerly important, is no longer attractive; it is in effect an indirect method of manufacture from ethylene.

The major use of ethyl chloride is in the manufacture of tetraethyl-lead:

$$\underset{\substack{\text{sodium-lead} \\ \text{alloy}}}{4\,NaPb} + 4C_2H_5Cl \longrightarrow Pb(C_2H_5)_4 + 4NaCl + 3Pb$$

It finds a number of other uses as an ethylating agent. USA production in 1967 was 276 000 tons.

Vinyl Chloride

Vinyl chloride, practically all of which is used to make poly(vinyl chloride) and vinyl chloride copolymers, is the largest tonnage organo-chlorine compound made, with the exception of its precursor, ethylene dichloride. UK capacity for vinyl chloride in 1970 was 430 000 tons per annum.

ACETYLENE-BASED PROCESS

Vinyl chloride has been made since the early 1930s by the process quoted in most organic chemistry text-books, the addition of hydrogen chloride to acetylene:

$$CH \equiv CH + HCl \longrightarrow CH_2 = CHCl$$

The reaction is carried out in the vapour phase, over a catalyst consisting of mercuric chloride supported on active carbon, at temperatures between 370 and 450 K depending on the age of the catalyst. (As the catalyst ages it loses activity, and to maintain the rate of reaction the temperature is raised.) It has been suggested that the reaction proceeds through an intermediate vinyl mercuric compound as shown below:

$$CH \equiv CH + HgCl_2 \longrightarrow \underset{\underset{H}{|}}{\overset{\overset{Cl}{|}}{C}} = \underset{\underset{HgCl}{|}}{\overset{\overset{H}{|}}{C}} \xrightarrow{\ HCl\ } CH_2 = CHCl + HgCl_2$$

Yields are variously quoted from 80 to 95%: probably the higher figure is realistic. The main by-product is 1,1-dichloroethane.

This process has a number of characteristics which the reader will recognise as being highly desirable: it has only one stage, the reaction conditions are mild, no heat need be supplied since the reaction is exothermic, and the stoichiometry is favourable, i.e. at 100% yield, all the raw materials appear in the product. Further, as it uses hydrogen chloride as raw material, this process can very conveniently be operated alongside other chlorohydrocarbon processes which give hydrogen chloride as a by-product. It does, however, have one major drawback—it is based on acetylene, a relatively expensive hydrocarbon. The high cost of acetylene provides a considerable economic incentive to base vinyl chloride manufacture on a cheaper two-carbon raw material; the most obvious such material is ethylene.

ETHYLENE DICHLORIDE PROCESS

The first process used for the manufacture of vinyl chloride from ethylene is generally known as the 'ethylene dichloride process'; as we shall see later, there are now other processes that involve ethylene dichloride as an intermediate.

In the ethylene dichloride process, ethylene and chlorine are allowed to react to give ethylene dichloride (1,2-dichloroethane), which is then pyrolised to give vinyl chloride and hydrogen chloride:

$$CH_2 = CH_2 + Cl_2 \longrightarrow CH_2ClCH_2Cl \longrightarrow CH_2 = CHCl + HCl$$

The chlorination may be carried out in the gas phase, at temperatures from 360 to 400 K. Under these conditions it is considered to be a free-radical chain reaction involving the following propagation steps:

$$CH_2 = CH_2 + Cl\cdot \longrightarrow CH_2ClCH_2\cdot$$

$$CH_2ClCH_2\cdot + Cl_2 \longrightarrow CH_2ClCH_2Cl + Cl\cdot$$

Homolysis of chlorine would not be expected to occur to any appreciable extent at the temperatures used, and it is suggested that initiation involves reaction of chlorine at the reactor wall, or on the catalyst if one is used:

$$Cl_2 \xrightarrow{\text{surface}} 2\,Cl\cdot$$

A variety of materials, e.g. calcium chloride, iron, magnesium oxide, have been suggested as catalysts for this reaction, but a catalyst is not essential.

The chlorination may also be carried out in the liquid phase, generally in solution in ethylene dichloride, with, e.g. ferric chloride as a catalyst. Temperatures of up to about 340 K are used. Under these conditions the reaction probably proceeds mainly by an ionic mechanism:

$$CH_2 = CH_2 + Cl_2 \longrightarrow \underset{\substack{Cl^+ \\ + \\ Cl^-}}{CH_2 - CH_2} \longrightarrow CH_2ClCH_2Cl$$

The catalyst is presumed to function by polarising the chlorine molecule and thereby increasing its electrophilicity.

Yields of about 95% are obtained by both gas-phase and liquid-phase chlorination.

The second stage of the process involves pyrolysis of the ethylene dichloride at about 770 K. This may be carried out in empty tubes or in tubes packed with active carbon. The reaction is a chain reaction, the following initiation and propagation steps being suggested:

initiation:

$$CH_2ClCH_2Cl \longrightarrow Cl\cdot + \cdot CH_2CH_2Cl$$

propagation:

$$Cl\cdot \; + \; CH_2ClCH_2Cl \longrightarrow CH_2Cl\dot{C}HCl \; + \; HCl$$

$$CH_2Cl\dot{C}HCl \longrightarrow CH_2 = CHCl + Cl\cdot$$

The initiation reaction probably occurs at least partly at the reactor wall, or on the catalyst surface.

Vinyl chloride is also produced in one of the termination reactions:

$$Cl\cdot \; + \; \cdot CH_2CH_2Cl \longrightarrow CH_2 = CHCl \; + \; HCl$$

Vinyl chloride can be further dehydrochlorinated to acetylene. To avoid this, this stage is operated at about 50% conversion; yields of up to 99% are claimed.

Although it has the advantage of being based on a cheaper hydrocarbon than the acetylene process, this route has the major drawback of converting half the chlorine fed to it to hydrogen chloride, and for many years this severely limited the use of ethylene in vinyl chloride manufacture.

THE BALANCED ROUTE

One at first sight elegant way round this disadvantage of the ethylene dichloride process is to use the so-called balanced route. In this, an acetylene-based and an ethylene-based process are run in parallel, the hydrogen chloride produced by ethylene dichloride pyrolysis being fed to the acetylene-based process:

$$C_2H_4 + Cl_2 \longrightarrow C_2H_4Cl_2 \longrightarrow CH_2 = CHCl + HCl$$

$$C_2H_2 + HCl \longrightarrow CH_2 = CHCl$$

The balanced route has been operated, e.g. by British Geon Ltd. (now BP) in the UK. However, this solution to the problem of hydrogen chloride by-product is not as attractive as it appears at first sight, for two reasons. Firstly, it requires three reaction stages, so that capital and operating costs are increased. Secondly, half the hydrocarbon fed is still the expensive acetylene. In the form described above, it is not now a favoured way of making vinyl chloride.

Modified forms of this process have recently been developed. In these, a mixture of acetylene and ethylene, produced by hydrocarbon cracking, is reacted first with hydrogen chloride, under conditions such that only the acetylene reacts. The resulting vinyl chloride is removed from the gas stream. The remaining ethylene is then reacted with chlorine to give ethylene dichloride. The hydrogen chloride produced by pyrolysis of the ethylene dichloride is fed to the first reaction stage:

$$C_n H_{2n+2} \longrightarrow C_2 H_2 + C_2 H_4 \xrightarrow{\;\;HCl\;\;} CH_2 = CHCl + C_2 H_4$$

$$C_2 H_4 \xrightarrow{\;\;Cl_2\;\;} C_2 H_4 Cl_2 \longrightarrow CH_2 = CHCl + HCl$$

The essential difference between this process and the balanced route described above is that substantial savings in the separation costs of ethylene and acetylene are made. However, the high capital and energy costs of the high-temperature hydrocarbon pyrolysis are still incurred, and the conversion of the acetylene/ethylene mixture to vinyl chloride requires three stages. This route is probably not now economically attractive compared with the type of process described below.

OXYCHLORINATION PROCESSES

The problem of the disposal of by-product hydrogen chloride is not a new one; in the nineteenth century large amounts of hydrogen chloride were produced as a by-product of alkali manufacture by the Leblanc process. In 1868, Henry Deacon patented a process for the conversion of hydrogen chloride to chlorine by oxidation with air over a copper chloride catalyst at about 720 K. This process was the major source of chlorine until the introduction of the electrolytic process for sodium hydroxide and chlorine, when it fell into disuse. In recent years, with the re-occurrence of a hydrogen chloride disposal problem, interest in it has revived, and a number of improved versions have been developed.

One way of avoiding wastage of chlorine would be to use the Deacon process to convert the hydrogen chloride by-product from ethylene dichloride pyrolysis back to chlorine:

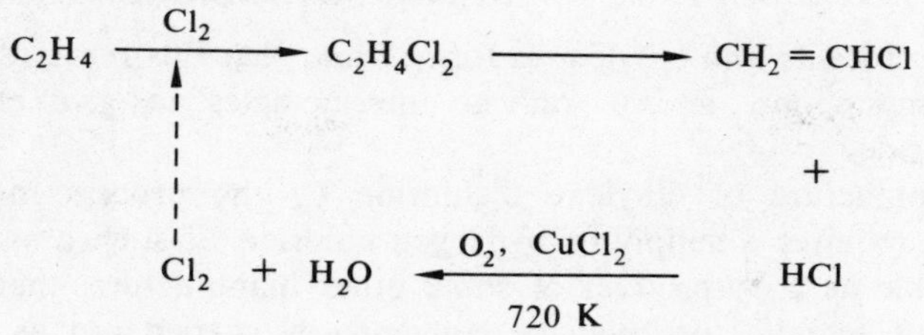

$$C_2H_4 \xrightarrow{\quad Cl_2 \quad} C_2H_4Cl_2 \longrightarrow CH_2=CHCl$$

$$+$$

$$Cl_2 + H_2O \xleftarrow[720\ K]{O_2,\ CuCl_2} \qquad HCl$$

However, this involves adding an extra stage to the process, with the resultant increase in capital and operating costs.

A more elegant approach is to carry out the oxidation of the hydrogen chloride in the presence of ethylene, i.e. to carry out an *oxychlorination*. Although oxychlorination has been used since 1934 for the production of chlorobenzene from benzene and hydrogen chloride, in the Raschig process for phenol (see page 185), processes for the oxychlorination of ethylene have been developed only relatively recently.

Considerably lower temperatures than those used in the Deacon process are used in oxychlorination of ethylene. The catalyst is normally a mixture of cupric chloride and potassium chloride on a solid support. It is thought that under operating conditions the salt mixture is at least partly in the liquid phase, absorbed on the support:

$$C_2H_4 + 2HCl \xrightarrow[520-620\ K]{O_2,\ CuCl_2/KCl} CH_2ClCH_2Cl + H_2O$$

Yields of over 90% at about 95% conversion are obtainable.

It is suggested that ethylene dichloride is formed by reaction of ethylene and cupric chloride:

$$C_2H_4 + 2CuCl_2 \longrightarrow CH_2ClCH_2Cl + Cu_2Cl_2$$

Cupric chloride is regenerated by the reaction of oxygen and hydrogen chloride with cuprous chloride:

$$Cu_2Cl_2 + \tfrac{1}{2}O_2 \longrightarrow CuO.CuCl_2$$

$$CuO.CuCl_2 + 2HCl \longrightarrow 2CuCl_2 + H_2O$$

The mechanism of the reaction between cupric chloride and ethylene is not known with certainty at present. It is suggested that a complex of the chloride and ethylene is formed and that this reacts through chloronium ion and/or radical intermediates to give ethylene dichloride.

Manufacture of ethylene dichloride by the process indicated above requires a supply of hydrogen chloride. If such a supply is available as a by-product of some other manufacture, then vinyl chloride manufacture may be conveniently carried out as shown below:

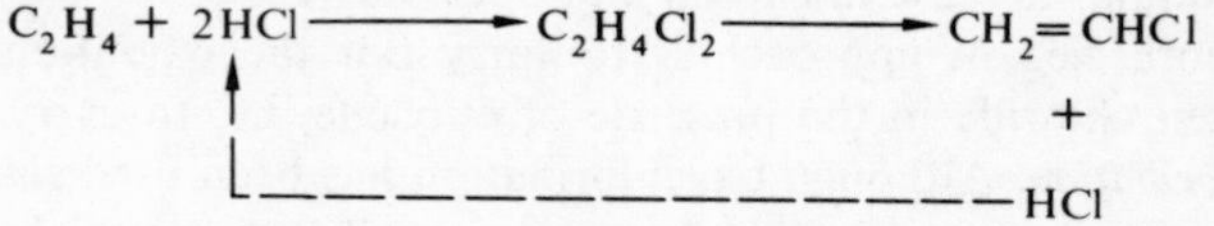

$$C_2H_4 + 2HCl \longrightarrow C_2H_4Cl_2 \longrightarrow CH_2{=}CHCl$$

$$+ $$

$$\text{---------------------- } HCl$$

If such a supply is not available, then to avoid the necessity of making hydrogen chloride from hydrogen and chlorine, a chlorination and an oxychlorination stage may be run in parallel, and the hydrogen chloride from the pyrolysis stage recycled to the oxychlorination:

$$C_2H_4 + Cl_2$$

$$\searrow$$

$$C_2H_4Cl_2 \longrightarrow CH_2{=}CHCl$$

$$C_2H_4 + 2HCl \nearrow$$

$$+$$

$$\text{-------------------- } HCl$$

This has the obvious disadvantage of requiring an extra reaction stage, with the consequent effects on capital and operating costs.

A more recently developed process involves carrying out the oxychlorination in the liquid phase in an aqueous solution of cupric chloride. This has what would appear to be the very considerable advantage that a mixture of chlorine and hydrogen chloride can be fed to the oxychlorination stage, so that if desired the process can be operated with only chlorine and ethylene as raw materials, with only two stages:

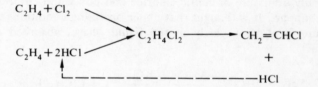

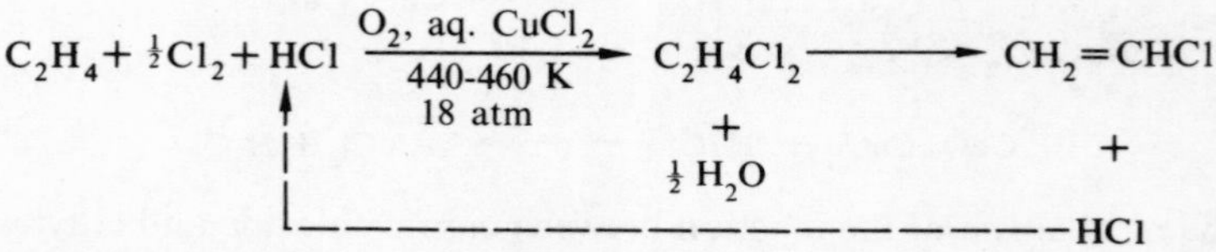

$$C_2H_4 + \tfrac{1}{2}Cl_2 + HCl \xrightarrow[\substack{440\text{-}460\ K \\ 18\ atm}]{O_2,\ aq.\ CuCl_2} C_2H_4Cl_2 \longrightarrow CH_2{=}CHCl$$

$$+$$

$$\tfrac{1}{2}H_2O$$

$$+$$

$$\text{----------------------- } HCl$$

This process has a considerable degree of flexibility in that if by-product hydrogen chloride is available, this can be used to replace some or all of the chlorine feed.

A further development, which has not yet reached the stage of commercial operation, but which will be attractive if good yields can be achieved, is to carry out the oxychlorination at higher temperatures, when vinyl chloride can be produced in one stage:

$$C_2H_4 + HCl \xrightarrow[750\ K]{O_2, CuCl_2/KCl} CH_2=CHCl + H_2O$$

CURRENT SITUATION IN VINYL CHLORIDE MANUFACTURE
The fact that the conventional ethylene dichloride process converts half the chlorine feed to it into hydrogen chloride has considerably limited its application, and until very recently most vinyl chloride was made from acetylene. In 1965, for example, the proportions of vinyl chloride manufacture based on acetylene in the USA and the UK respectively were about 55 and 80%. The development of oxychlorination processes and, in the UK at least, the disappointing performance of cracking processes for the manufacture of acetylene, have now made ethylene the favoured raw material for vinyl chloride manufacture. In the UK, ICI currently operate an oxychlorination process in two plants of capacity 150 000 tons per annum each, and BP have a plant of capacity 260 000 tons per annum under construction. When these plants are fully operational, acetylene-based manufacture of vinyl chloride will probably be discontinued in the UK.

Vinylidine Chloride

Vinylidine chloride is the monomer for poly(vinylidine chloride), sold under the trade name 'Saran'. This polymer is made in relatively small tonnages, and finds its main outlets in certain fibre applications, and in film for packaging.

Vinylidine chloride is made by dehydrochlorination of 1,1,2-trichloroethane by treatment with an aqueous suspension of calcium hydroxide at about 320 K. The trichloroethane may be made in a number of ways, e.g. by chlorination of vinyl chloride in the liquid phase at 300 to 320 K:

$$CH_2=CHCl \xrightarrow{Cl_2} CH_2ClCHCl_2 \xrightarrow{Ca(OH)_2} CH_2=CCl_2 + CaCl_2$$

Trichloroethylene

1,1,2-Trichloroethylene was the first organic chemical to be manufactured from acetylene. Manufacture was started in 1908 in Bosnia
(now Yugoslavia), and in England.

Manufacture from acetylene has continued to the present day,
and still accounts for much production. The process involves
addition of chlorine to acetylene to give 1,1,2,2-tetrachloroethane,
followed by elimination of hydrogen chloride from the tetrachloroethane:

$$CH\equiv CH + Cl_2 \longrightarrow CHCl_2CHCl_2 \xrightarrow{-HCl} CHCl = CCl_2$$

The reaction between acetylene and chlorine is carried out in the
liquid phase in solution in tetrachloroethane, at about 350 K and in
the presence of a catalyst, e.g. ferric chloride. Elimination of
hydrogen chloride was originally brought about by treatment with
a slurry of calcium hydroxide at 370 K. When operated in this way
the process is obviously wasteful of raw materials in that it converts
chlorine into valueless calcium chloride. More recently, processes
in which hydrogen chloride is eliminated by pyrolysis have also
been used. The reaction is carried out at 570 to 770 K in the presence
of a catalyst, e.g. barium chloride on silica gel or active carbon, or
at about 870 K in the absence of a catalyst.

The uncatalysed dehydrochlorination probably proceeds mainly
by a free-radical chain reaction involving steps of the following
type:

initiation:

$$CHCl_2CHCl_2 \longrightarrow CHCl_2\dot{C}HCl + Cl\cdot$$

propagation:

$$Cl\cdot + CHCl_2 CHCl_2 \longrightarrow CHCl_2\dot{C}Cl_2 + HCl$$

$$CHCl_2\dot{C}Cl_2 \longrightarrow CHCl = CCl_2 + Cl\cdot$$

It has been suggested that in the presence of a catalyst the reaction
is a *beta*-elimination in which the role of the catalyst is to polarise
the reactant molecule:

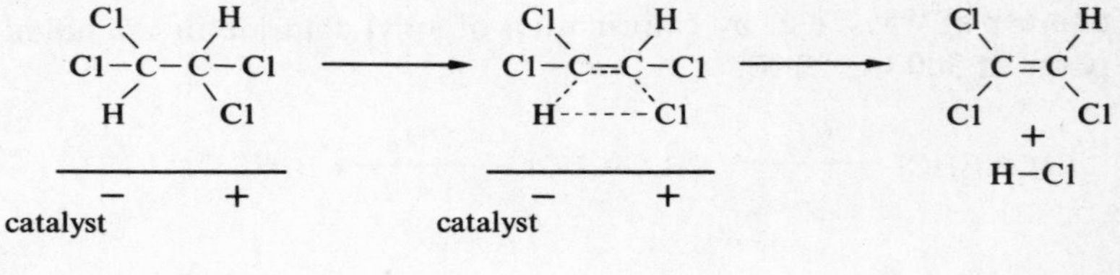

Trichloroethylene can also be made from ethylene, for example by chlorination firstly to ethylene dichloride and then to 1,1,1,2-tetrachloroethane, followed by pyrolysis:

$$CH_2=CH_2 \xrightarrow{\text{Cl}_2} CH_2ClCH_2Cl \xrightarrow{\text{Cl}_2} CH_2ClCCl_3 \longrightarrow CHCl=CCl_2$$
$$+ \qquad\qquad\qquad +$$
$$2HCl \qquad\qquad\quad HCl$$

The heavy chlorine consumption of such processes discouraged manufacture based on ethylene for many years. Recently, however, the development of outlets for large amounts of hydrogen chloride in vinyl chloride manufacture, and also of a number of oxychlorination routes to trichloroethylene, coupled with the decrease in ethylene costs compared with acetylene costs, have made ethylene-based manufacture economically attractive. An increasing proportion of trichloroethylene is now derived from ethylene.

Total USA production in 1967 was about 218 700 tons.

APPLICATIONS

The major use of trichloroethylene is in solvent applications, of which the most important is metal degreasing. For example, in the USA, about 90% of the trichloroethylene consumed is used in this application. In engineering manufacture, metal parts are often greasy or oily after forming, and require cleaning before passing to the finishing operations. A very convenient way of carrying this out is by 'vapour degreasing'. This involves immersing the part to be cleaned in an atmosphere of solvent vapour, so that the hot vapours condense on the metal, dissolving the grease and flushing it from the work. The article rapidly reaches vapour temperature, and condensation ceases, leaving the work dry and ready for finishing. The operation is conveniently carried out in a tank in which solvent is boiled by a heater at the bottom of the tank, and vapour is condensed and prevented from escaping to the atmosphere by condenser coils round the top of the tank (see Fig. 4-2).

The requirements for a solvent for this process are more exacting than would appear at first sight. It must, of course, be a good solvent for oil and grease. It is highly desirable that it should be non-flammable, in that a tank of flammable vapour presents a very substantial hazard. Its boiling point should be high enough to give rapid, efficient cleaning, but not so high that the work cannot be handled when it is removed from the bath. A low latent heat of evaporation is desirable to minimise energy requirements. It should preferably be non-toxic, and should certainly not be highly toxic. It should be stable under the operating conditions, and should not, for instance, give rise to any corrosive decomposition products. Finally, it should if possible be cheap.

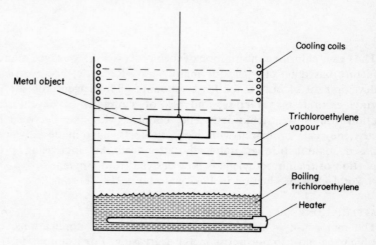

Fig. 4-2. *Vapour degreasing bath.*

Trichloroethylene fulfils most of these requirements, and is the major solvent used for vapour degreasing. It has excellent solvent powers, is non-flammable, and has a boiling point of 360 K, which is about the optimum. It is one of the least toxic of the chlorohydrocarbons, and given proper design of equipment and control of working conditions, can readily be safely handled. Since it contains only chlorine atoms attached to doubly-bonded carbon, it is highly resistant to hydrolysis (cf. vinyl chloride and chlorobenzene). This is important in that moisture may well be present under the operating conditions, and the production of the highly corrosive hydrogen chloride is highly undesirable.

In the presence of air, however, oxidation occurs, leading to the formation of a variety of products of which the most important are dichloroacetyl chloride, phosgene, carbon monoxide, and hydrogen chloride:

$$CHCl = CCl_2 \xrightarrow{O_2} CHCl_2COCl, \ COCl_2, \ HCl, \ CO.$$

The precise mechanism of this reaction does not appear to be very well established.

Other degradation reactions can occur under the catalytic effect of metal chlorides formed by the attack of hydrogen chloride on parts being cleaned.

It is thus necessary to add stabilisers to trichloroethylene which is to be used in vapour degreasing. A very large number of compounds have been proposed for this application: their mode of action is not well documented. Some, e.g., amines and phenols, presumably act as antioxidants (see Chapter 8), others, e.g., epoxides, may act as hydrogen chloride acceptors:

$$\underset{O}{C-C} + HCl \longrightarrow -\underset{OH}{C}-\underset{Cl}{C}-$$

Normally, a mixture of stabilisers is used.

Trichloroethylene has a number of other relatively minor uses. It is used to some extent in dry-cleaning, although to a large extent it has been displaced from this application by perchloroethylene, which has less tendency to cause dyes to 'bleed', and it has a number of other uses as a solvent. It is used to some extent as an anaesthetic, particularly in childbirth.

Perchloroethylene (1,1,2,2-tetrachloroethylene)
The first process used for the manufacture of perchloroethylene was based on acetylene, and involved trichloroethylene as an intermediate:

$$CH \equiv CH \longrightarrow CHCl = CCl_2 \xrightarrow{Cl_2} CHCl_2CCl_3 \xrightarrow{-HCl} CCl_2 = CCl_2$$

This process is still used for some manufacture of perchloroethylene. Elimination of hydrogen chloride from the pentachloroethane is normally brought about by treatment with a suspension of calcium hydroxide; it can also be carried out catalytically, e.g. over a copper chloride catalyst at 570 K.

Most manufacture (more than two-thirds of the total in the USA) is now by processes based on raw materials other than acetylene. For example, exhaustive chlorination of a range of compounds under forcing conditions gives a mixture of perchloroethylene and carbon tetrachloride. Thus, Progil, in France, produce carbon tetrachloride and perchloroethylene by chlorinolysis of propene:

$$C_3H_6 \xrightarrow{Cl_2, 870 \text{ K}} C_2Cl_4 + CCl_4 + 6HCl$$

At the temperatures involved, perchloroethylene and carbon tetrachloride are rapidly interconverted, and exist in an equilibrium mixture:

$$2CCl_4 \rightleftharpoons CCl_2 = CCl_2 + 2Cl_2$$

The proportions of the products obtained can be controlled to some extent by varying the operating conditions. If desired, unwanted product can be recycled to the reactor.

USA production of perchloroethylene in 1967 was 238 000 tons.

APPLICATIONS

The properties of perchloroethylene are generally similar to those of trichloroethylene. The main differences are that it has a higher boiling point (394 K), and a lower solubility in water (0·015 g/100 g compared with 0·11 g/100 g for trichloroethylene). It undergoes oxidative degradation in a very similar way to trichloroethylene, but in this case to give trichloroacetyl chloride and phosgene:

$$CCl_2 = CCl_2 \xrightarrow{O_2} CCl_3COCl, \; COCl_2.$$

As in the case of trichloroethylene, this reaction is retarded by the use of stabilisers.

The major use of perchloroethylene is in dry-cleaning. It has advantages in this application over carbon tetrachloride and trichloroethylene in that it causes far fewer dyes to bleed, i.e., to migrate from the fabric, than these solvents, and it is much less toxic than carbon tetrachloride. It finds some use in metal cleaning. However, in vapour degreasing its high boiling point can be a disadvantage in that the work is too hot to handle direct from the bath. It has a variety of other solvent applications.

Allyl Chloride

Allyl chloride was not made commercially until 1945, there being no method suitable for the manufacture of large quantities avail-

able until just before World War II. This is a case where manufacture of an evidently useful and versatile intermediate had to
await a fundamental chemical discovery. This discovery, made in
the laboratories of the Shell Development Company in the USA
was that at temperatures above about 570 K, chlorine reacts with
propene to give substitution on the 3-carbon atom rather than by
addition to the double bond.

Substitution occurs by a free radical chain reaction mechanism
involving the initiation and propagation steps indicated below:

$$Cl_2 \longrightarrow 2Cl \cdot \qquad (1)$$

$$Cl \cdot + CH_3CH = CH_2 \longrightarrow \cdot CH_2CH = CH_2 + HCl \qquad (2)$$

$$Cl_2 + \cdot CH_2CH = CH_2 \longrightarrow ClCH_2CH = CH_2 + Cl \cdot \qquad (3)$$

Addition to the double bond also proceeds by a free-radical
mechanism:

$$Cl \cdot + CH_3CH = CH_2 \longrightarrow CH_3\overset{\cdot}{C}HCH_2Cl \qquad (4)$$

$$Cl_2 + CH_3\overset{\cdot}{C}HCH_2Cl \longrightarrow CH_3CHClCH_2Cl + Cl \cdot \qquad (5)$$

Reaction (2) has a higher activation energy than reaction (4), and is
therefore more temperature-sensitive. Thus, whereas addition is the
faster reaction at low temperatures, as the temperature is increased
substitution becomes more favoured. A further factor is that at
high temperatures hydrogen atom abstraction from the radical
formed in reaction (4) becomes important:

$$Cl \cdot + CH_3\overset{\cdot}{C}HCH_2Cl \longrightarrow CH_2 = CHCH_2Cl + HCl$$

In commercial operation a reaction temperature of about 770 K
is used, and the reaction time is about 2 seconds. About 4 moles of
propene per mole of chlorine are used; the contacting pattern is of
extreme importance, and has major effects on the yield. Under the
correct conditions yields of about 85% are obtained, the major by-
products being cis- and trans-1,2-dichloropropene, 1,2-dichloro-
propane and 2-chloropropene.

The major uses of allyl chloride are in the manufacture of glycerol,
epichlorohydrin and allyl alcohol; it has a number of other small-
tonnage uses as an intermediate. Allyl chloride is not made in the
UK. Capacity in the USA in 1968 was about 135 000 tons per
annum.

Allyl alcohol is obtained by hydrolysis of allyl chloride under alkaline conditions:

$$CH_2{=}CHCH_2Cl \xrightarrow{\text{NaOHaq., 430 K}} CH_2{=}CHCH_2OH + NaCl$$

It has a number of relatively small-tonnage uses, e.g. in polymer applications.

Epichlorohydrin is made by reacting allyl chloride with hypochlorous acid and then treating the resulting dichlorohydrins with alkali under carefully controlled conditions:

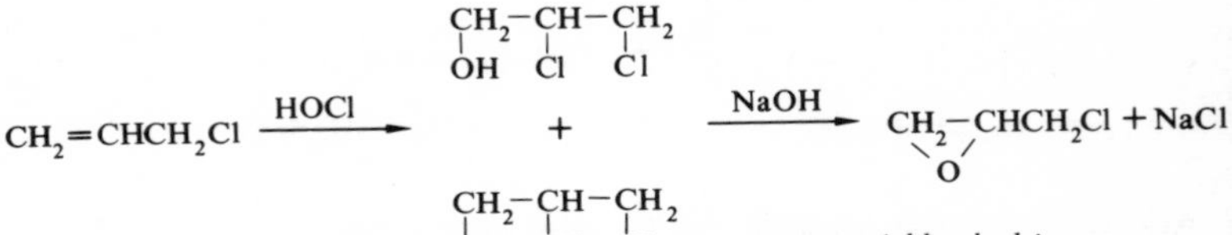

epichlorohydrin

Epichlorohydrin is used in the manufacture of epoxy resins (see Chapter 7).

Glycerol can be made from allyl chloride by a number of routes. One involves the hydrolysis of propylene dichlorohydrins, made as described above, with alkali:

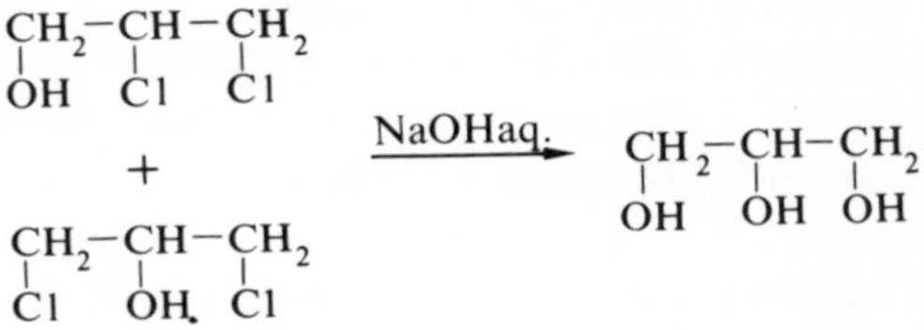

It should be noted that in many countries soap manufacture still provides the main or only source of glycerol.

Chloroprene

Chloroprene (2-chlorobuta-1,3-diene) is the monomer for neoprene, one of the earliest synthetic rubbers (see page 254). The original, and still the major, method of manufacture is by addition of hydrogen chloride to vinylacetylene, the vinylacetylene being made by dimerisation of acetylene:

$$2CH{\equiv}CH \longrightarrow CH{\equiv}C{-}CH{=}CH_2 \xrightarrow{\text{HCl}} CH_2{=}\underset{}{\overset{\text{Cl}}{\overset{|}{C}}}{-}CH{=}CH_2$$

This process is used by duPont at Maydown in Northern Ireland, at a scale of about 30 000 tons per annum.

The dimerisation of acetylene is carried out by contacting acetylene with an aqueous solution containing cuprous chloride and other salts, e.g. ammonium chloride, at about 290 K. Conversion is limited to about 10% to limit the formation of higher polymers of acetylene. Hydrochlorination of the vinylacetylene is brought about by treatment with concentrated hydrochloric acid containing cuprous chloride, at about 300 to 330 K. The overall yield is about 88%.

This process is now subject to competition from processes based on raw materials other than acetylene. For example, in the mid-1950s the Distillers Company Ltd. introduced a process using butadiene as raw material. In this process, butadiene is chlorinated in the gas phase at about 570 K to give a mixture of *cis*- and *trans*-1,4-dichlorobut-2-enes and 3,4-dichlorobut-1-ene:

$$CH_2=CH-CH=CH_2 \xrightarrow{\;Cl_2\;} \begin{array}{c} \underset{\underset{Cl}{|}}{CH_2}-CH=CH-\underset{\underset{Cl}{|}}{CH_2} \\ + \\ CH_2=CH-\underset{\underset{Cl}{|}}{CH}-\underset{\underset{Cl}{|}}{CH_2} \end{array}$$

The reaction is a free-radical chain reaction, and conditions must be carefully designed to avoid the occurrence of a variety of side reactions.

The 3,4-dichlorobut-1-ene, b.p. 396 K, which is the product required for the final stage of the process, is separated from the mixture of 1,4-dichlorobut-2-enes, b.p. 428 K, by distillation. The 1,4-dichlorobut-2-enes are subjected to isomerisation in the presence of copper and other salts to produce further 3,4-dichlorobut-1-ene:

$$\underset{\underset{Cl}{|}}{CH_2}-CH=CH-\underset{\underset{Cl}{|}}{CH_2} \qquad\qquad \underset{\underset{Cl}{|}}{CH_2}-\underset{\underset{Cl}{|}}{CH}-CH=CH_2$$

$$\Big\updownarrow \tfrac{1}{2}Cu_2Cl_2 \qquad\qquad -\tfrac{1}{2}Cu_2Cl_2 \Big\updownarrow$$

$$\underset{\underset{Cl}{|}}{CH_2}-CH=CH-CH_2^+ \; CuCl_2^- \longleftrightarrow \underset{\underset{Cl}{|}}{CH_2}-\overset{+}{CH}-CH=CH_2 \; CuCl_2^-$$

In the final stage of the process, the 3,4-dichlorobut-1-ene is dehydrochlorinated by heating with aqueous alkali:

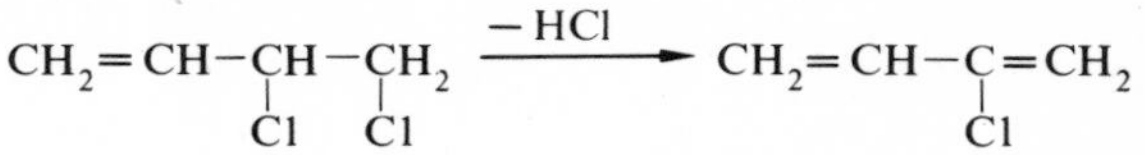

$$CH_2=CH-\underset{\underset{Cl}{|}}{CH}-\underset{\underset{Cl}{|}}{CH_2} \xrightarrow{-HCl} CH_2=CH-\underset{\underset{Cl}{|}}{C}=CH_2$$

A small amount of 1-chlorobuta-1,3-diene is produced, and is separated by fractional distillation.

This process is operated in France at a scale of 20 000 tons per annum, and a plant of the same capacity is under construction in the USA.

Ethylene and Propylene Chlorohydrins

Chlorohydrins are obtained by the reaction of alkenes with chlorine and water. The reaction can be considered to be an addition of hypochlorous acid to the alkene, e.g.:

$$Cl_2 + H_2O \rightleftharpoons HOCl + HCl$$

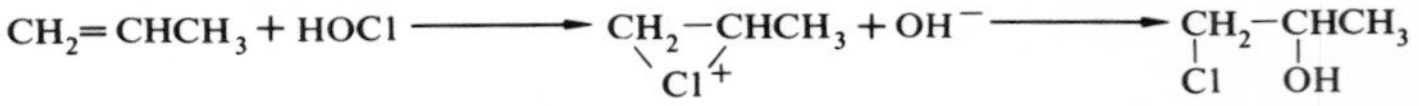

$$CH_2=CHCH_3 + HOCl \longrightarrow \underset{Cl^+}{CH_2-CHCH_3} + OH^- \longrightarrow \underset{Cl}{\underset{|}{CH_2}}-\underset{OH}{\underset{|}{CHCH_3}}$$

Alternatively, a chloronium ion may be formed by reaction of propene with chlorine, and subsequently attacked by water:

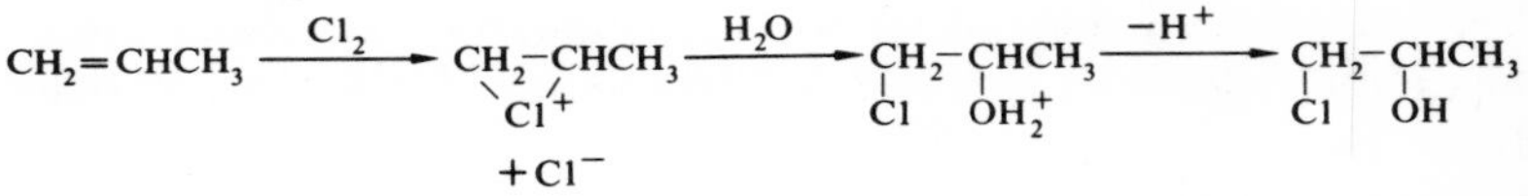

$$CH_2=CHCH_3 \xrightarrow{Cl_2} \underset{\underset{+Cl^-}{Cl^+}}{CH_2-CHCH_3} \xrightarrow{H_2O} \underset{Cl\ \ OH_2^+}{\underset{|\ \ \ |}{CH_2-CHCH_3}} \xrightarrow{-H^+} \underset{Cl\ \ OH}{\underset{|\ \ \ |}{CH_2-CHCH_3}}$$

There seems no reason to suppose that both mechanisms do not operate.

In either case, the chloronium ion can be attacked by chloride ion as well as by water or hydroxide ion, and dichlorides are by-products of the reaction:

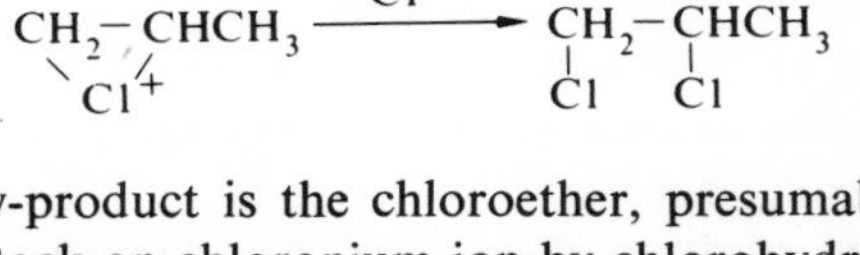

$$\underset{Cl^+}{CH_2-CHCH_3} \xrightarrow{Cl^-} \underset{Cl\ \ Cl}{\underset{|\ \ |}{CH_2-CHCH_3}}$$

A further by-product is the chloroether, presumably formed by nucleophilic attack on chloronium ion by chlorohydrin:

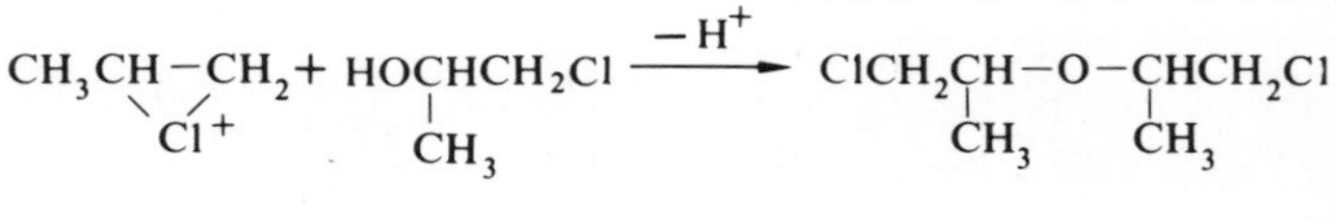

$$\underset{Cl^+}{CH_3CH-CH_2} + \underset{CH_3}{\underset{|}{HOCHCH_2Cl}} \xrightarrow{-H^+} \underset{CH_3\ \ \ CH_3}{\underset{|\ \ \ \ \ \ |}{ClCH_2CH-O-CHCH_2Cl}}$$

The yield of chlorohydrin obtained is very much dependent on the conditions under which the reaction is carried out. Formation of dichloride is minimised by the use of a low concentration of chlorine, and that of ether by ensuring that the concentration of chlorohydrin remains low. Temperatures ranging from 280 to 320 K are used. Yields of both ethylene and propylene chlorohydrin by this process are about 85%.

The only substantial use of these products is as intermediates in the preparation of the corresponding alkene oxides. These are made by treatment of the chlorohydrin/hydrochloric acid solution obtained as described above, with calcium hydroxide. (Calcium hydroxide is used because it is the cheapest base available.) The reaction involves an intramolecular nucleophilic displacement of chloride ion:

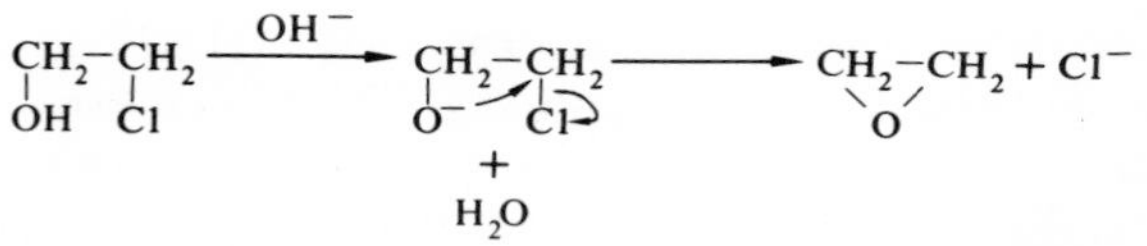

$$CH_2-CH_2 \xrightarrow{\ OH^- \ } CH_2-CH_2 \longrightarrow CH_2-CH_2 + Cl^- $$

Thus, ethylene and propylene oxides may be prepared by the following overall routes:

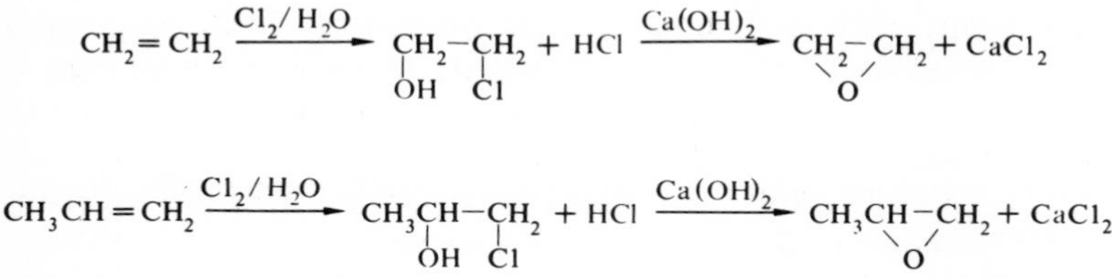

$$CH_2=CH_2 \xrightarrow{Cl_2/H_2O} CH_2-CH_2 + HCl \xrightarrow{Ca(OH)_2} CH_2-CH_2 + CaCl_2$$

$$CH_3CH=CH_2 \xrightarrow{Cl_2/H_2O} CH_3CH-CH_2 + HCl \xrightarrow{Ca(OH)_2} CH_3CH-CH_2 + CaCl_2$$

In both cases, yields of about 80% based on the alkene are obtained.

The overriding factor in the economics of these processes is the fact that they convert chlorine into valueless calcium chloride. Thus, the manufacture of 1 ton of ethylene oxide consumes 2·35 tons of chlorine, which at 1970 UK prices costs about £75. There is consequently a large economic incentive to manufacture these materials in some way which does not involve such high raw material costs. As we have seen, most ethylene oxide manufacture (all manufacture in the UK) is by the direct oxidation process. Although this gives a lower yield of ethylene oxide, it uses only ethylene and air or oxygen as raw materials. In the case of propylene oxide, economically viable alternatives to the chlorohydrin route have only recently been developed, and at the time of writing, most propylene oxide is still made by the chlorohydrin process.

Insecticides

Insecticides are an important group of products of the organic chemical industry, and make a major contribution to the quality of life by increasing agricultural efficiency and output, and by controlling insect-borne disease. Chlorine-containing compounds at present make up the largest group of synthetic organic insecticides. However, they have the characteristic of being extremely persistent, and in recent years concern has developed about the long-term effects of these materials on the environment. As a consequence, their use is now beginning to dwindle.

DDT

The discovery of the insecticidal properties of DDT in the Swiss laboratories of the Geigy Company in 1939 was an event of major importance. Prior to this discovery, the only insecticides available were naturally occurring products, e.g. derris, and a limited range of synthetic insecticides, all of which had major technical disadvantages, e.g. in some cases high toxicity to warm-blooded animals, low persistence, low potency. DDT was found to be a highly effective insecticide, with a combination of properties which apparently approached the ideal. It is highly potent, and kills a wide range of insects by both contact and ingestion. It is of low acute toxicity to warm-blooded animals. It has little smell, and does not taint food. It is chemically stable, and therefore persistent. (This persistence is now considered to be a major disadvantage of DDT.) Finally, since it is made by a simple process from low-cost raw materials, it is cheap.

DDT is manufactured by the condensation of chloral with chlorobenzene in the presence of concentrated sulphuric acid:

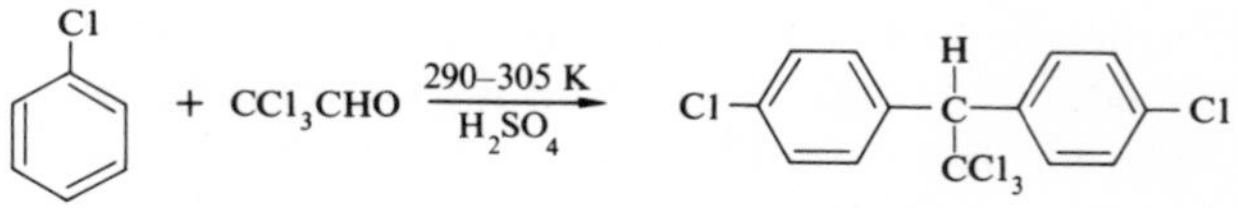

1,1−bis(p− chlorophenyl)−2, 2, 2−
trichloroethane (main product)

DDT

The mechanism of the reaction is presumed to be as shown below:

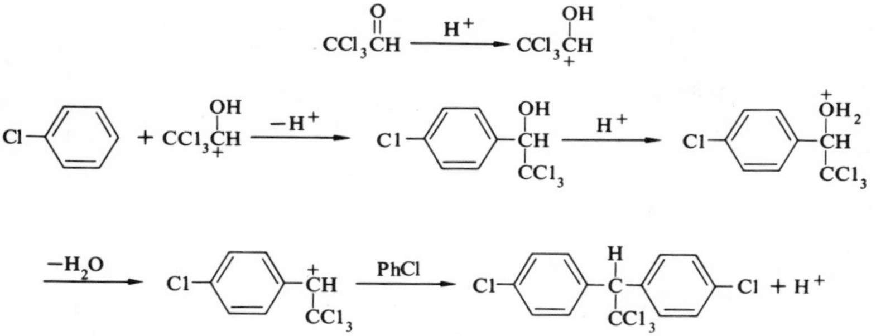

This type of reaction between aldehydes and ketones and aromatic compounds is of general applicability, and is discussed in Chapter 5.

As would be expected, isomers in which *ortho-* substitution has occurred are also obtained. Technical DDT consists of rather more than 70% of the *p,p'*-isomer, and about 20% of the less active *o,p*-isomer, together with a number of other by-products, including a trace of the *o,o'*-isomer. For most applications these impurities have no deleterious effect, and the product is used without purification.

Chloral is prepared by chlorination of either ethanol or acetaldehyde.

Capacity for DDT manufacture in the USA in 1965 was about 90 000 tons per annum.

BENZENE HEXACHLORIDE

Benzene hexachloride, or more correctly, 1,2,3,4,5,6-hexachlorocyclohexane, was found to have strong insecticidal properties in about 1942, by ICI workers in the UK, and by Lebas in France. Its properties as an insecticide are broadly similar to those of DDT, and it has achieved substantial importance, though by no means to the same extent as DDT.

Benzene hexachloride is prepared by the addition of chlorine to benzene under the influence of light. The reaction undoubtedly involves free radicals, but its exact mechanism is not known. One obvious candidate mechanism is a chain reaction of the type shown below:

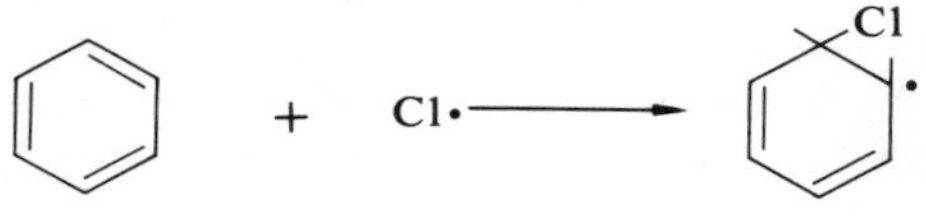

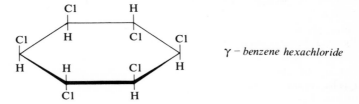

etc.

However, the situation may well be more complicated than this.

There are eight possible geometrical isomers of benzene hexachloride, of which only one, the γ-isomer, has insecticidal properties:

γ – *benzene hexachloride*

The γ-isomer makes up about 12% of benzene hexachloride as normally prepared. As the other isomers are inactive, their formation represents a considerable wastage of raw materials, and much effort has been expended in attempts to vary the isomer ratio. Limited control over the γ-isomer content can be achieved by varying the concentration of chlorine used in the process, and by the use of solvents.

Crude benzene hexachloride has a musty smell and causes tainting of foods with which it comes in contact. Purified product, 'lindane', of about 99% γ-isomer content is considerably better in this respect. It is prepared by fractional crystallisation of the crude product.

Benzene hexachloride is used on a much smaller scale than DDT.

OTHER CHLORINE-CONTAINING INSECTICIDES

There are a number of other commercially important chlorine-containing insecticides. A very potent group of compounds are those based on the Diels-Alder reaction of hexachlorocyclopentadiene with certain alkenes, the reaction being of the general type:

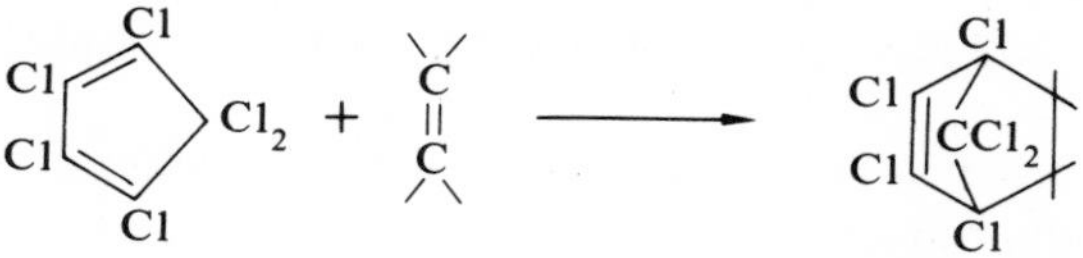

Thus, reaction of hexachlorocyclopentadiene with cyclopentadiene, followed by chlorination of the product, gives 'chlordane':

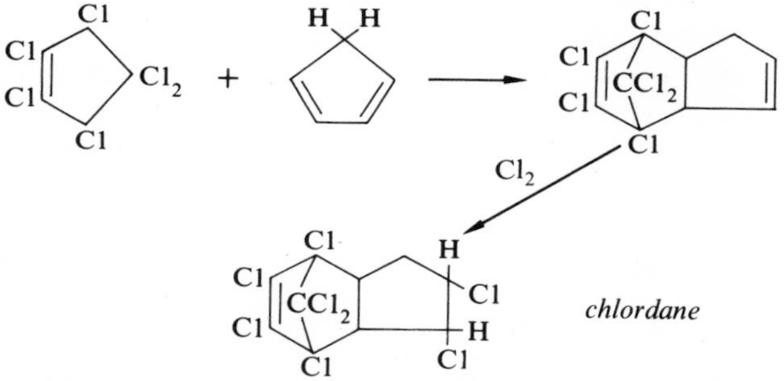

Cl_2

chlordane

The well-known insecticides 'aldrin' and 'dieldrin' are of this class.

Many of these products have considerable toxicity to warm-blooded animals, and their use is now severely restricted.

Herbicides

Herbicides have wide and important use in agriculture, for killing weeds, for killing some crops prior to harvesting, and in a number of other applications. A number of important herbicides are chlorine-containing compounds, the most well-known being the selective weed killers 2,4-D and MCPA. Since these have similar applications and are made by very similar processes, we shall consider them together.

2,4-D AND MCPA

2,4-D (2,4-dichlorophenoxyacetic acid) and MCPA (2-methyl-4-chlorophenoxyacetic acid) have a much greater herbicidal effect on broad-leafed than on narrow-leafed plants (i.e. cereals and grasses). They can consequently be used to kill weeds amongst cereal crops, or, on the domestic scale, in lawns. They act as plant hormones and cause excessive, disorganised growth of the plant which leads to death by drought. They are of very major importance in modern agricultural practice. 2,4-D is currently the most important herbicide made in the USA. In the UK MCPA is the more important product of the two.

Both products are made from the corresponding chlorophenol and sodium monochloroacetate by the Williamson synthesis:

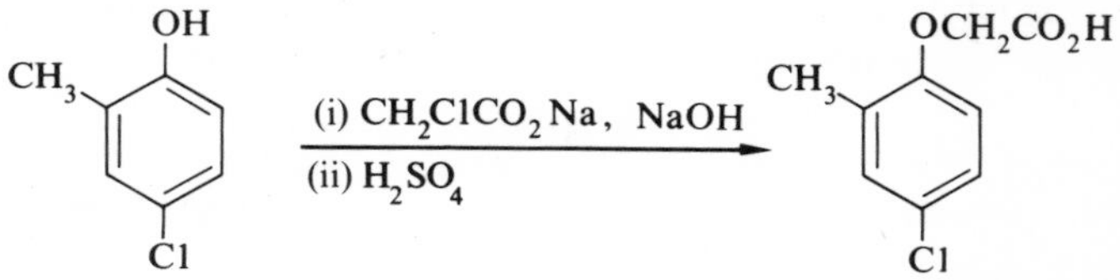

CH$_3$—C$_6$H$_3$(OH)(Cl) $\xrightarrow[\text{(ii) } H_2SO_4]{\text{(i) } CH_2ClCO_2Na,\ NaOH}$ CH$_3$—C$_6$H$_3$(OCH$_2$CO$_2$H)(Cl)

MCPA

Cl—C$_6$H$_3$(OH)(Cl) $\xrightarrow[\text{(ii) } H_2SO_4]{\text{(i) } CH_2ClCO_2Na,\ NaOH}$ Cl—C$_6$H$_3$(OCH$_2$CO$_2$H)(Cl)

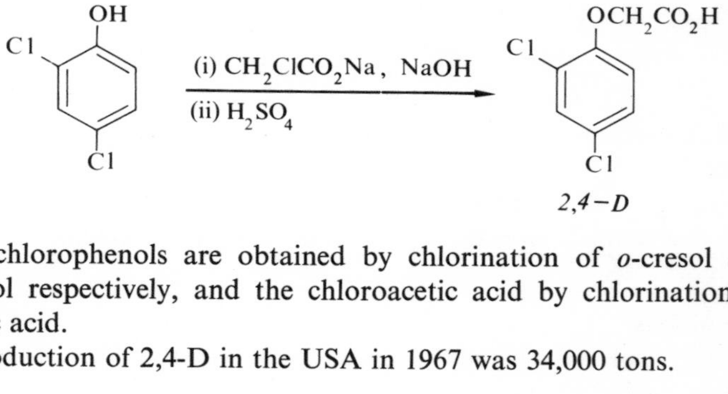

2,4-D

The chlorophenols are obtained by chlorination of *o*-cresol and phenol respectively, and the chloroacetic acid by chlorination of acetic acid.

Production of 2,4-D in the USA in 1967 was 34,000 tons.

Dyestuffs and Pigments

As was indicated at the beginning of this chapter, a great deal of use is made of chlorine-containing compounds as synthetic intermediates in the dyestuffs industry. It is not proposed to discuss these specialised and relatively small tonnage applications here.

There are many cases where incorporation of chlorine in a dyestuff or pigment molecule has important effects on the colour and on other technological properties of the product. For example, introduction of a chlorine atom into the molecule of Malachite Green (I) at the *ortho*-position of the unsubstituted benzene ring gives a blue dye (II):

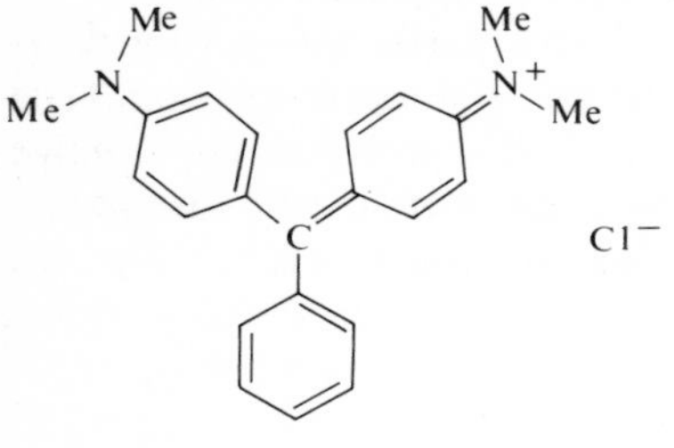

I 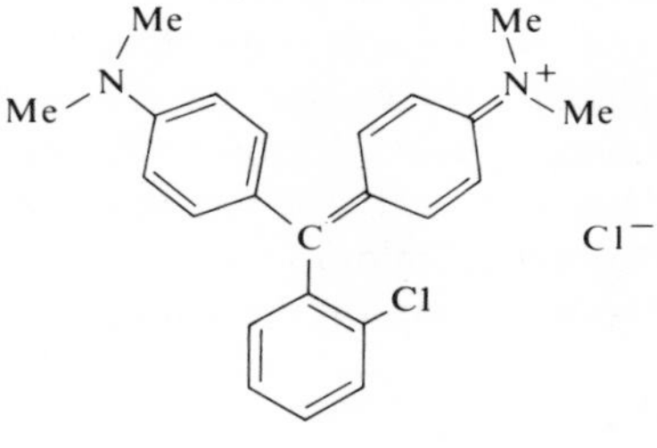   II

It is suggested that the chlorine prevents co-planarity of the chlorine-substituted ring with the rest of the molecule and thus limits the participation of its π-electrons in resonance.

An example of much more technical importance is provided by the phthalocyanine pigments. Copper phthalocyanine is a blue pigment with outstandingly good properties and a wide range of applications:

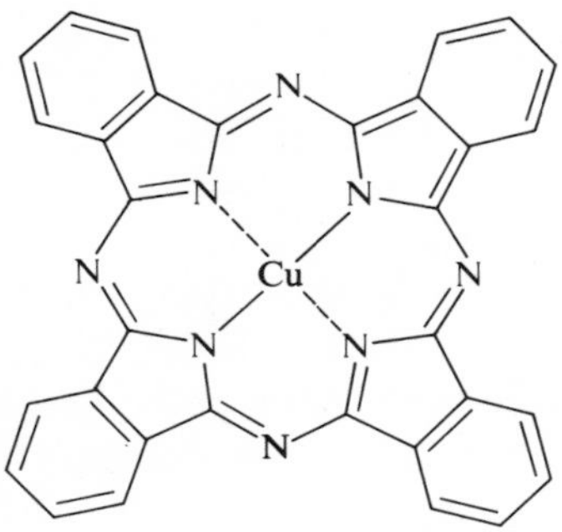

copper phthalocyanine

Chlorination, which leads to the replacement of approximately fourteen of the hydrogen atoms by chlorine, converts it into a green pigment, which is also of considerable commercial importance.

Chlorine plays an important part in the technology of the *reactive dyes*. Reactive dyes were first developed by ICI in the 1950s, and are of particular importance in dyeing cellulosic fibres such as cotton, although they are now used to some extent in dyeing wool and nylon. The distinguishing feature of this type of dye is that they react chemically with the fibre and become fixed to it by covalent bonds. They give fast dyeing on cotton by a more convenient process, and at a lower cost than had previously been possible.

Reactive dye molecules consist of two parts, the portion that imparts the colour, the chromophore, and the reactive system. The most important group of reactive dyes is that in which the reactive system is a chlorotriazine group. These are made by the reaction of cyanuric chloride with water-soluble dyestuffs which bear an amino group on the molecule, e.g.:

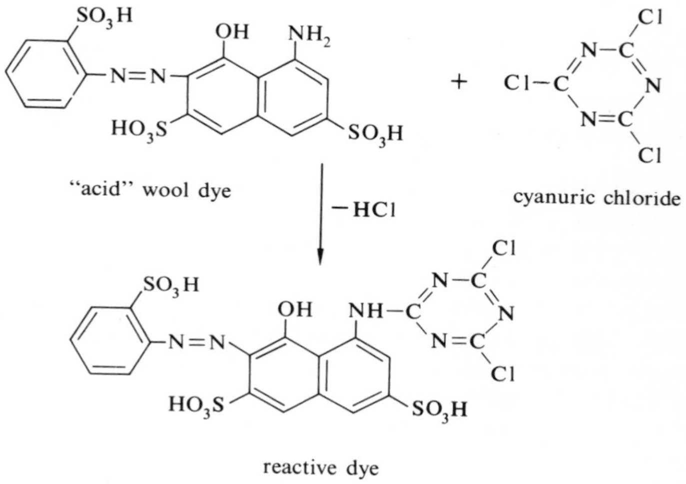

"acid" wool dye   cyanuric chloride

reactive dye

When cellulosic fibres are dyed with this type of dye, reaction occurs between chlorine atoms on the triazine ring and hydroxyl groups on the cellulose molecule (see page 200 for the structure of cellulose). Thus, the fabric is immersed in a neutral solution of the dyestuff, and the dyestuff becomes adsorbed on the fibres. The pH is then adjusted to about 10·5. At this pH a proportion of the cellulose hydroxyl groups are ionised, and nucleophilic displacement of chlorine by cellulose-O^- occurs:

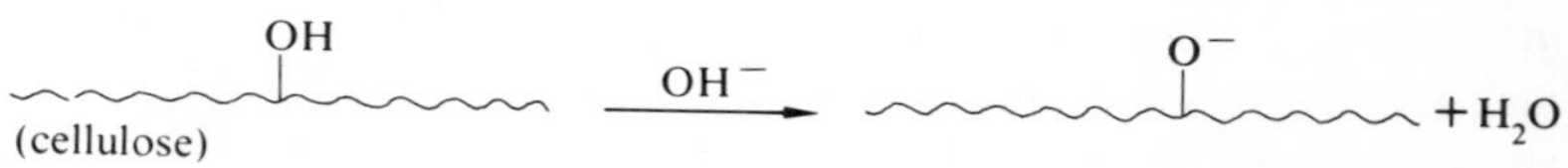

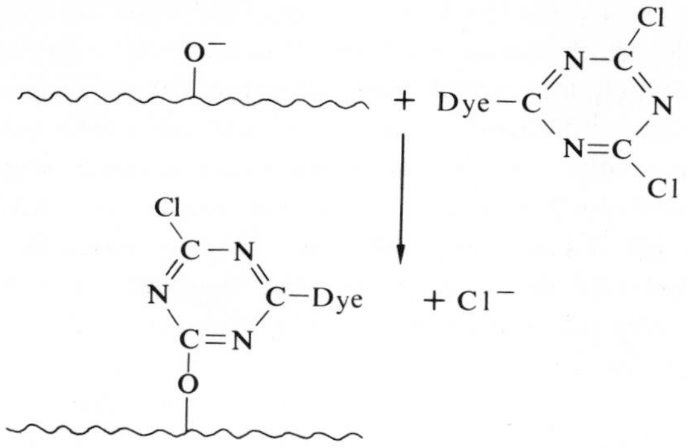

$$\text{O}^- \quad + \quad \text{Dye} - \text{C} \underset{\text{N}=\text{C}}{\overset{\text{N}-\text{C}}{\left\langle \begin{array}{c} \text{Cl} \\ \end{array} \right.}} \text{N} \quad \text{Cl}$$

$$\overset{\text{Cl}}{\underset{\text{N}}{\text{C}}} - \text{N} \\ \text{C} - \text{Dye} \quad + \text{Cl}^-$$

The dye is thus attached to the fibre by a stable ether linkage.

FLUORINE COMPOUNDS

The high strength of the carbon–fluorine bond, the high electro-negativity of fluorine, and the relatively small size of the fluorine atom all combine to make the organic chemistry of fluorine quite different from the organic chemistry of the other halogens, and consequently the technological applications of fluorine-containing organic compounds are quite distinct from those of compounds of chlorine, bromine and iodine. Fluorine compounds are not subject to nucleophilic aliphatic substitution, or to elimination reactions of the type undergone by compounds of chlorine, bromine and iodine, so that they do not find uses in synthetic applications analogous to those of compounds of the other halogens.

On the other hand, incorporation of fluorine into an organic molecule can produce technological effects not produced by other halogens, the most important being thermal and chemical stability, low toxicity, and low free surface energy, and a number of fluoro-chemicals have achieved substantial importance in various end-uses.

The consumption of fluorine-containing organic chemicals in the USA in 1963 was 184 000 tons. More than 85% of this was made up of chlorofluoromethanes and chlorofluoroethanes, the main uses of which are as aerosol propellants and as refrigerant fluids. These applications require materials of low boiling point, which should preferably be non-flammable and non-toxic; the chlorofluoromethanes and chlorofluoroethanes are pre-eminently suitable for these applications.

Chlorofluoromethanes

Dichlorodifluoromethane, CCl_2F_2, b.p. 243 K, is the largest-tonnage fluorochemical made, and accounts for more than half the output of organic fluorine compounds. Its main use is as a refrigerant fluid, the other major use being as an aerosol propellant, in which application it is generally used as a mixture with trichlorofluoromethane, CCl_3F, b.p. 297 K. These materials are also used as blowing agents in polyurethane foam manufacture (see Chapter 7), and trichlorofluoromethane finds some use as a solvent. Chlorodifluoromethane, $CHClF_2$, b.p. 229 K, chlorotrifluoromethane, $CClF_3$, b.p. 192 K, and trifluoromethane, CHF_3, b.p. 191 K, are also used as refrigerant fluids.

The most important method of making the chlorofluoromethanes and other chlorofluorohydrocarbons involves the replacement of chlorine by fluorine by treatment of a chlorocompound with anhydrous hydrogen fluoride in the presence of antimony pentachloride. Thus, dichlorodifluoromethane is made by the reaction of carbon tetrachloride with a slight excess over the stoichiometric amount of hydrogen fluoride, and antimony pentachloride at about 320 K:

$$CCl_4 + 2HF \xrightarrow{\quad SbCl_5 \quad} CCl_2F_2 + 2HCl$$

Other chlorofluoromethanes are obtained by similar methods, the products obtained depending on starting materials, temperature and pressure, and the ratio of hydrogen fluoride to chlorinated starting material.

Chlorofluoroethanes

The most important chlorofluoroethane is 1,2-dichlorotetrafluoroethane, $CClF_2CClF_2$, b.p. 269 K, which is used mainly as an aerosol propellant. It is less readily hydrolysed than trichlorofluoromethane, and so is more suitable for aerosol packs of preparations containing water than is a mixture of trichlorofluoromethane and dichlorodifluoromethane. It is made by the reaction of anhydrous hydrogen fluoride with hexachloroethane, or with perchloroethylene and chlorine, in both cases in the presence of antimony pentachloride:

$$CCl_3CCl_3 + 4HF \longrightarrow CClF_2CClF_2 + 4HCl$$

$$CCl_2{=}CCl_2 + Cl_2 + 4HF \longrightarrow CClF_2CClF_2 + 4HCl$$

Other chlorofluoroethanes have some small tonnage uses as solvents.

Fluorine-containing Monomers and Polymers

A number of fluorine-containing polymers are produced, the most well known being polytetrafluoroethylene (see Chapter 7). These polymers have outstanding chemical and thermal resistance. They are, however, expensive, and their manufacture is consequently on a much smaller scale than that of the major general purpose commercial polymers such as polyethylene and poly(vinyl chloride). For example, total world consumption of polytetrafluoroethylene in 1969 was about 13 000 tons. The more important of these monomers and their polymers are discussed briefly below.

TETRAFLUOROETHYLENE

The polymerisation of tetrafluoroethylene is discussed in Chapter 7. The monomer is prepared by the pyrolysis of chlorodifluoromethane at 870 to 1 070 K, the chlorodifluoromethane being made by the reaction of chloroform and anhydrous hydrogen fluoride under the conditions discussed above:

$$CHCl_3 \xrightarrow{\text{HF, SbCl}_5} CHClF_2 \longrightarrow CF_2{=}CF_2$$
$$+ \qquad\qquad +$$
$$2HCl \qquad\qquad 2HCl$$

The pyrolysis is thought to involve difluorocarbene as an intermediate:

$$CHClF_2 \rightleftharpoons CF_2 + HCl$$

$$2CF_2 \rightleftharpoons CF_2{=}CF_2$$

Polytetrafluoroethylene has outstanding resistance to thermal degradation and chemical attack.

CHLOROTRIFLUOROETHYLENE

Polychlorotrifluoroethylene has thermal and chemical properties similar to, though not quite so good as, those of polytetrafluoroethylene, and has the advantage of being easier to fabricate. The monomer is made as shown below:

$$CCl_3CCl_3 + 3HF \longrightarrow CClF_2CCl_2F + 3HCl$$

or

$$CCl_2 = CCl_2 + Cl_2 + 3HF \longrightarrow CClF_2CCl_2F + 3HCl$$

$$CClF_2CCl_2F + Zn \longrightarrow CF_2 = CFCl + ZnCl_2$$

VINYL FLUORIDE

Poly(vinyl fluoride) is the basis for the manufacture of 'Tedlar' film. It has very good resistance to weathering and chemicals, and is used as protective coating for constructional materials such as metals, wood, cement–asbestos, and plastics, for use in the cladding of buildings.

The monomer is made from acetylene and hydrogen fluoride, either by a one-stage process analogous to that used for vinyl chloride manufacture, or by a two-stage process:

$$CH \equiv CH + HF \longrightarrow CH_2 = CHF$$

$$CH \equiv CH + 2HF \longrightarrow CH_3CHF_2 \xrightarrow{\text{pyrolysis}} CH_2 = CHF + HF$$

VINYLIDINE FLUORIDE

Poly(vinylidine fluoride) also has very good resistance to weathering and chemical attack. It is used in applications similar to those of poly(vinyl fluoride), as a coating for constructional materials, and in a number of other applications where its resistance to chemical attack is of value, e.g. as linings and coatings for chemical processing equipment.

Vinylidine fluoride is made by pyrolysis of 1,1,1-chlorodifluoroethane, itself made by the reaction of 1,1,1-trichloroethane and hydrogen fluoride:

$$CCl_3CH_3 + 2HF \longrightarrow CClF_2CH_3 \longrightarrow CF_2 = CH_2 + HCl$$

BROMINE COMPOUNDS

We have seen that bromine is more than four times as expensive as chlorine. Since its atomic weight is about twice that of chlorine, the halogen cost involved in the use of bromine compounds as intermediates in synthetic applications is likely to be about nine times as much as that involved in using the corresponding chlorine compounds. It is therefore not surprising that bromine is only used in industrial synthesis where it has some distinct advantage over chlorine, e.g. in respect of its higher reactivity. Similarly, bromine is only used in end-use products where the advantages gained compensate for the relatively high cost. For example, the incorporation of bromine into dyestuffs molecules can produce desirable shade changes, the effect being usually greater than when chlorine is used.

Bromine compounds are made on a much smaller scale than chlorine compounds; we shall consider a number of the more important products.

Methyl Bromide

The major use of methyl bromide is as a fumigant for insect and rodent extermination. There is some use as a fire extinguisher, and a small amount is used in synthetic applications as a methylating agent. Production in the USA in 1967 was 8 800 tons.

Methyl bromide is made from methanol, either by treatment with an alkali metal bromide and sulphuric acid, or with bromine and a reducing agent, e.g. sulphur dioxide:

$$2CH_3OH + 2NaBr + H_2SO_4 \longrightarrow 2CH_3Br + Na_2SO_4 + 2H_2O$$

$$2CH_3OH + Br_2 + SO_2 \longrightarrow 2CH_3Br + H_2SO_4$$

It will be evident that in these processes, the primary objective is to obtain efficient utilisation of the bromine source rather than of the methanol.

Bromochloromethane

Bromochloromethane is used as a fire-extinguishing fluid. It is more than twice as effective on a weight basis as carbon tetrachloride, and thus has advantages over this latter compound for use in aircraft, and in portable extinguishers. On the other hand, it is substantially more expensive than carbon tetrachloride.

It is made by the reaction of methylene chloride with bromine and aluminium filings:

$$CH_2Cl_2 \xrightarrow{\text{Br}_2, \text{Al}} CH_2BrCl + AlCl_3$$

Methylene dibromide is obtained as a by-product.

The effectiveness of bromochloromethane and other halogen-containing compounds as flame-extinguishing agents does not depend only on exclusion of oxygen from the flame. These agents act as flame inhibitors by becoming involved in the chemical reactions involved in combustion, and reducing the concentration of free radicals in the flame. Bromine compounds are considerably more active than chlorine compounds as inhibiting agents.

A number of other bromine-containing compounds are used as flame-extinguishing agents to some extent. Examples are dibromodifluoromethane, bromochlorodifluoromethane and, as we have already seen, methyl bromide. The latter compound is highly effective, but is also highly toxic, and this limits its use.

Ethylene Dibromide

Ethylene dibromide is by a very large margin the largest-tonnage organic bromine compound made, and about 80% of world bromine output is used in its manufacture. Total world production of ethylene dibromide in the early 1960s was about 120 000 tons per annum.

Ethylene dibromide is made, as would be expected, by the addition of bromine to ethylene. A variety of conditions are used, and both batch and continuous processing are employed:

$$C_2H_4 \xrightarrow{\text{Br}_2} CH_2BrCH_2Br$$

The major use of ethylene dibromide is in leaded gasoline. As we have seen, tetraethyl-lead is added to gasoline to improve its octane rating. To prevent fouling of the engine by lead compounds produced in the combustion, a mixture of ethylene dichloride and ethylene dibromide is also added to the gasoline. On combustion the lead is converted into lead chloride and lead bromide, and these pass out of the engine in the exhaust gases. Ethylene dichloride, which is of course very much cheaper than ethylene dibromide, is not effective if used alone. It seems quite likely that this application of ethylene dibromide will diminish in importance, since there is mounting pressure to reduce or eliminate the use of tetraethyl-lead in gasoline.

Some ethylene dibromide is used as a grain fumigant, for killing insects, and as a soil fumigant, for killing wireworms and nematodes, and much smaller amounts are used as an intermediate.

Halothane (2-bromo-2-chloro-1,1,1-trifluoroethane)
Halothane (ICI trade name 'Fluothane') has achieved very considerable importance as an anaesthetic since its introduction by ICI in 1954. It approaches being a perfect anaesthetic in that it is non-flammable, potent, has a wide margin between anaesthetic and lethal dosages, and produces few side-effects. It may be made by the following route:

$$CCl_2 = CHCl \xrightarrow{HF} CF_3CH_2Cl \xrightarrow{Br_2} CF_3CHBrCl$$

$$+ \qquad\qquad +$$

$$2\,HCl \qquad\qquad HBr$$

IODINE COMPOUNDS

Iodine is very much more expensive than the other halogens, and it is made and used on a much smaller scale. In 1966, total world production of iodine was estimated to be about 8 600 tons. USA consumption of iodine in 1964 was 1 400 tons of which about 340 tons was used in organic chemical manufacture. The high price of iodine restricts its use in the organic chemical industry to the manufacture of high-price, low-tonnage products in cases where the use of an iodine-containing compound offers particular advantages.

One example of an iodine-containing end-use product is the red dyestuff *erythrosin* which is widely used in colouring foodstuffs, for example, tinned cherries. Erythrosin (the disodium salt of 2,4,5,7-tetraiodofluorescein) is made by the treatment of fluorescein with iodine under oxidising conditions:

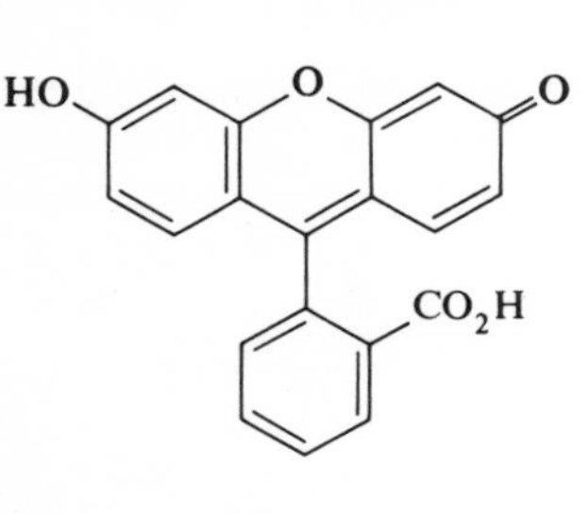

fluorescein 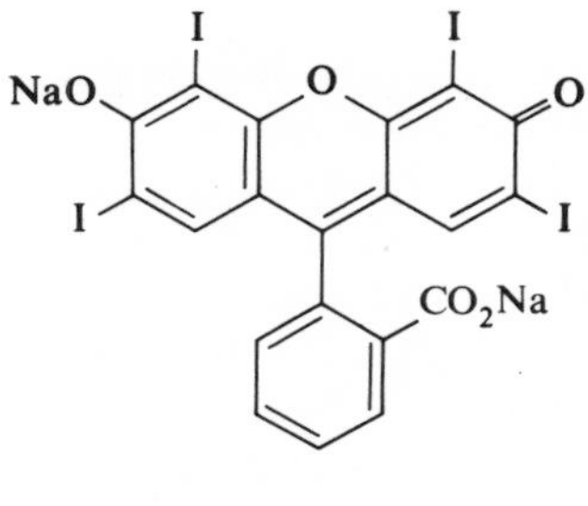   *erythrosin*

Chapter 5

Aromatic Substitution and Related Reactions

AROMATIC substitution reactions are of major importance in industrial chemistry in that they provide the means by which aromatic hydrocarbons can be converted into a large number of useful products.

ELECTROPHILIC AROMATIC SUBSTITUTION

The theory of electrophilic aromatic substitution is one of the cornerstones of modern organic chemistry, and it is not proposed to discuss general aspects of this theory in detail here. As the reader will probably know, the generally accepted mechanism is one in which an electrophile, which is often, but not necessarily, positively charged, adds to the aromatic ring to give an intermediate complex:

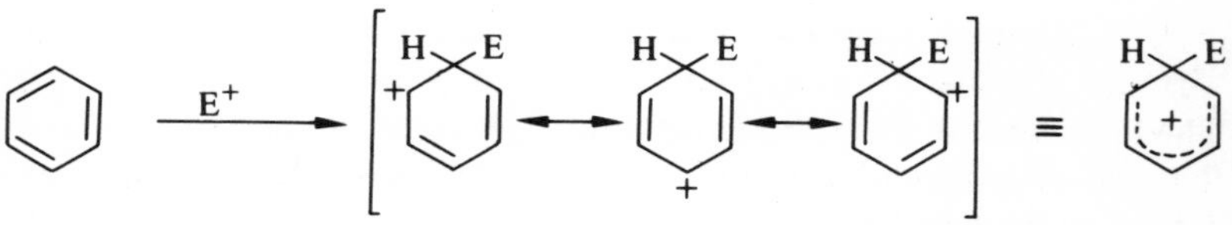

Loss of a proton from the complex regenerates the aromatic system:

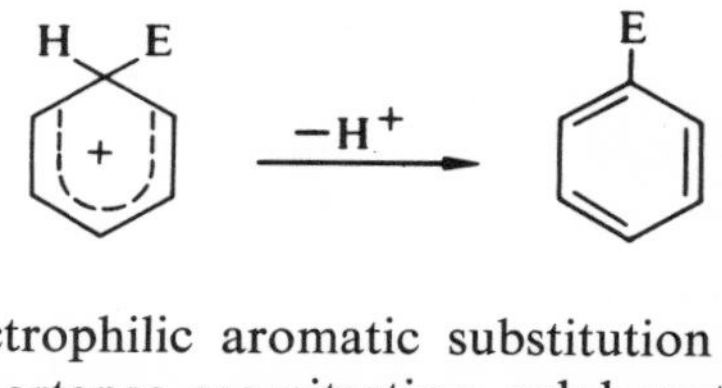

Types of electrophilic aromatic substitution that are of major commercial importance are nitration, sulphonation, Friedel–Crafts and related reactions, chlorination and diazonium coupling reactions.

Nitration

Nitration is one of the oldest processes used in the organic chemical industry and has been carried out commercially for over 100 years. The original applications of aromatic nitro compounds were as intermediates in the manufacture of dyestuffs, and this is still an important outlet, although there are now a number of other major applications, e.g. in explosives and in the preparation of isocyanates. In many cases, the nitro compound is an intermediate in the preparation of an amino compound.

It is generally considered that in most nitrations the electrophile is the nitronium ion, NO_2^+. This is present to a small extent in concentrated nitric acid, as a result of the following equilibria:

$$2\,HNO_3 \rightleftharpoons H_2NO_3^+ + NO_3^-$$

$$H_2NO_3^+ \rightleftharpoons NO_2^+ + H_2O$$

About 3% of pure nitric acid is ionised to nitronium ion. As would be expected, the more dilute the nitric acid, the lower the concentration of nitronium ion.

In the presence of other strong acids, the extent of ionisation to nitronium ion is substantially increased, and nitric acid in solution in sulphuric acid of concentration 94% or over is almost completely ionised to nitronium ion:

$$HNO_3 + 2\,H_2SO_4 \rightleftharpoons NO_2^+ + H_3O^+ + 2HSO_4^-$$

The lower the sulphuric acid concentration, or conversely, the higher the water concentration, the less this ionisation takes place; in sulphuric acid of less than 86% concentration it occurs to only a small extent. Hydrogen fluoride, boron trifluoride, perchloric acid and other strong acids produce the same effect. However, sulphuric acid is cheaper than these materials and most commercial nitrations are carried out with mixtures of nitric and sulphuric acids, so-called 'mixed acid' or 'nitrating mixture'. The composition of the mixed acid used is selected to suit the particular substrate involved. Thus, in the nitration of an unreactive substrate a mixed acid of high sulphuric acid concentration, and therefore of high nitronium ion concentration, is used.

Attack on the substrate by the nitronium ion proceeds by the mechanism already discussed:

Introduction of the strongly electron-withdrawing nitro group into the substrate deactivates it against further attack; it has been shown that under one particular set of conditions nitrobenzene is nitrated about ten thousand times more slowly than benzene. Consequently it is relatively easy to control the degree of substitution attained in nitrations, by adjustment of the reaction temperature and the constitution and amount of nitrating mixture used. This contrasts with Friedel–Crafts alkylations, where the group being introduced activates the substrate towards further attack, and where polysubstitution is difficult to avoid.

The products formed in many nitrations are capable of decomposing explosively, and care has to be taken in designing nitration processes so that conditions which could lead to explosive decomposition are avoided. One important factor is that the sensitivity of the product to explosive decomposition tends to be increased by the presence of by-products which can be formed by oxidation of the substrate by the nitrating mixture. For this reason, and to avoid excessive loss of yield by oxidation, nitrations are normally carried out under the mildest conditions, in terms of temperature, and composition of nitrating mixture, which will bring about the desired reaction at an acceptable rate. Careful attention to cooling and agitation are also necessary.

NITRATION OF BENZENE

The mononitration of benzene is carried out on a fairly substantial scale; production of nitrobenzene in the USA in 1965 was about 140 000 tons. The main use of nitrobenzene is in the manufacture of aniline, and its production is normally carried out as part of an overall operation for the manufacture of aniline from benzene.

In batchwise operation, nitrating mixture with a composition 60–53% sulphuric acid, 32–39% nitric acid and 8% water is added to benzene at about 320 K during two to four hours. The product is separated from the spent acid by decantation, and may be used as it stands for aniline manufacture. If pure nitrobenzene is required, it is washed acid-free and distilled. As we have already seen, nitrobenzene is much less readily nitrated than benzene, and it is not attacked to

any appreciable extent under these conditions; a yield of 95–98 %
nitrobenzene is obtained.

m-Dinitrobenzene may be made by nitrating the crude nitrobenzene, made as described above, at about 365 K with a nitrating
mixture of composition 75 % sulphuric acid, 20 % nitric acid, and 5 %
water.

Most large modern nitrobenzene plants use continuous processing,
typically with a number of stirred flow reactors in series; yields in
continuous processes are about the same as in batch processes.

Aniline Manufacture

Reduction of nitrobenzene to aniline may be brought about by
catalytic hydrogenation in the vapour phase (see Chapter 6), or by
the older method of refluxing with dilute hydrochloric acid and iron
borings. Most modern plants use hydrogenation. However, reduction
of nitro compounds with iron and hydrochloric acid still remains
important for the production of amines which are made only on the
small scale:

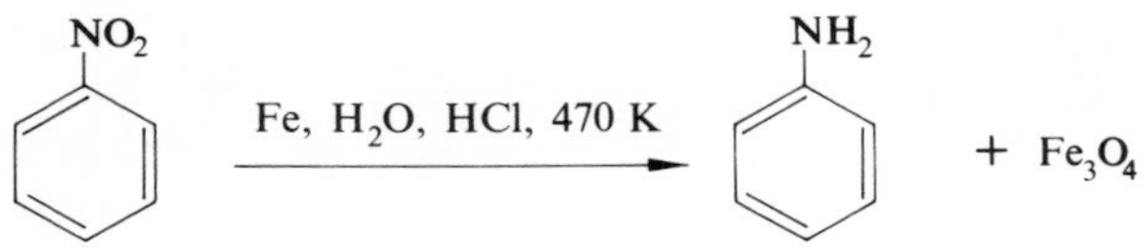

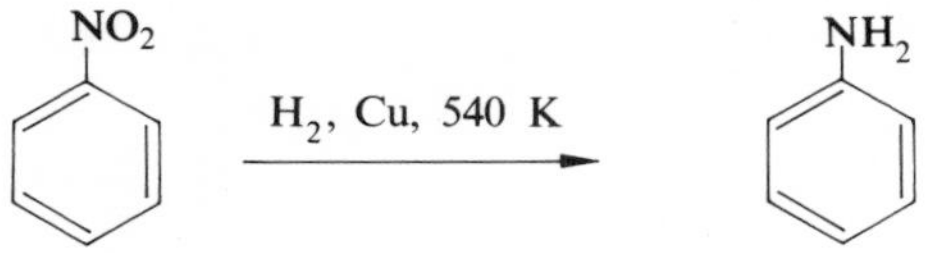

The yield given by iron reduction is about 95 %, while hydrogenation
gives about 98 %. The overall yield of aniline from benzene is
therefore high.

The main uses of aniline are in the production of dyes and rubber
chemicals.

NITRATION OF TOLUENE

Nitration of toluene is operated commercially for the production of
mono-, di-, and trinitrotoluenes. Trinitrotoluene, as is well known,

is a military explosive, and mono-, and dinitrotoluenes are intermediates in the manufacture of toluidines and dyestuffs intermediates, and tolylene di-isocyanates respectively. Because of the activating effect of the methyl group, toluene is more readily attacked than benzene, and mono-nitration is carried out under milder conditions than those used for benzene. For example, a nitrating mixture of composition 58% sulphuric acid, 19% nitric acid, and 23% water, and a reaction temperature of 328 K may be used. The product obtained contains about 60% *ortho*-isomer, 36% *para*-isomer, and 4% *meta*-isomer. This product distribution is not much affected by the conditions used, so that the relative production of the *ortho*- and *para*-isomers cannot be adjusted to any appreciable extent to suit market requirements.

Separation of *o*- and *p*-nitrotoluenes can be brought about by a series of distillation and crystallisation steps. However, if trinitrotoluene or some grades of dinitrotoluene mixtures are to be prepared, the mixture is nitrated further without separation.

Further nitration of *o*-nitrotoluene gives mainly 2,4- and 2,6-dinitrotoluenes in the approximate ratio 65:35:

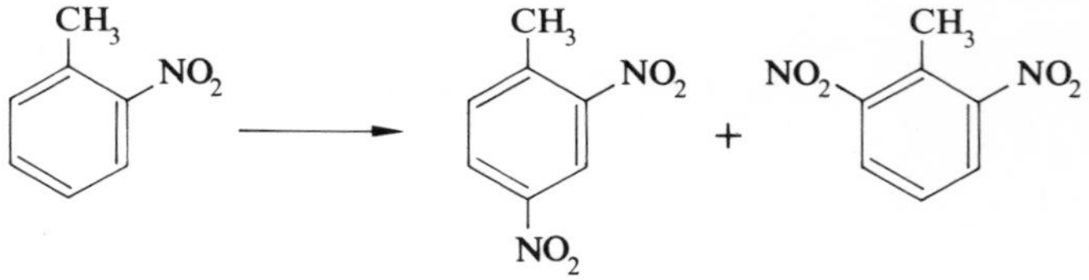

Further nitration of *p*-nitrotoluene gives mainly 2,4-dinitrotoluene:

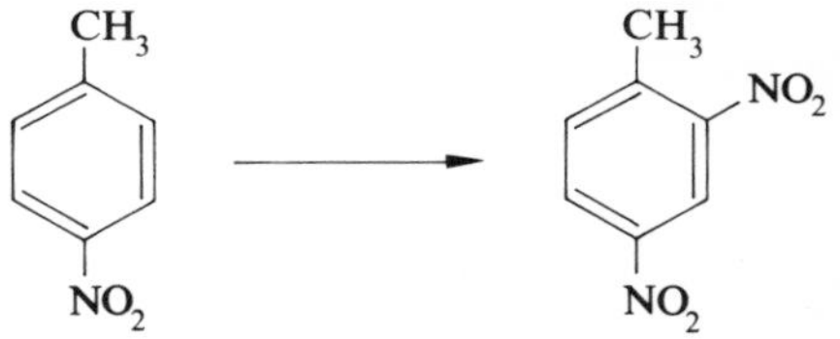

Nitration of the mixture of *o*- and *p*-nitrotoluenes gives 2,4- and 2,6-isomers in the ratio 80:20. Mixtures of dinitrotoluene isomers are intermediates in the manufacture of tolylene di-isocyanates (see below).

Nitration of the dinitrotoluenes, which requires forcing conditions because of the combined deactivating effect of two nitro groups, gives mainly 2,4,6-trinitrotoluene.

Tolylene Di-isocyanate Manufacture

Mixtures of 2,4- and 2,6-tolylene di-isocyanates, commonly known as TDI, are of considerable importance in polyurethane manufacture (see Chapter 7). The most commonly used material is an 80:20 mixture of isomers, but a 65:35 mixture, which is more expensive, has some technical advantages and also finds some application. These products are made from the corresponding mixtures of dinitrotoluenes, obtained as described above, by reduction to tolylenediamines followed by reaction with phosgene:

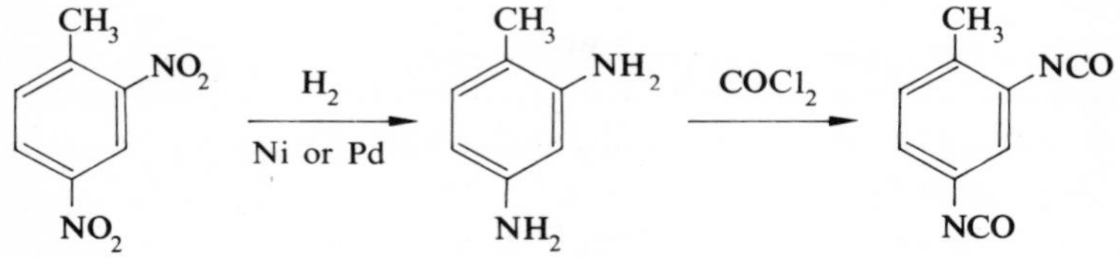

+ 2,6-isomer   + 2,6-isomer   + 2,6-isomer

The scale of operation is normally fairly large, typically of the order of 20 000 tons per annum, and production is generally carried out as an integrated operation starting from toluene. Continuous processing tends to be used in the more modern plants.

The reduction of dinitrotoluene is carried out by hydrogenation in the liquid phase in the presence of, e.g., a palladium or nickel catalyst (see Chapter 6). The phosgenation can be carried out under a variety of conditions. Commonly, it is done in solution in *o*-dichlorobenzene in a series of reactors at temperatures increasing from about 290 to 460 K. The reaction involves the formation of, and subsequent dehydrochlorination of, a carbamoyl chloride:

$$\text{RNH}_2 + \text{ClCCl}_{\overset{\text{O}}{\|}} \longrightarrow \text{RNHCCl}_{\overset{\text{O}}{\|}} + \text{HCl}$$

$$\text{RNHCCl}_{\overset{\text{O}}{\|}} \longrightarrow \text{RN=C=O} + \text{HCl}$$

OTHER NITRATIONS

A number of other nitrations are important in preparing intermediates for dyestuffs and other fine chemicals. These are mainly carried out on a relatively small scale, and there would be no point

156 INTRODUCTION TO INDUSTRIAL ORGANIC CHEMISTRY

in discussing all of them. We shall briefly consider two of the more important ones.

Nitration of Chlorobenzene

As the reader will probably know, chlorine deactivates the aromatic ring towards electrophilic substitution, but directs the entering group into the *ortho-* and *para-* positions. Nitration of chlorobenzene at 310 to 340 K with a nitrating mixture containing 52·5% sulphuric acid, 35·5% nitric acid, and 12% water gives a mixture of nitro-chlorobenzene isomers containing about 34% *ortho*-isomer, 65% *para*-isomer, and 1% *meta*-isomer. This can be separated into its components by a combination of fractional distillation and fractional crystallisation. No economically viable way of substantially altering the *ortho–para* ratio appears to have been found.

The nitrochlorobenzenes find their main applications as dyestuffs intermediates.

Nitration of Naphthalene

Naphthalene is more readily nitrated than benzene; the nitro group enter predominantly in the 1-position. Thus, nitration at 320 to 340 K with a nitrating mixture containing 60% sulphuric acid, 16% nitric acid, and 24% water gives a product containing about 95% 1-nitronaphthalene:

(main product)

Preferential attack on the 1-position may be rationalised by considering the resonance structures of the intermediate ions. Two of the canonical forms of the intermediate ion in attack at the 1-position retain the low-energy benzenoid structure compared with only one in the case of the ion formed by attack at the 2-position:

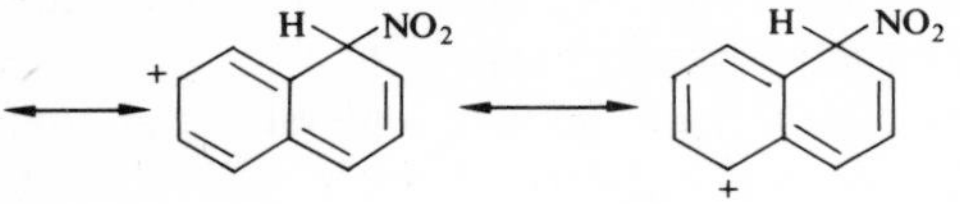

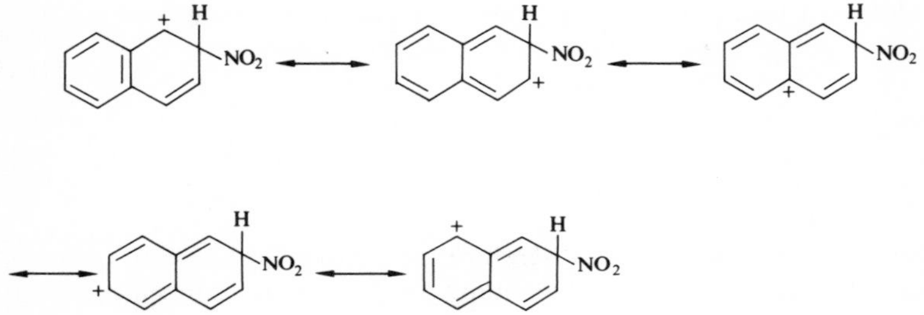

It is thus reasonable to suppose that the ion produced by attack at the 1-position is more strongly resonance-stabilised. If we assume that the transition states are similar to the intermediate ions, then this leads us to conclude that the transition state in substitution at the 1-position is of lower energy than the transition state in substitution at the 2-position.

One major use of 1-nitronaphthalene has been in the production of 1-aminonaphthalene, an important dyestuffs intermediate, by reduction. However, the product so obtained contains up to 4% of the 2-isomer, which is a potent carcinogen, and manufacture has been discontinued in a number of countries, including the UK.

Sulphonation

Aromatic compounds are sulphonated for two main reasons, either to obtain products in which the highly polar sulphonate group confers desirable technological properties, or in order that the sulphonate group may be subjected to further reactions in a synthetic route to a desired product. The dyestuffs industry provides an important example of the first type of application, the incorporation of sulphonate groups being very widely used to achieve water solubility of the dyestuff. In dyes for some fibres, e.g. wool, the group also plays a part in fixing the dyestuff to the fibre by forming salt linkages with amino groups on the fibre. The largest-tonnage synthetic use of aromatic sulphonates is in the manufacture of phenols, by alkaline fusion, though it should be noted that the manufacture of phenol itself in this way has been largely discontinued.

The mechanism of sulphonation is not so well understood as that of nitration. The major agents used industrially are sulphuric acid, oleum (a solution of sulphur trioxide in sulphuric acid), and sulphur trioxide. In each case, the consensus of opinion is that the electrophile involved is sulphur trioxide or a sulphur trioxide complex:

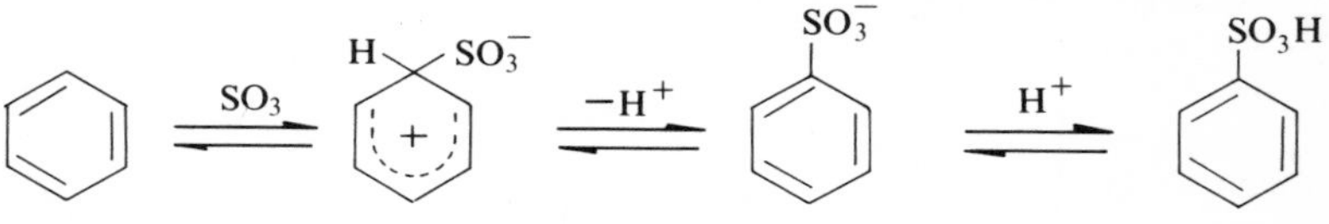

Sulphonating agents act, therefore, as sources of sulphur trioxide, and they vary in strength depending on the effective concentration of sulphur trioxide that they provide. Sulphur trioxide itself is, as would be expected, an extremely active sulphonating agent.

In sulphuric acid, free sulphur trioxide is in equilibrium with H_2SO_4:

$$H_2SO_4 \rightleftharpoons SO_3 + H_2O$$

$$H_2O + H_2SO_4 \rightleftharpoons H_3O^+ + HSO_4^-$$

overall:
$$2H_2SO_4 \rightleftharpoons SO_3 + H_3O^+ + HSO_4^-$$

Evidently the more dilute the acid, the less active a sulphonating agent it is, and for any particular aromatic compound there is a minimum concentration of acid below which sulphonation does not occur. The more reactive the compound is towards electrophiles, the lower this concentration is.

The isomer distribution obtained in sulphonation is very much affected by reaction conditions. For example, depending on the conditions used, sulphonation of toluene can be made to give a mixture mainly of *ortho*- and *para*-isomers, mainly *para*-isomer, or a mixture of mainly *meta*- and *para*-isomers. Two major factors are involved. Firstly, the ratio of isomers produced in the initial attack on the ring varies considerably with the reagent and conditions, the variation being attributed to changes in the nature of the sulphonating species, and in the degree of stabilisation of the transition states by solvation. The effects of acid concentration and reaction temperature in the sulphonation of toluene are illustrated by Table 5-1. Secondly, sulphonation is reversible and under favourable conditions the kinetically controlled mixture of isomers initially formed is converted by a process of desulphonation and sulphonation into the most thermodynamically stable mixture of isomers. Thus, sulphonation of

TABLE 5-1

MONOSULPHONATION OF TOLUENE BY H_2SO_4

Acid concn. (%)	Temp. (K)	Isomer distribution (%)		
		ortho	meta	para
77·6	298	21·2	2·1	76·7
	338	16·4	4·3	79·3
85·5	298	38·8	2·6	58·6
	338	30·5	4·4	65·1
95·8	278	51·6	4·7	43·7
	298	50·2	4·9	44·9
	338	42·4	7·0	50·6

Source of data: H. Cerfontain, F.L.J. Sixma, and L. Vollbracht, *Rec. Trav. Chim.*, **82**, 659 (1963). By permission of The Royal Netherlands Chemical Society and H. Cerfontain.

toluene with 74% sulphuric acid at 414 K gave 3·2% *ortho*-, 59·6% *meta*-, and 37·2% *para*-toluenesulphonic acid (compare Table 5-1). The flexibility of product distribution in sulphonation is of considerable commercial importance.

Since the sulphonic acid group deactivates the ring against further attack, there is in most cases no difficulty in avoiding disubstitution.

SULPHONATING AGENTS

Water is produced when sulphonations are carried out with sulphuric acid, and consequently the acid becomes more dilute as reaction proceeds:

$$RH \ + \ H_2SO_4 \longrightarrow RSO_3H \ + \ H_2O$$

As the acid is diluted the reaction will slow down and at some concentration depending on the compound being sulphonated, will stop. Thus, unless some means is used of removing the water, it is necessary to use an excess of sulphuric acid in order to achieve complete conversion of the aromatic starting material. For example, in the sulphonation of benzene with monohydrate (100% H_2SO_4), a 100% excess of sulphuric acid has to be used. This tends to be wasteful of raw materials in that separation of the excess of acid from the product is difficult, and may involve, e.g., neutralisation with lime to give virtually valueless calcium sulphate.

Sulphur trioxide produces no water on reaction, so that difficulties of the type discussed above do not arise:

$$RH + SO_3 \longrightarrow RSO_3H$$

The major problem in using sulphur trioxide as a sulphonating agent is that its extremely high reactivity can lead to difficulties in control and excessive formation of by-products. However, in a number of cases these difficulties have been overcome by correct design of equipment and operating conditions, and sulphonation using sulphur trioxide is commercially important, particularly in the manufacture of detergents. The fact that specialised equipment is required for sulphonation with sulphur trioxide tends to limit its use to large-tonnage applications.

The use of oleum enables sulphonation to be carried out under the same conditions as those used with sulphuric acid, but with a smaller excess of acid.

Chlorosulphonic acid, which may be regarded as an adduct of sulphur trioxide and hydrogen chloride, is used mainly for the preparation of sulphonyl chlorides:

$$RH + ClSO_3H \longrightarrow RSO_3H + HCl$$

$$RSO_3H + ClSO_3H \longrightarrow RSO_2Cl + H_2SO_4$$

SULPHONATION OF BENZENE

Benzenesulphonic acid has been of major importance as an intermediate in the so-called sulphonation process for phenol (see page 182). However, this process is now obsolescent. Benzenesulphonic acid has some other small tonnage uses as an intermediate. Benzene-1,3-disulphonic acid is an intermediate in the manufacture of resorcinol:

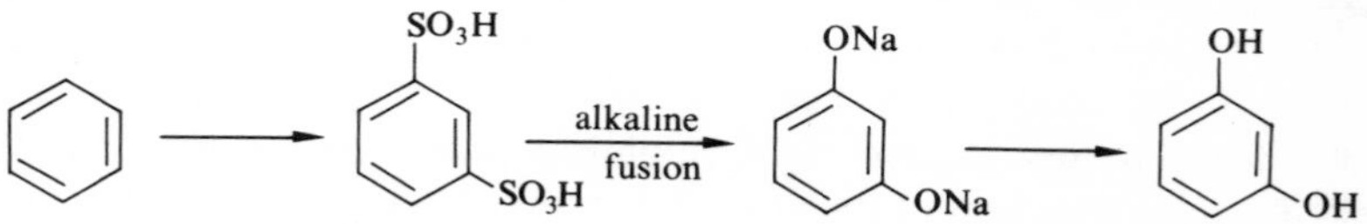

Various conditions have been used for the manufacture of benzenesulphonic acid. For example, the sulphonation can be carried out with a 100% excess of monohydrate at a temperature progressing from 340 to 380 K. As has been pointed out, the use of such a large excess of acid is wasteful, and a number of processes have been developed to avoid the necessity for it. In one process, water is continually removed from the reaction mixture by co-distillation with benzene, thus avoiding dilution of the sulphuric acid, and considerably reducing the amount required. In other processes oleum is used. Sulphonation with sulphur trioxide gives large amounts of diphenylsulphone as by-product, and is apparently not commercially important.

Introduction of a second sulphonic acid group requires considerably more forcing conditions. Thus, if monohydrate is used, a reaction temperature of about 470 K is required, and even so the reaction is slow. Under these conditions the reaction products tend to be the thermodynamically controlled products, and up to 34% of the *para*-isomer, for which there is little commercial demand, is obtained. If oleum is used, the reaction may be carried out at about 350 K in a relatively short time, and high yields of the initially formed *meta*-isomer obtained.

SULPHONATION OF TOLUENE

Sulphonation of toluene is not of very great commercial importance. A relatively small amount of cresol is made from toluenesulphonic acids, and these acids have some other small tonnage uses. Sulphonation can be carried out under a variety of conditions, and as we have seen, the isomer distribution can be controlled to a substantial extent.

Chlorosulphonation of toluene is used in the manufacture of saccharin:

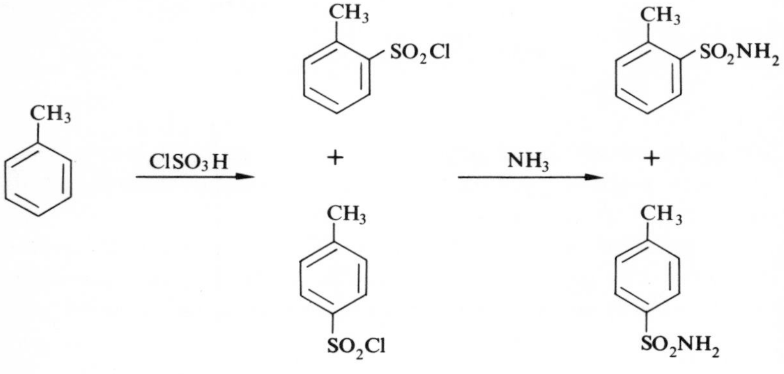

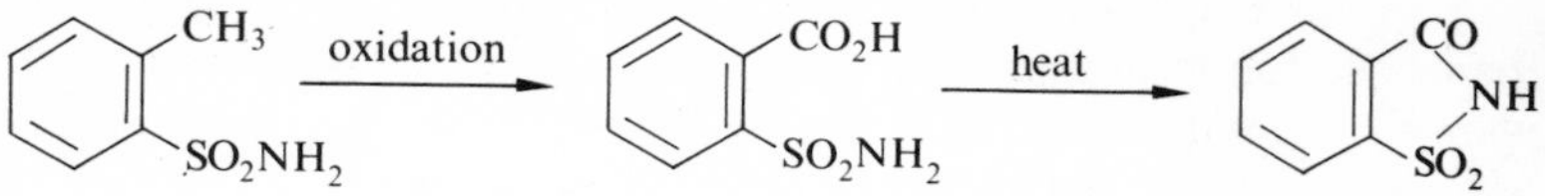

saccharin

The chlorosulphonation is generally carried out at 273 to 278 K, with an excess of chlorosulphonic acid. The toluenesulphonyl chloride obtained contains about 40% of the *ortho*-isomer, the rest being mainly *para*-isomer. It does not appear to be possible to obtain a higher *ortho*- to *para*- ratio than this.

SULPHONATION OF NAPHTHALENE

Naphthalenesulphonic acids and their derivatives are of major importance as intermediates in the dyestuffs industry.

As we have seen, the 1-position in naphthalene is more readily attacked by electrophiles than the 2-position, and monosulphonation of naphthalene at low temperatures gives mainly naphthalene-1-sulphonic acid:

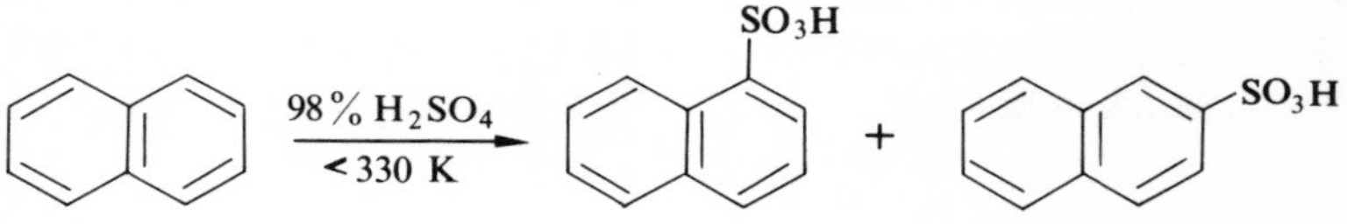

Naphthalene-1-sulphonic acid is used for making 1-naphthol and a number of other dyestuffs intermediates.

Sulphonation at higher temperatures (*ca.* 430 K) gives mainly the thermodynamically more stable naphthalene-2-sulphonic acid:

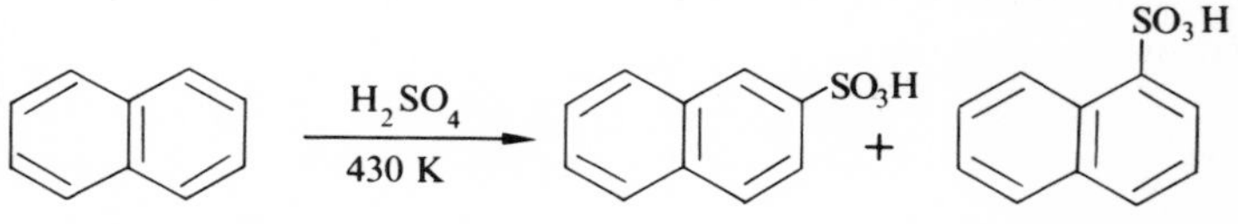

The crude naphthalene-2-sulphonic acid is treated with steam, when the more reactive naphthalene-1-sulphonic acid is hydrolysed to naphthalene, which is removed by steam distillation.

Further sulphonation becomes complex, and we shall not discuss it. A number of the di-, tri,- and tetra-sulphonic acids have applications in the dyestuffs industry. Thus, the important dyestuffs intermediate

H-acid is made from naphthalene-1,3,6-trisulphonic acid by the following route:

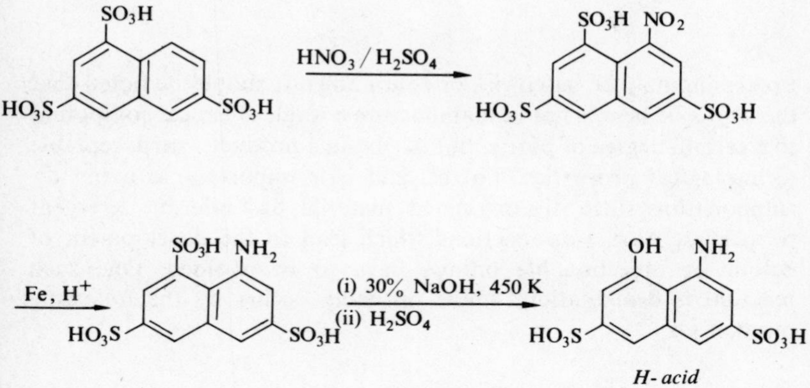

$$SO_3H \text{ (naphthalene with } SO_3H, HO_3S, SO_3H) \xrightarrow{HNO_3 / H_2SO_4} SO_3H \ NO_2 \ (HO_3S, SO_3H)$$

$$\xrightarrow{Fe,\ H^+} SO_3H \ NH_2 \ (HO_3S, SO_3H) \xrightarrow[\text{(ii) } H_2SO_4]{\text{(i) } 30\% \text{ NaOH, 450 K}} OH \ NH_2 \ (HO_3S, SO_3H)$$

H- acid

Production of H-acid in the USA in 1964 was about 1 600 tons.

ALKYLBENZENESULPHONATE DETERGENTS

The sodium alkylbenzenesulphonates form the most important class of synthetic detergents at the present time. Production in the USA in 1966 was about 270 000 tons.

The basic requirement for a detergent is that its molecules should contain both hydrophilic and lipophilic portions, that is, portions which have high affinities for water and for oil respectively. Thus, in the oldest-established detergent, soap, the polar carboxylate group is the hydrophilic portion and the long alkyl chain is the lipophilic portion. For good detergent properties there has to be a balance between the two types of group, and in soap, for instance, the alkyl group has to contain between 12 and 18 carbon atoms for optimum detergent properties. The sulphonate group is of course hydrophilic, and sodium monosulphonates of monoalkylbenzenes with about 12 carbon atoms in the alkyl group have good detergent properties. The preparation of alkylbenzenes for detergent manufacture will be discussed in the section on Friedel–Crafts reactions.

Sulphonation of alkylbenzenes can be carried out with sulphuric acid or oleum at temperatures in the range 300 to 330 K depending on the strength of the sulphonating agent used:

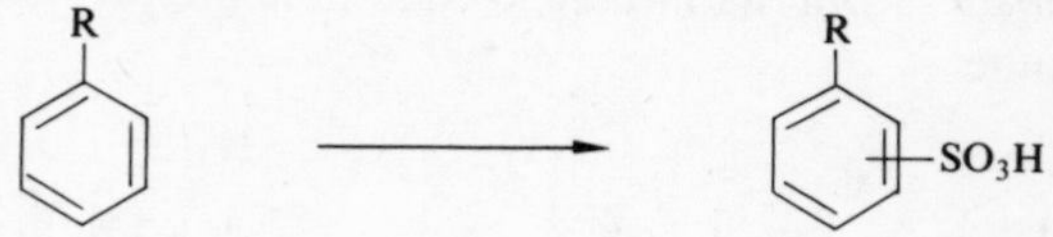

Processing may be batchwise or continuous. It should be noted that the objective here is not to manufacture a single chemical compound to a certain degree of purity, but to obtain a product with acceptable technological properties. To this end it is important to avoid disulphonation since disulphonated material has inferior detergent properties. Also, side reactions which lead to the development of colour or objectionable odours have to be avoided. One such reaction is dealkylation, which probably occurs by the following mechanism:

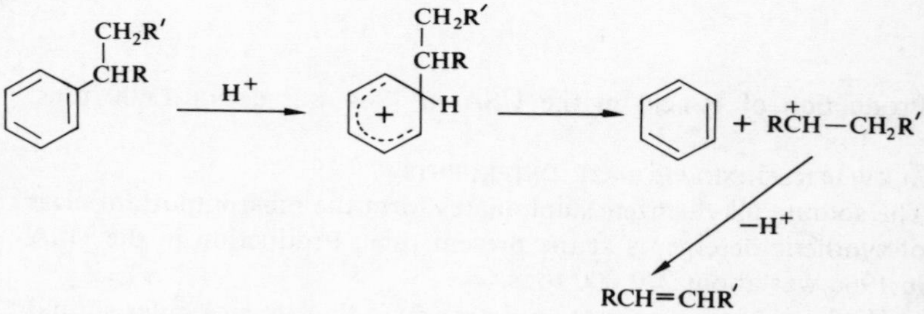

It is thought that the formation of long-chain alkenes leads to the production of odours.

The sulphonic acid is insoluble in the spent acid and can be separated by decantation. After separation sodium hydroxide is added to the sulphonic acid, and the resulting sodium alkylbenzene-sulphonate solution is spray-dried.

In recent years a number of processes have been developed for sulphonation with sulphur trioxide. These have the advantages that no spent acid is produced, and that a higher quality product than that produced by sulphonation with acid is obtained. Sulphur trioxide is normally used in the gas phase, diluted with air, and is contacted with the alkylbenzene either in a conventional reactor, or in a reactor in which a thin film flows over a cooled surface. The tendency to sulphone formation is much less than with benzene.

Friedel–Crafts and Related Reactions

Friedel–Crafts reactions find a number of applications in industrial operations. Alkylation is the most important such reaction, and is used, for example, in the manufacture of styrene, phenol by the cumene process, and alkylbenzenesulphonate detergents, as well as in a number of smaller-scale applications, but acylation is also applied to some extent. In addition there are some other reactions related to Friedel–Crafts reactions which are of substantial commercial importance, for example, the reaction of aldehydes and ketones with phenols.

In a Friedel–Crafts reaction, the alkylating or acylating agent and the catalyst react to form either a carbonium or acylium ion or a polarised complex, and the ion or the complex then attacks the aromatic ring. Thus, in alkylation with an alkene in the presence of aluminium chloride and hydrogen chloride, the following reactions occur:

$$RCH=CH_2 + AlCl_3 + HCl \rightleftharpoons [R\overset{+}{C}HCH_3 \quad AlCl_4^-]$$

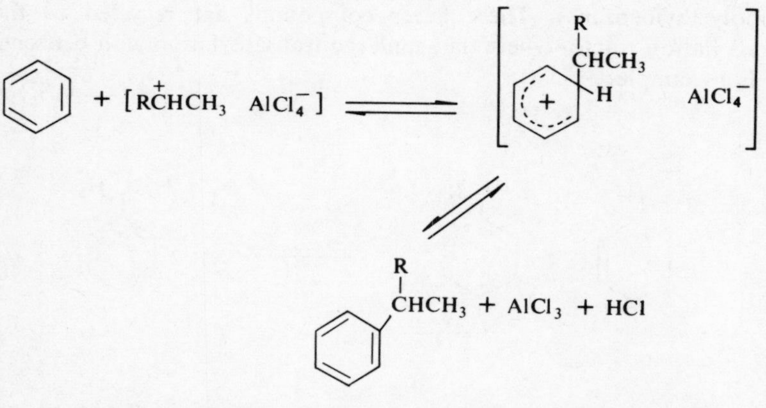

Friedel–Crafts reactions are often complicated by the occurrence of rearrangement of the attacking agent and also, in some cases, of the aromatic starting material. Since alkyl groups activate the ring towards further attack, there is a marked tendency to polysubstitution in alkylations, and measures have to be taken to control this.

ETHYLBENZENE

The preparation of ethylbenzene is of very major importance, as a step in the manufacture of styrene (see Chapter 6).

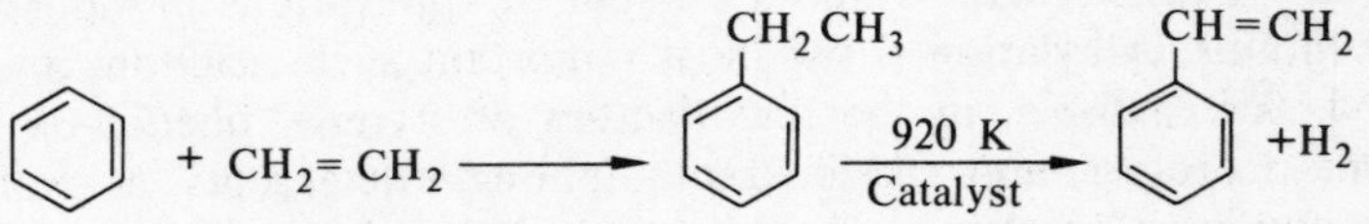

$$C_6H_6 + CH_2{=}CH_2 \longrightarrow C_6H_5{-}CH_2CH_3 \xrightarrow[\text{Catalyst}]{920\ K} C_6H_5{-}CH{=}CH_2 + H_2$$

The alkylation may be carried out in the liquid phase or in the gas phase; liquid-phase operation appears to be most common. In the liquid-phase process, benzene, and ethylene containing a small amount of ethyl chloride (as a source of hydrogen chloride), are allowed to react at about 370 K in the presence of aluminium chloride. To keep down the amount of polyalkylation, an excess of benzene is used (mole ratio of benzene to ethylene 1:0·6), but even so substantial amounts of polyethylbenzenes are formed. The aluminium chloride is present as a liquid catalyst complex which is insoluble in the benzene/ethylbenzene mixture, and so can be separated by decantation and re-used. The reaction mix, after separation of the catalyst layer, is distilled to give benzene, ethylbenzene, and polyethylbenzenes. These latter compounds are recycled to the alkylation reactor where they undergo transalkylation with benzene. For example:

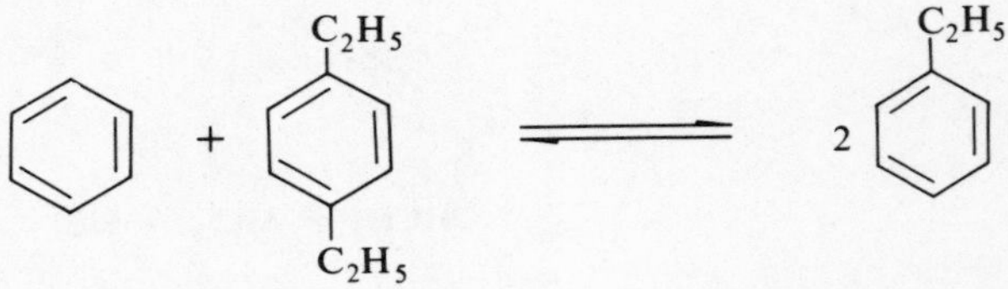

$$C_6H_6 + C_6H_4(C_2H_5)_2 \rightleftharpoons 2\,C_6H_5C_2H_5$$

The overall formation of polyethylbenzenes is thus suppressed. The yield of ethylbenzene obtained is about 95% calculated on both benzene and ethylene.

In vapour-phase processes the most usual catalyst is a 'solid phosphoric acid' catalyst, made by impregnating a porous support, normally a diatomaceous earth, with aqueous phosphoric acid and then removing most of the water by drying at an elevated temperature. Silica–alumina catalysts have also been used. In either case, protonation of the ethylene occurs on the catalyst to give an adsorbed ethyl carbonium ion which then attacks the benzene. When a phosphoric acid catalyst is being used, the reaction is carried out at about 570 K and 40 atm.

In most vapour-phase processes it is not possible to recycle polyethylbenzenes to the alkylation reactor because the catalyst does not promote transalkylation. It is thus necessary to operate at a higher benzene ratio than in the liquid-phase process, or provide a separate dealkylation stage. Typically, a molar ratio of benzene to ethylene of $1:0.2$ is used. Yields similar to those in the liquid-phase process can be achieved.

The major advantage of the vapour-phase processes is that the process streams involved are not very corrosive. This contrasts with the liquid-phase process where the reaction mix, containing aluminium chloride and hydrogen chloride, is extremely corrosive, making it necessary to use relatively expensive materials of construction.

CUMENE

As we saw in Chapter 3, cumene is an intermediate in what is currently the most important process for the manufacter of phenol. It is made by alkylation of benzene with propene:

$$\text{(benzene)} + CH_3CH=CH_2 \longrightarrow (CH_3)_2CH\text{-(benzene)}$$

Most manufacture is by vapour-phase alkylation over a solid phosphoric acid catalyst, but liquid-phase operation in the presence of aluminium chloride or sulphuric acid is also used. Milder conditions can be used than in alkylation with ethylene, since propene is more readily protonated. Thus, vapour-phase processes use a temperature of about 520 K and a pressure of about 25 atm. As in the ethylation of benzene, polyalkylation tends to occur, but there is in this case a further side-reaction in that propene can polymerise to low polymers under the reaction conditions (see below). Both polyalkylation and polymerisation are suppressed by the use of an excess of benzene.

Yields of more than 90% calculated on both benzene and propene are obtained.

LINEAR ALKYLBENZENES

Alkylbenzenes for use in the manufacture of sodium alkylbenzene-sulphonate detergents were formerly made by alkylation of benzene with a C_{12} alkene mixture, 'propylene tetramer', obtained, together with 'propylene trimer', by polymerisation of propene in the presence of an acid catalyst:

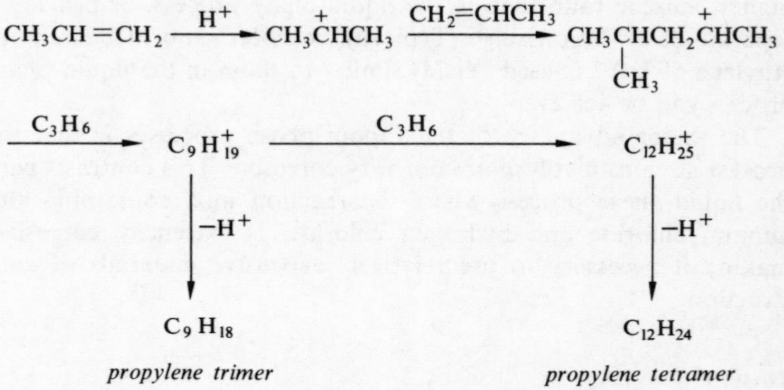

$$CH_3CH=CH_2 \xrightarrow{H^+} CH_3\overset{+}{C}HCH_3 \xrightarrow{CH_2=CHCH_3} CH_3\overset{+}{C}HCH_2\overset{+}{C}HCH_3$$
$$\overset{|}{CH_3}$$

$$\xrightarrow{C_3H_6} C_9H_{19}^+ \xrightarrow{C_3H_6} C_{12}H_{25}^+$$

$$\downarrow{-H^+} \qquad\qquad \downarrow{-H^+}$$

$$C_9H_{18} \qquad\qquad C_{12}H_{24}$$

propylene trimer   propylene tetramer

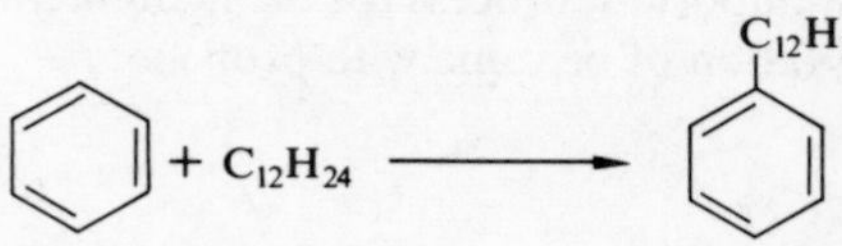

$$\text{benzene} + C_{12}H_{24} \longrightarrow \text{C}_{12}H_{25}\text{-benzene}$$

However, the detergents made from alkylbenzenes of this type have very poor biodegradability as a consequence of the high degree of branching of the alkyl groups, and their extensive use in domestic detergents led to severe pollution of rivers, and to problems of foam formation in sewage works. Their use has now been discontinued in many countries, including the UK; in these countries they have been replaced by alkylbenzenesulphonates with straight chains, these being much more readily broken down by bacteria.

There are two major sources of alkylating agents for the preparation of linear alkylbenzenes. As we saw in Chapter 2 linear alkenes of the appropriate molecular weight can be made by wax cracking. Alternatively, linear alkanes may be separated from a petroleum fraction of the appropriate boiling range by treatment with a molecular sieve, or by complex formation with urea, and then treated with chlorine to give a mixture of monochloroalkanes. This mixture is then used as the alkylating agent, or in some cases is first de-hydrochlorinated to a mixture of linear alkenes, which is then used in the alkylation. In either case, one molecule of chlorine is converted into hydrogen chloride for every molecule of alkylate made, and unless there is an outlet for this hydrogen chloride, this represents a considerable wastage of raw materials. Linear 1-alkenes can be synthesised from ethylene (see page 245), but this process does not appear to be commercially attractive compared with wax cracking at the time of writing.

In alkylation with chloroalkanes, the catalyst used is usually aluminium chloride:

$$\text{C}_6\text{H}_6 + \text{C}_{12}\text{H}_{25}\text{Cl} \xrightarrow{\ \text{AlCl}_3\ } \text{C}_6\text{H}_5\text{C}_{12}\text{H}_{25} + \text{HCl}$$

The reaction temperature is usually about 310 K. To avoid poly-alkylation, which gives material which cannot be used in detergent manufacture, an excess of benzene is used.

In alkylation with alkenes, hydrogen fluoride is often used as catalyst; reaction temperatures of 270 to 280 K are employed.

ALKYLPHENOLS

As would be expected, phenols are readily alkylated, under mild conditions. Catalysts commonly used commercially for such alkylations are sulphuric acid, phosphoric acid, boron trifluoride, and aluminium chloride. Solid catalysts such as silica–alumina or acidic ion-exchange resins can also be used, and have the advantage that there is no difficulty in separating the catalyst from the reaction mixture. The alkylating agents normally used are alkenes.

The alkyl group initially enters at positions *ortho-* and *para-* to the hydroxyl group, but since migration of alkyl groups occurs rather readily in the presence of acidic catalysts, isomerisation of the initially formed *ortho*-isomer to the more stable *para*-isomer may occur, and the product distribution finally obtained depends markedly on the conditions used. For example, alkylation of phenol with isobutene at 313 to 318 K in the presence of 0·1 % of concentrated sulphuric acid gives both *ortho-* and *para*-t-butylphenol, while at 343 K in the presence of 5 % of concentrated sulphuric acid the product is practically all *para*-t-butylphenol.

If alkylation of phenol is carried out in the presence of aluminium phenoxide as catalyst, almost exclusive *ortho-* substitution can be achieved. It has been suggested that phenol forms a complex with the aluminium phenoxide, and that this complex undergoes a concerted reaction with the alkene as shown below:

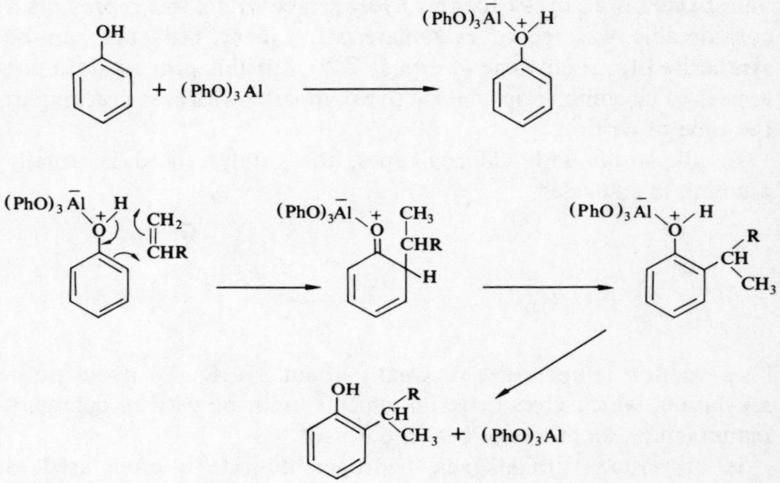

Major uses of alkylphenols are as intermediates in the manufacture of non-ionic surface-active agents (see page 258), and as antioxidants (see Chapter 8) or as intermediates in the manufacture of antioxidants (see page 174). These are moderate to small tonnage uses, so that these alkylations are carried out on a smaller scale than those discussed previously.

Octyl-, nonyl-, and dodecyl-phenols, made by alkylation of phenol with di-isobutylene, propylene trimer, and propylene tetramer respectively, are the main products used in surface-active agent manufacture. (Di-isobutylene is made by acid-catalysed dimerisation of isobutene; cf. propylene trimer and tetramer.) Reaction conditions are adjusted to give the maximum yields of the *para*-isomers, which are the required products. Thus:

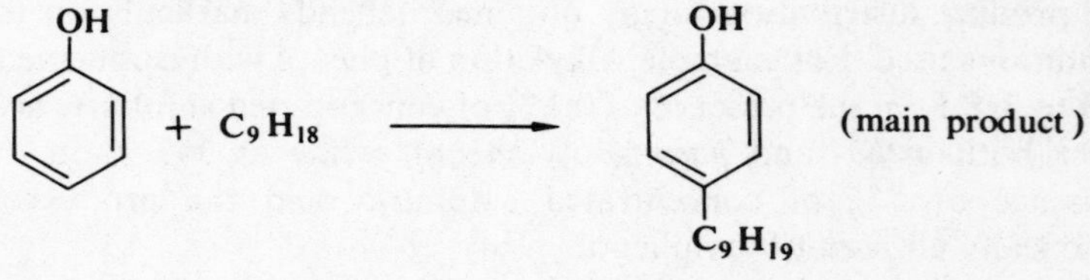

nonylphenol

Small amounts of *ortho*- and di-substituted products may also be obtained, and are separated by distillation.

p-t-Butylphenol, obtained by alkylation of phenol with isobutene under suitable conditions, is used in preparing oil-soluble phenolic resins (see Chapter 7):

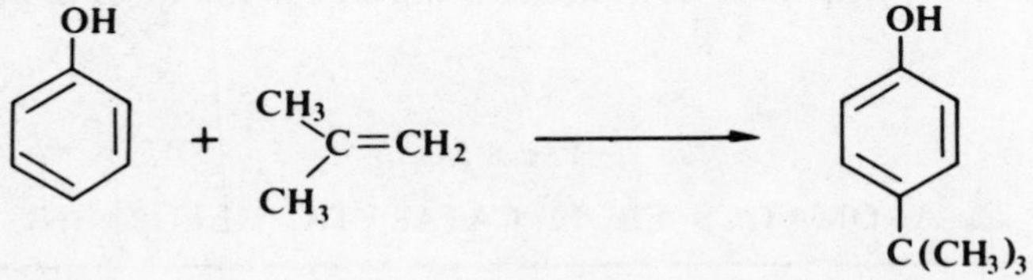

2,6-Di-t-butylphenol is obtained by alkylation of phenol with isobutene using aluminium phenoxide as catalyst:

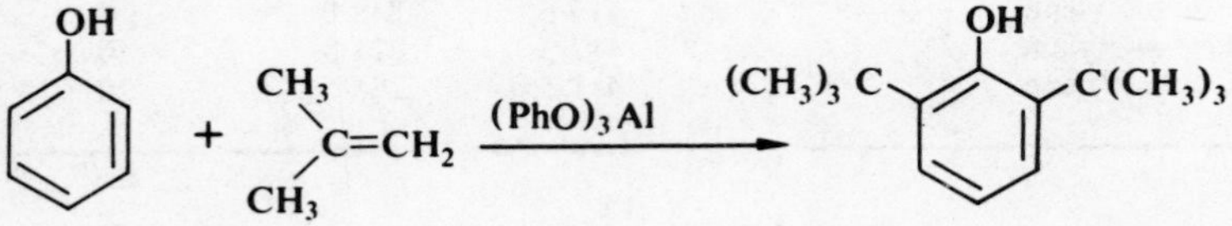

It is used as an antioxidant for gasoline and as an intermediate in the manufacture of other antioxidants.

A number of t-butyl derivatives of cresols and xylenols are used as antioxidants. One such product which is of substantial importance is 4-methyl-2,6-di-t-butylphenol, obtained by alkylation of *p*-cresol:

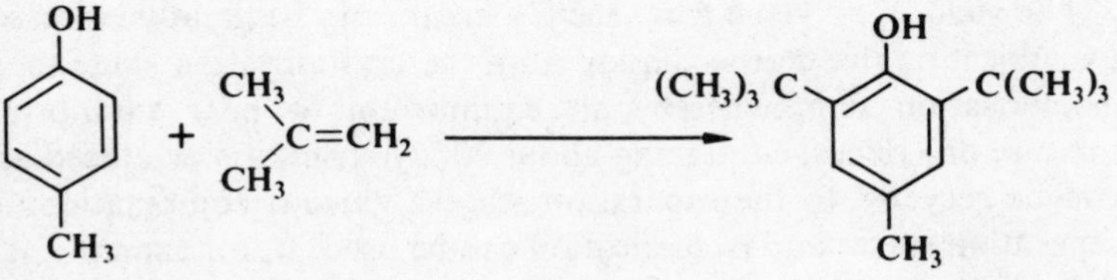

ISOMERISATION OF ALKYLBENZENES

In the presence of acidic catalysts, alkylbenzenes can undergo various isomerisation and disproportionation reactions involving migration and rearrangement of alkyl groups. One such reaction of commercial importance is the isomerisation of C_8 aromatics.

The C_8 stream produced by catalytic reforming of naphtha contains all the C_8 alkylbenzenes, in the approximate proportions shown in Table 5-2. *o*-Xylene and *p*-xylene are required for the manufacture of phthalic anhydride and terephthalic acid respectively, but there is little demand for *m*-xylene. Ethylbenzene is an intermediate in the manufacture of styrene, but isolation from C_8 reformate fractions is not in general an important method of production, alkylation of benzene being by far the most important source.

TABLE 5-2

C_8 AROMATICS FROM CATALYTIC REFORMING

Compound	b.p. (K)	m.p. (K)	Approximate distribution (%)
Ethylbenzene	409·4	178·2	20
o-Xylene	417·6	248·0	20
m-Xylene	412·3	225·3	40
p-Xylene	411·5	286·5	20

o-Xylene can fairly readily be separated from the other components by fractional distillation, but separation of *p*-xylene in this way is impracticable because of the small difference between its boiling point and that of *m*-xylene. However, *p*-xylene has a much higher melting point than the other components of the C_8 fraction, and about 60 % of it can be separated by fractional crystallisation from the mixture at low temperatures (*ca.* 240 K).

The yield of *p*-xylene from the C_8 stream can be greatly increased by subjecting the mother liquor from the crystallisation stage to an isomerisation step, whereby an equilibrium or near equilibrium mixture of xylenes, containing about 20 % *p*-xylene, is produced and can be recycled to the separation stages. Various combinations of separation stages and isomerisation can be used. If, for example, it is desired to maximise the production of *p*-xylene from the C_8 feed available, the separation of *o*-xylene may be omitted and the whole C_8 stream fed to the crystalliser.

The isomerisation may be carried out in the liquid phase in the presence of, e.g., BF_3/HF or $AlCl_3/HCl$, or, as is more common, in the vapour phase over cracking-type catalysts, e.g. silica–alumina, at about 770 K. In both cases, the reaction involves protonation of the hydrocarbon, followed by a series of 1,2-shifts of methyl groups:

CH$_3$ CH$_3$ CH$_3$ CH$_3$ CH$_3$

CH$_3$

$-$H$^+$ $\|$ H$^+$ $-$H$^+$ $\|$ H$^+$ $-$H$^+$ $\|$ H$^+$

H CH$_3$ CH$_3$ H   CH$_3$
CH$_3$ +   + CH$_3$ + CH$_3$
                                        H

In addition to isomerisation, a substantial amount of disproportionation to toluene and polymethylbenzenes can occur.

Separation of ethylbenzene from the C$_8$ stream by distillation is
possible, but expensive, since because of the small difference between
its boiling point and that of *p*-xylene, a very large column is required
and the energy consumption is high. Separation of ethylbenzene from
C$_8$ reformate is consequently not a very attractive method of
manufacture of this material. Ethylbenzene is not isomerised to
xylenes under the conditions discussed above, but undergoes cracking
and disproportionation reactions leading to the production of
benzene and polyethylbenzenes. Processes which do bring about the
isomerisation of ethylbenzene to xylenes are available. They involve
reaction of the C$_8$ mixture in the vapour phase over a dual function
catalyst similar to those used in catalytic reforming (see Chapter 2),
in the presence of hydrogen. The reactions involved are assumed to
be similar to those which occur in catalytic reforming. Thus, the
ethylbenzene is partially hydrogenated to a cycloalkene, which
isomerises and is then dehydrogenated, regenerating an isomeric
aromatic hydrocarbon.

Whichever procedure is used, the operating costs for manufacture
of *p*-xylene are relatively high, in that fractional crystallisation at low
temperatures is an expensive method of isolation, requiring as it does
the use of refrigeration and solids handling equipment. Consequently,
the price of *p*-xylene is high compared with prices of other simple
aromatic hydrocarbons [*ca.* £56/ton; cf. benzene *ca.* £30/ton
(1970)].

PHENOL-ALDEHYDE CONDENSATIONS

The reaction of phenols with aldehydes or ketones under acidic or, less often, basic conditions is of substantial commercial importance. The use of such reactions in the preparation of DDT and phenol-formaldehyde resins is discussed in Chapters 4 and 7, respectively. These reactions are also important in the preparation of bisphenol A and a number of antioxidants.

Bisphenol A [2,2-bis(p-hydroxyphenyl)propane] is used in making polycarbonates and epoxy resins (see Chapter 7). It is made by the reaction of acetone with phenol in the presence of acidic catalysts, hydrochloric acid or anhydrous hydrogen chloride being most commonly used. The mechanism is analogous to that involved in the formation of DDT:

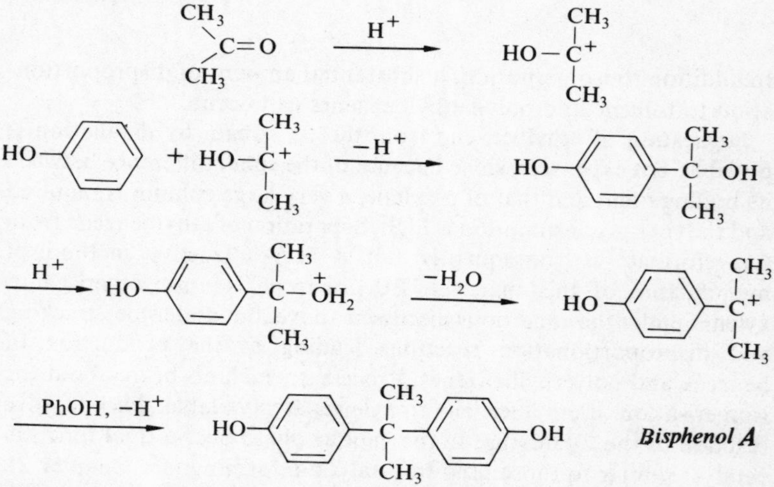

Bisphenol A

The desired reaction is subject to a number of side-reactions, for example, attack in the *ortho*-position, or further attack of the initial product, and careful design of the reaction conditions is necessary in order to obtain good yields and a product that does not require extensive, costly purification. Typically, a 3:1 molar ratio of phenol to acetone is used, and a reaction temperature of about 320 K employed. A thiol, e.g. methyl mercaptan or thioglycollic acid, is generally used as co-catalyst. Yields of up to 98% are claimed, though under unfavourable conditions much lower yields are obtained.

Typical of the antioxidants made by phenol-aldehyde condensations is 2,2'-methylenebis(4-methyl-6-t-butylphenol), an important rubber antioxidant:

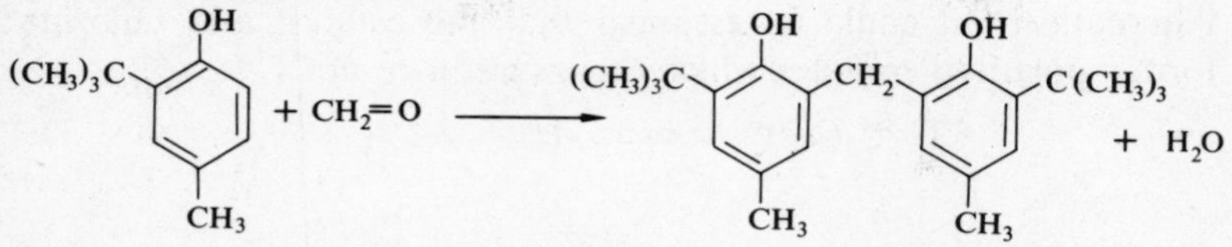

The aldehydes most used in the manufacture of antioxidants of this type are formaldehyde and n-butyraldehyde.

Halogenation

CHLORINATION OF BENZENE

The chlorination of benzene is the only nuclear aromatic halogenation that has been carried out on a large-tonnage scale. Chlorobenzene was formerly of great importance as an intermediate in the manufacture of phenol by the so-called monochlorobenzene process, but this process is now being superseded by more modern methods of phenol manufacture. Ammonolysis of chlorobenzene provides one route to aniline, and there are a number of other uses for chlorobenzene, dichlorobenzenes and, to a lesser extent, the higher chlorobenzenes.

Chlorination of benzene is carried out in the liquid phase at about 330 K, ferric chloride normally being present as catalyst. One suggestion is that attack is by molecular chlorine and that the catalyst functions by removing a chloride ion from the intermediate complex:

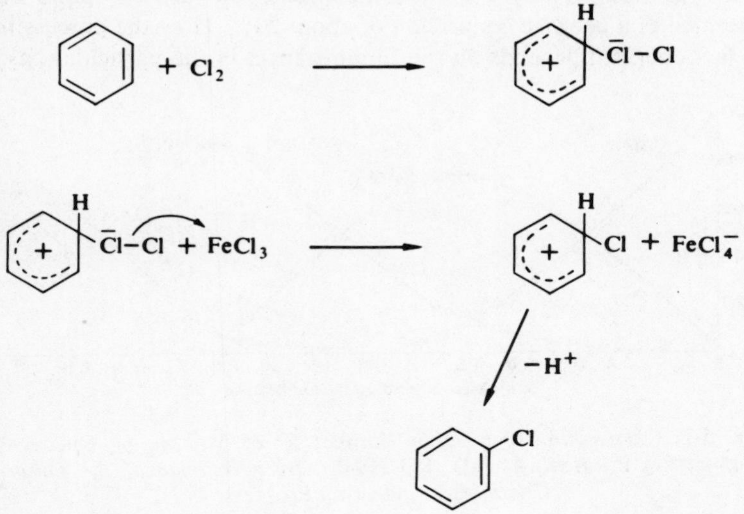

Alternatively, it could be assumed that the catalyst and chlorine form a polarised complex which acts as a source of Cl^+:

$$Cl_2 + FeCl_3 \rightleftharpoons [Cl^+ \ FeCl_4^-]$$

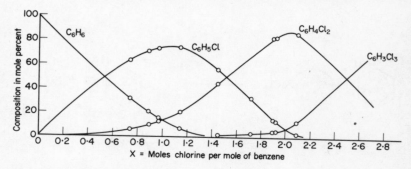

Although the introduction of a chlorine atom deactivates the ring somewhat towards further attack, this is not sufficient to prevent the occurrence of polysubstitution, and depending on the extent to which the reaction is taken, di-, tri-, and higher chlorobenzenes are formed:

The change of product composition with extent of reaction in a batch chlorination of benzene is shown in Fig. 5-1. It can be seen that in order to obtain only monochlorobenzene, the process must be operated at a benzene conversion of about 20%. How the process is in fact operated depends on the circumstances in the particular case

Fig. 5-1. *Chlorination of benzene.* Source: *R. B. MacMullin,* Chemical Engineering Progress, **44**, (3), 183 (1948). *By permission of the Editor,* Chemical Engineering Progress.

involved. If only monochlorobenzene is required, the recycling costs involved in working at low conversion have to be balanced against the extra raw material costs incurred by operating at higher conversion and producing some dichlorobenzenes. Typically, about 0·6 moles chlorine per mole benzene are used in these circumstances.

Apart from its use in phenol manufacture, which as has been indicated is becoming less important, and a limited amount of use for making aniline, the major use of chlorobenzene at present is in making DDT. It finds a number of other small uses, e.g. as a solvent, and as a dyestuffs intermediate.

p-Dichlorobenzene is used in quite substantial tonnages as a moth repellent, and de-odouriser, and o-dichlorobenzene is used as a solvent, e.g. in the manufacture of tolylene di-isocyanate; both are used to some extent as dyestuffs intermediates.

Diazonium Coupling

Substitution of aromatic compounds by aromatic diazonium salts, generally called 'coupling', is of prime importance in the dyestuffs industry.

Aromatic diazonium ions are weak electrophiles and will attack only highly activated aromatic rings; the reactions of importance in dyestuffs manufacture are with phenols and amines. Coupling with phenols is normally brought about by reaction of the diazonium salt with a solution of the sodium salt of the phenol, the negatively charged phenoxide ion being much more susceptible to electrophilic attack than the phenol itself. Activation of the substrate in this way is not possible with amines, and coupling with amines is substantially slower than with phenols.

The electrophilicity of diazonium compounds is enhanced by electron-withdrawing groups on the aromatic ring, and p-nitro-benzenediazonium chloride, for example, is much more reactive than benzenediazonium chloride.

The major application of diazonium compounds is in the manufacture of *azo dyes*, one of the simplest of which is Orange II, made by coupling diazotised sulphanilic acid with 2-naphthol:

Orange II

A large number of phenols and amines can be coupled with a variety
of diazonium compounds, and the range of azo dyes made is very
wide; the total number of commercial azo dyes has been estimated
to be about 1 600. They fall into a number of classes, based on the
method of application and the fibres for which they are used.
Orange II is one of the class of 'acid dyes', used for dyeing wool. The
'direct dyes', of which Congo Red was the first made, are used for
dyeing cotton:

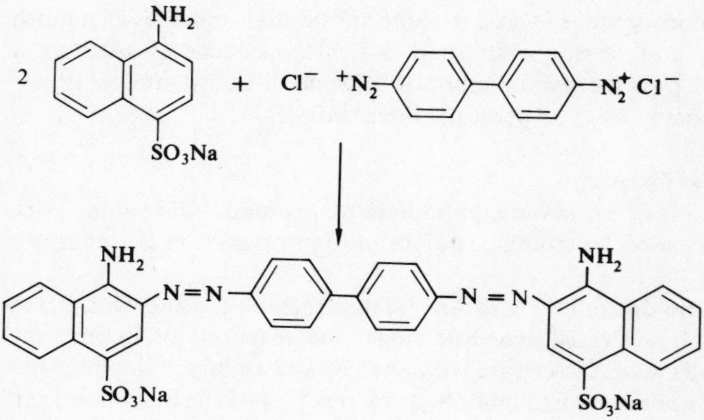

Congo Red

Another class of dyes, generally used on cellulosic fibres, in which
diazonium coupling is used, is that of the *azoic dyes*. In this case, the
coupling is carried out on the fibre to produce an insoluble coloured
molecule, and thus give a dyeing with very good fastness to washing.
If, for example, the fabric is impregnated first with a solution of
2-naphthol in aqueous sodium hydroxide, and then with a solution of
p-nitrobenzenediazonium chloride, the dye Para Red is produced on
the fibre:

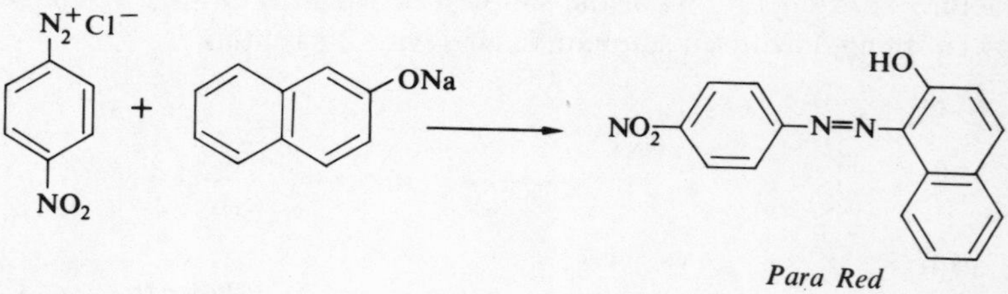

Para Red

Note that the replacement of the sulphonate group in Orange II by a nitro group, giving Para Red, leads to two technological changes, the loss of water solubility, and a change of colour.

Isomerisation and Disproportionation of Benzenecarboxylic Acids

In the early 1950s it was shown by workers at Henkel and Cie in Germany that good yields of potassium terephthalate could be obtained by heating the potassium salts of various benzenecarboxylic acids, e.g. phthalic, isophthalic and benzoic acids, at about 670 K in the presence of cadmium or zinc salts, and under carbon dioxide pressure:

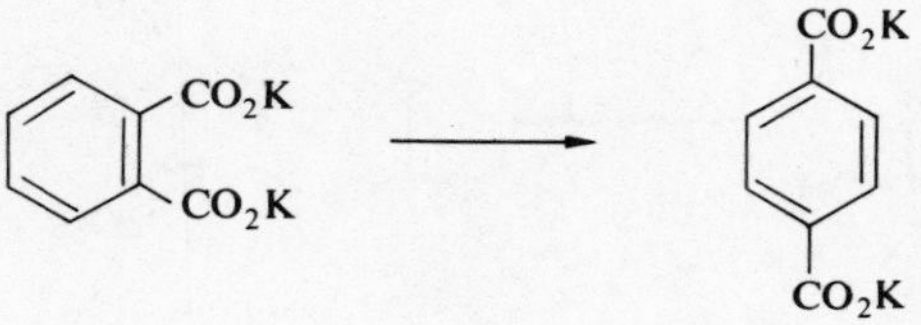

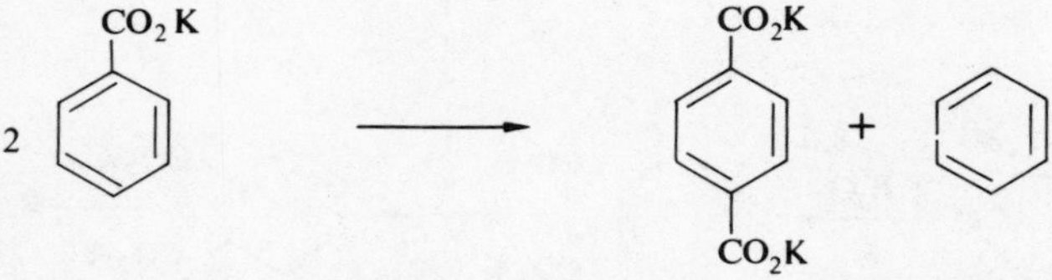

These reactions provide the basis of routes to terephthalic acid from lower cost hydrocarbons than the relatively expensive p-xylene, since benzoic acid can be made by liquid-phase free-radical oxidation of toluene, and phthalic anhydride by gas-phase oxidation of o-xylene over a vanadium pentoxide catalyst (see Chapter 3). Processes using these routes are operated in Japan. Thus, Mitsubishi manufacture terephthalic acid from toluene as indicated below:

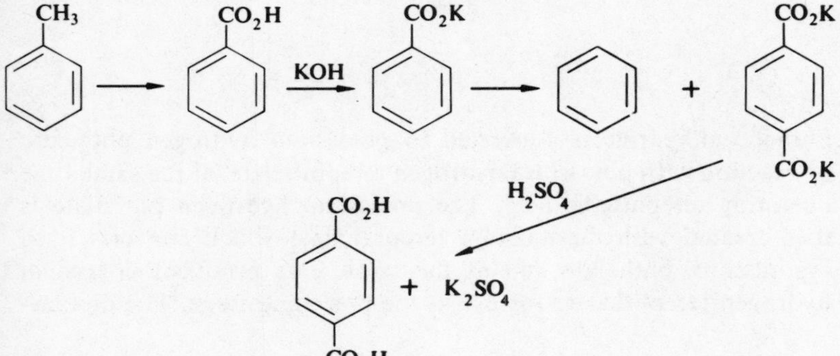

Yields in the disproportion stage are greater than 90%. It can be seen that in addition to toluene, potassium hydroxide and sulphuric acid are required as raw materials, and potassium sulphate is given as a co-product. The economics of the process are markedly affected by the return obtained for the potassium sulphate; Mitsubishi sell it for use as fertiliser. As we have already seen, benzene, the other co-product, has a higher value than toluene, and can be readily disposed of.

In the process based on o-xylene, there is no overall consumption of potassium, except for small losses:

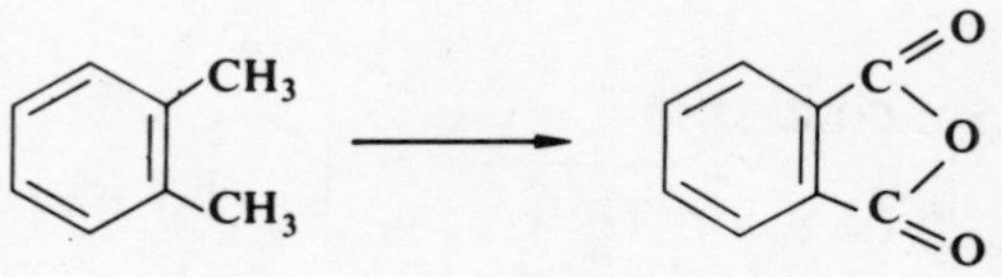

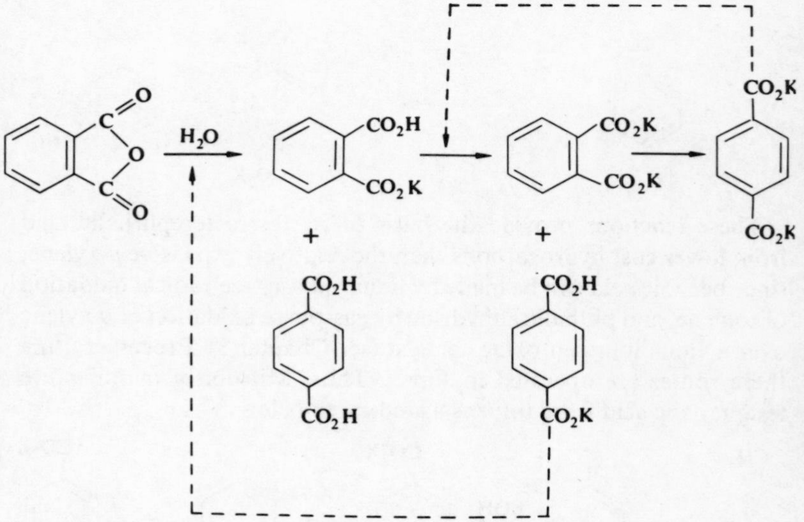

Phthalic anhydride is converted to potassium hydrogen phthalate by reaction with potassium hydrogen terephthalate, at the same time liberating terephthalic acid. The potassium hydrogen phthalate is then treated with dipotassium terephthalate which converts it to dipotassium phthalate and at the same time produces potassium hydrogen terephthalate for use in the preceding stage. The dipotassium phthalate solution is evaporated to dryness, and the dry potassium salt is passed to the isomerisation stage. The yield in the isomerisation stage is greater than 90%.

In both variants the isomerisation or disproportionation reaction is carried out with the dry potassium salt, in the solid phase, and there are some difficulties in the engineering of this stage of the processes.

The isomerisation/disproportionation reaction has not received a great deal of attention from research workers outside industry, and there is considerable uncertainty as to its mechanism. One suggested scheme is shown below:

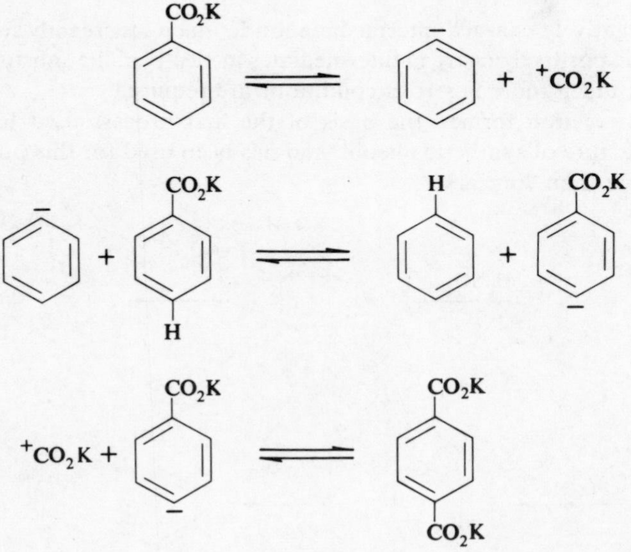

$$C_6H_5CO_2K \rightleftharpoons C_6H_5^- + {}^+CO_2K$$

$$C_6H_5^- + \text{para-}C_6H_4(CO_2K)(H) \rightleftharpoons C_6H_6 + \text{para-}C_6H_4(CO_2K)^-$$

$${}^+CO_2K + \text{ortho-}C_6H_4(CO_2K)^- \rightleftharpoons \text{para-}C_6H_4(CO_2K)_2$$

The mode of action of the catalyst is not clear.

NUCLEOPHILIC AROMATIC SUBSTITUTION

Although not so important as electrophilic substitution, nucleophilic aromatic substitution finds a number of commercial applications. In general, nucleophilic displacement of a substituent from an aromatic ring is much less facile than displacement of the same substituent from an aliphatic compound, and it will be noted that in a number of the processes discussed below very severe conditions are required.

Alkaline Fusion of Sodium Arylsulphonates

The sulphonate group of sodium arylsulphonates can be replaced by hydroxyl by reaction with sodium hydroxide at about 570 to 590 K. The reaction, which is conveniently brought about in a melt of sodium hydroxide and the sulphonate, is thought to involve displacement of a sulphite ion by a two-step mechanism analogous to that involved in electrophilic substitution:

$$C_6H_5SO_3^- + OH^- \rightleftharpoons [\text{intermediate: } HO, SO_3^-] \longrightarrow C_6H_5OH + SO_3^{2-}$$

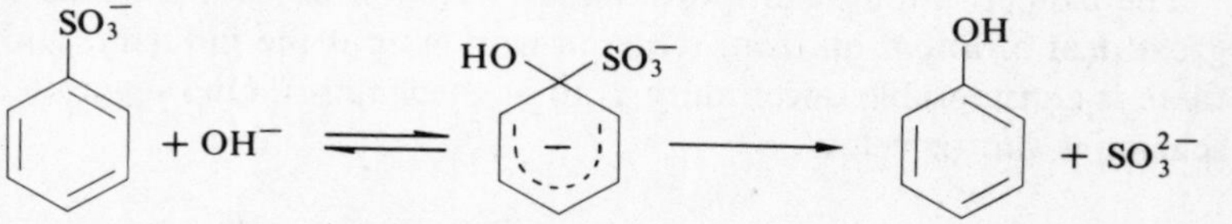

The negatively charged intermediate ion is much less readily formed than the positively charged intermediates in electrophilic substitution so that much more vigorous conditions are required.

This reaction formed the basis of the first process used for the manufacture of synthetic phenol, and has been used for this purpose for more than 70 years:

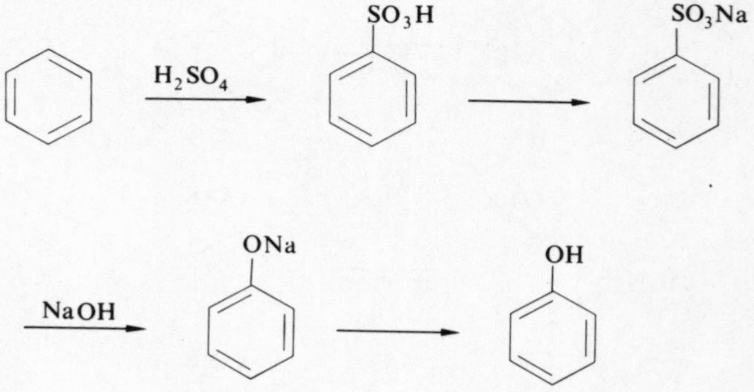

$$C_6H_6 \xrightarrow{H_2SO_4} C_6H_5SO_3H \longrightarrow C_6H_5SO_3Na$$

$$\xrightarrow{NaOH} C_6H_5ONa \longrightarrow C_6H_5OH$$

A yield of about 85 to 90% based on benzene is given.

The major disadvantage of this process is its very heavy consumption of sulphuric acid and sodium hydroxide. An excess of sulphuric acid is required in the sulphonation stage for reasons discussed earlier in this chapter, and an excess of sodium hydroxide is required in the fusion stage; despite a number of ingenious methods of minimising the usage of these reagents, typically about $1\frac{3}{4}$ tons each of sulphuric acid and sodium hydroxide are required per ton of phenol made. In addition, the four-stage process has fairly high capital and operating costs. Consequently, the process compares unfavourably with a number of newer phenol processes, and many plants in which it was operated have been shut down. In the UK, for instance, the plant that was operated by Monsanto at a scale of 10 000 tons per annum was shut down in 1964.

Alkaline fusion of sulphonates still provides the major synthetic route to a number of other phenols, made on a smaller scale than phenol itself, e.g. 2-naphthol, resorcinol, *p*-cresol.*

Hydrolysis of Aryl Chlorides
The hydrolysis of aryl chlorides provides another method of introducing a hydroxyl group into an aromatic ring.

CHLOROBENZENE
Hydrolysis of chlorobenzene can be brought about by treatment with 10–15% aqueous sodium hydroxide at about 640 K, a pressure of about 300 atm being required to keep the reactants in the liquid phase. This reaction forms the basis of a phenol process developed by Dow in the early 1920s:

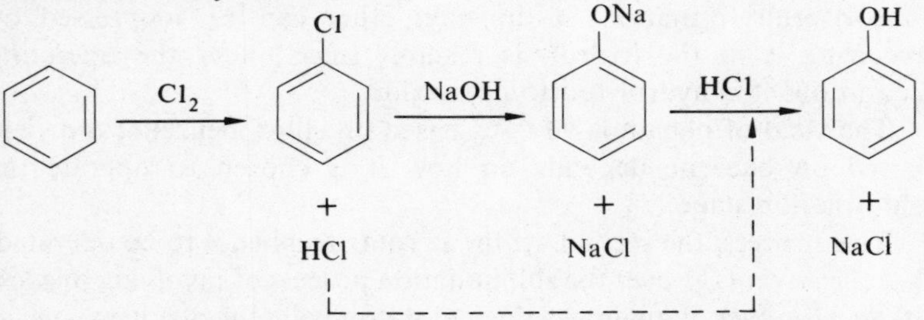

It is thought that the main path of the hydrolysis is *via* dehydrochlorination of the chlorobenzene to the reactive intermediate benzyne, followed by the addition to this of water:

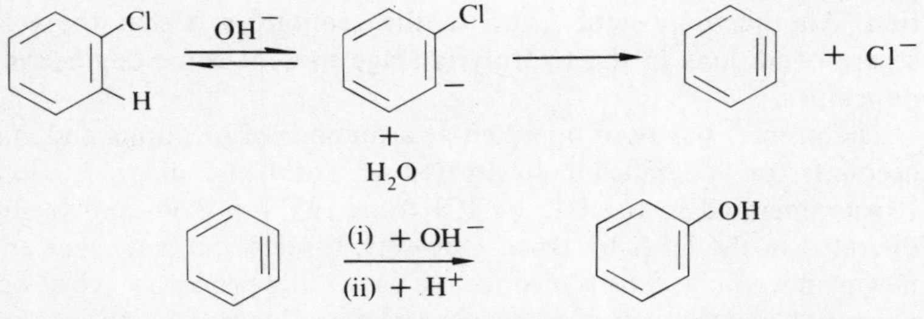

---

* Most cresol made is so-called natural cresol, isolated from coal carbonisation by-products. Only a minor proportion is made synthetically at present.

The main evidence for this is that hydrolysis of chlorobenzene labelled with carbon-14 at the 1-position gives a substantial proportion of phenol with the hydroxyl group in a position *ortho-* to that originally occupied by the chlorine atom.

Diphenyl ether and phenylphenols are produced as by-products, possibly by reactions of the type indicated below:

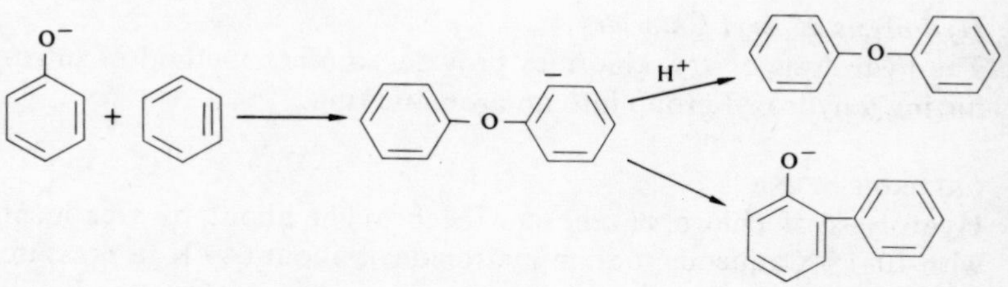

The overall formation of diphenyl ether can be suppressed by recycling it to the hydrolysis reactor, since under the operating conditions it is hydrolysed to phenoxide.

The yield of phenol is 90-95% based on chlorobenzene; the yield based on benzene depends on how it is chosen to operate the chlorination stage.

This process, the second synthetic route to phenol to be operated, has the advantage over the sulphonation process of involving one less stage. However, it again has the major disadvantage of a heavy usage of ancillary raw materials, in this case chlorine and sodium hydroxide. The sodium chloride formed can, if desired, be converted back to chlorine and sodium hydroxide by electrolysis, but in this case the process effectively has four stages, and a heavy electricity consumption. Another important factor in the economics is that the very severe conditions in the hydrolysis stage present some engineering difficulties.

The process has been operated at a number of locations and still accounts for a significant proportion of world phenol production. It was operated in the UK by ICI from 1951 to 1964, and is still operated in the USA by Dow. However, it seems unlikely that any new plants will be constructed for the use of this process, since newer processes, notably the cumene process (see Chapter 3), show considerable economic advantage over it.

Hydrolysis of chlorobenzene can also be carried out catalytically in the vapour phase and this reaction is used in the regenerative (Raschig) process for phenol, in conjunction with an oxychlorination stage for the production of chlorobenzene:

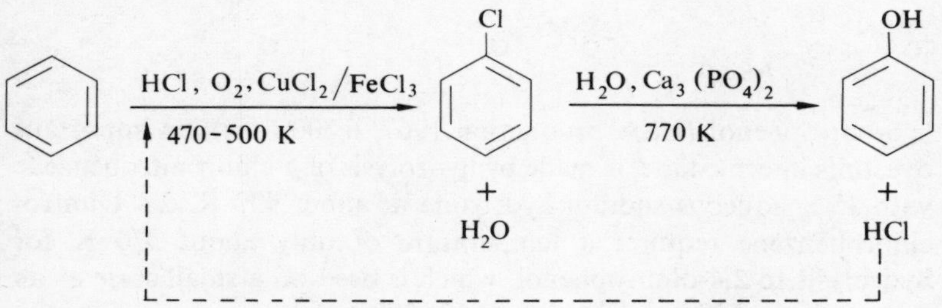

$$
\text{benzene} \xrightarrow[470-500\ K]{HCl,\ O_2,\ CuCl_2/FeCl_3} \text{chlorobenzene (Cl)} + H_2O \xrightarrow[770\ K]{H_2O,\ Ca_3\,(PO_4)_2} \text{phenol (OH)} + HCl
$$

Yields of up to 89% based on benzene are claimed.

This process, developed in Germany during the early 1930s, requires only benzene and a small amount of hydrogen chloride (to make up losses) as raw materials, and in this respect shows considerable advantage over the sulphonation process and the monochlorobenzene process described above. Further, it has only two stages. It has not, however, been very widely adopted. The reasons for this are not altogether clear. One factor may be that the process acquired a reputation for corrosion problems in the early days of its operation, though it is claimed that these problems are not basic to the process. Also, the economics of the process are adversely affected to some extent by the fact that both stages have to be operated at low conversion (*ca.* 10%) in order to obtain acceptable yields. In most circumstances, the process seems to compare unfavourably with the cumene process. However, it is operated in the USA, where in 1966 it accounted for about 17% of total phenol production, and it is reported that a plant is to be built in Spain to operate it.

CHLORONITROBENZENES

Chloronitrobenzenes with the nitro groups *ortho-* or *para-* to the chlorine atom are much more readily hydrolysed than chlorobenzene itself. The reaction in this case is thought to have a bimolecular mechanism of the type discussed under alkaline fusion of aryl sulphonates. The nitro groups facilitate reaction by stabilising the intermediate ions and the transition states involved in their formation:

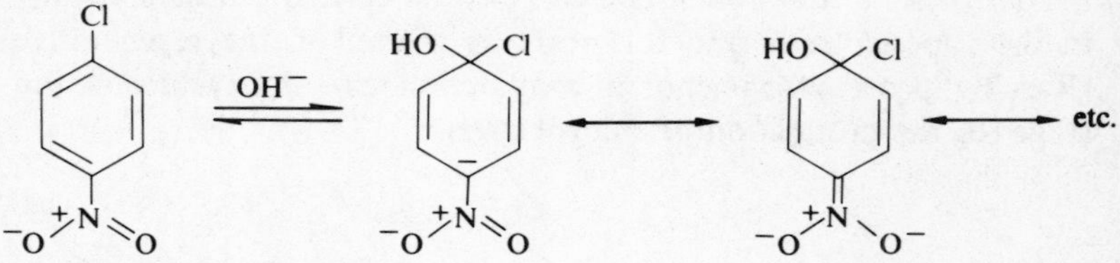

p-Nitrophenol (USA production 1967, 6 800 tons), an important dyestuffs intermediate, is made by hydrolysis of *p*-chloronitrobenzene with 15% aqueous sodium hydroxide at about 430 K. 2,4-Dinitro-chlorobenzene requires a temperature of only about 370 K for hydrolysis to 2,4-dinitrophenol, which is used on a small scale as an intermediate (USA production 1967, 350 tons).

Ammonolysis of Aryl Chlorides

Ammonolysis of aryl chlorides finds some application in the manufacture of aromatic amines. For example, reaction of chlorobenzene with aqueous ammonia provides one route to aniline. The reaction is carried out by treating chlorobenzene with about a four to five molar proportion of aqueous ammonia at 480 to 490 K and under about 60 atm pressure (to keep the reactants in the liquid phase). The large excess of ammonia is required to suppress the competing hydrolysis of chlorobenzene, but even with its use, about 5% of the chlorobenzene is converted into phenol. The yield of aniline is about 90% based on chlorobenzene:

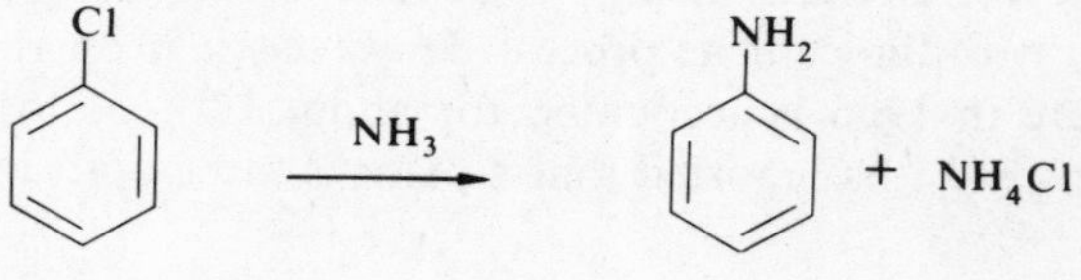

The reaction is thought to follow a path analogous to that involved in the hydrolysis of chlorobenzene, i.e. to involve benzyne as an intermediate.

This process accounts for only a small proportion of aniline manufacture, most being made by the reduction of nitrobenzene. It is operated in one plant in the USA, but not in the UK.

As with hydrolysis, strongly electron-withdrawing groups *ortho*- and *para*- to the chlorine atom render ammonolysis more facile. Thus, *p*-nitroaniline, an intermediate for dyestuffs and fine chemicals, is made by the reaction of *p*-chloronitrobenzene with aqueous ammonia at about 450 K and 40 atm:

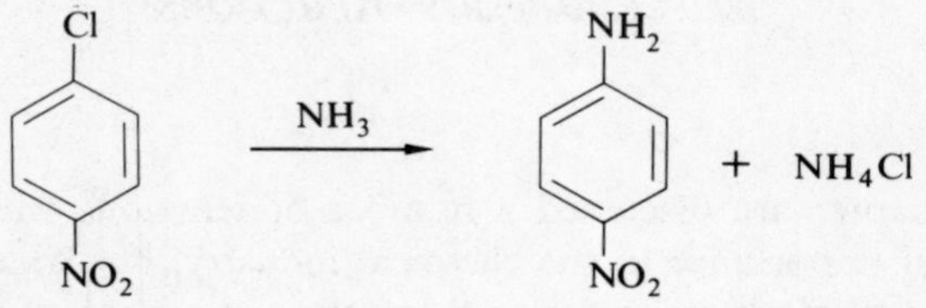

Chapter 6

Miscellaneous Reactions

IN this chapter are discussed a number of reactions which are of substantial importance in the chemical industry, but which are not considered to merit a chapter to themselves in a book of this size.

HYDRATION

Hydration, that is the addition of the elements of water to a compound, is of substantial commercial importance. Hydration of alkenes, for instance, provides one of the major routes to alcohols, although it has the important limitation that, since the addition obeys Markownikoff's rule, primary alcohols are not obtained except in the hydration of ethylene. Hydration of acetylene has been of considerable commercial importance in some countries, and hydration of ethylene oxide provides by far the major route to ethylene glycol.

Hydration of Ethylene

The hydration of ethylene, which is the major commercial source of ethanol, is carried out by two methods.

SULPHURIC ACID PROCESS

In this process, ethylene is absorbed in concentrated sulphuric acid to form a mixture of ethyl hydrogen sulphate and diethyl sulphate which is then diluted with water and the sulphates hydrolysed to ethanol:

$$CH_2{=}CH_2 + H_2SO_4 \longrightarrow CH_3CH_2SO_4H$$

$$CH_3CH_2SO_4H + CH_2{=}CH_2 \longrightarrow (CH_3CH_2)_2SO_4$$

188

$$CH_3CH_2SO_4H + H_2O \longrightarrow CH_3CH_2OH + H_2SO_4$$

$$(CH_3CH_2)_2SO_4 + 2H_2O \longrightarrow 2CH_3CH_2OH + H_2SO_4$$

The absorption is carried out in 94–98 % acid at 330 to 360 K and under a pressure of 17 to 35 atm. Higher temperatures, or more concentrated acid, give a higher rate of absorption, but also cause more side reactions. The product from the absorption stage contains 1·0–1·4 moles reacted ethylene/mole sulphuric acid, and consists mainly of ethyl hydrogen sulphate, diethyl sulphate, and sulphuric acid, together with polymeric oils and tars. It is diluted with water to give a final acid strength of about 50 % after hydrolysis, and held at 330 to 370 K while hydrolysis occurs. Some diethyl ether is formed as a by-product of hydrolysis:

$$(C_2H_5)_2SO_4 + C_2H_5OH \longrightarrow (C_2H_5)_2O + C_2H_5SO_4H$$

$$C_2H_5SO_4H + C_2H_5OH \longrightarrow (C_2H_5)_2O + H_2SO_4$$

The yield of ethanol is about 90 %.

This process was first operated in 1930, in the USA. Its major disadvantage is that large volumes of diluted sulphuric acid are produced, and the necessity to reconcentrate this adds a substantial amount to the capital and operating costs (particularly energy costs) of the process. The highly corrosive nature of the process streams also adds to costs.

The sulphuric acid process is used in the UK by Courtauld, at a scale of 40 000 tons per annum.

CATALYTIC HYDRATION

The direct catalytic hydration of ethylene in the vapour phase was first operated commercially by Shell, in 1947. This process would appear at first sight to have substantial advantages over the sulphuric acid process in that it has only one stage, and has no heavy energy requirements for acid reconcentration. However, for reasons we shall consider below, the economic choice between these two processes is not quite so clear-cut as might be expected.

The hydration is carried out over a solid phosphoric acid catalyst (see page 166), and is presumed to involve protonation of ethylene followed by reaction of water with the resulting ethyl carbonium ion:

$$CH_2{=}CH_2 \xrightarrow{\;H^+\;} CH_3CH_2^+ \xrightarrow{\;H_2O\;} CH_3CH_2\overset{+}{O}H_2 \xrightarrow{\;-H^+\;} CH_3CH_2OH$$

Side reactions can occur leading to the formation of diethyl ether and low polymers of ethylene:

$$CH_3CH_2^+ + CH_3CH_2OH \longrightarrow CH_3CH_2\overset{\overset{\displaystyle H}{|}}{\underset{+}{O}}CH_2CH_3 \xrightarrow{\;-H^+\;} CH_3CH_2OCH_2CH_3$$

$$CH_3CH_2^+ + CH_2{=}CH_2 \longrightarrow CH_3CH_2CH_2CH_2^+ \xrightarrow{\;-H^+\;} CH_3CH_2CH{=}CH_2$$

Hydration is reversible and exothermic:

$$C_2H_4 + H_2O \rightleftharpoons C_2H_5OH \qquad \Delta H = -46 \text{ kJ/mol}$$

High conversion to ethanol at equilibrium is favoured by low temperature, high pressure, and high water concentration. The effects of temperature and pressure on conversion at equilibrium in the reaction between equimolar quantities of ethylene and water are shown in Fig. 6-1.

In order to achieve acceptable rates of reaction over a solid

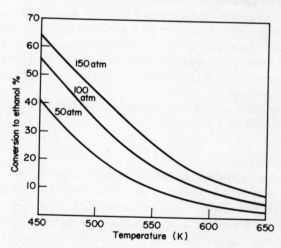

Fig. 6-1. *Hydration of ethylene-calculated equilibrium conversions (molar ratio $C_2H_4 : H_2O = 1:1$).*

Source of data: A. A. Vvedensky and L. F. Feldman, *Zhur. Obschei Khim.*, **15**, 37 (1945).

phosphoric acid catalyst, a reaction temperature of about 570 K has to be used; it can be seen from Fig. 6-1 that at this temperature and atmospheric pressure, the conversion to ethanol in the reaction between equimolar quantities of ethylene and water will be low. Increase in operating pressure has a marked effect on the position of the equilibrium. However, the extent to which pressure can be used to improve the conversion is limited by an increased tendency to polymer formation at high pressures, and by the extra capital and operating costs involved. Pressures of about 70 atm are generally used.

The other way of achieving a higher conversion of ethylene would be to use an excess of water. However, the use of a solid phosphoric acid catalyst places limits on the ratio of water to ethylene which can be used in the process, in that the composition of the phosphoric acid phase of the catalyst depends on the operating temperature and on the partial pressure of water in the gas stream. At high partial pressures of water, the catalyst takes up water and its activity drops, so that the reaction slows down. Also, as the phosphoric acid becomes diluted it tends to drain from the support. These factors limit the molar water/ethylene ratio which can be used to about 0·6:1.

Under the operating conditions indicated above conversions of about 5% are obtained, the yield being about 95%. The gas stream from the reactor is cooled to separate the liquid products, and the remaining ethylene is then recycled. The necessity to re-heat and re-circulate very large amounts of ethylene makes the process quite a heavy energy consumer, but even so, it probably still shows economic advantage over the sulphuric acid process under most conditions. It is operated in the UK by BP at a scale of 84 000 tons per annum (1970), and there are plans to substantially increase this scale by 1972.

It is evident that the development of a catalyst that would allow the reaction to be carried out at, say, 470 K would lead to a substantial improvement in the economics of the process. No such catalyst appears to be available at the time of writing.

The major use of ethanol has for many years been in the manufacture of acetaldehyde, but this outlet will dwindle as the Wacker Chemie process replaces ethanol dehydrogenation in acetaldehyde manufacture. There are a variety of other uses as an intermediate, e.g. in the manufacture of ethyl acetate, and a substantial amount of ethanol is used in solvent applications.

Hydration of Propene

Hydration of propene gives isopropanol, the formation of the secondary propyl carbonium ion occurring much more readily than formation of the primary carbonium ion. Thus, for direct hydration:

$$CH_3CH=CH_2 \xrightarrow{H^+} CH_3\overset{+}{C}HCH_3 \xrightarrow{H_2O} CH_3\overset{\overset{+}{O}H_2}{C}HCH_3 \xrightarrow{-H^+} CH_3\overset{OH}{C}HCH_3$$

Isopropanol has been manufactured by hydration of propene since 1920, and is one of the oldest-established petrochemicals. Until 1951, all manufacture was by the two-stage sulphuric acid process, but substantial quantities are now manufactured in Europe (though not in the USA) by direct hydration over solid catalysts.

About 60 % of the isopropanol made is converted to acetone. The rest goes to a variety of uses including solvent applications, and other chemical manufactures.

SULPHURIC ACID PROCESS

This process was first operated by the Standard Oil Company, and by the Carbide and Carbon Chemical Corporation, in the USA in 1920. The process is similar to that used for ethanol but, since propene is more reactive towards electrophilic reagents than ethylene, the conditions used in the absorption stage are milder than in the ethanol process, sulphuric acid of about 80 % concentration, and temperatures of about 300 K normally being used. The use of lower concentration acid substantially reduces reconcentration costs compared with those in the ethanol process. Yields of over 90 % can be achieved, the main by-products being di-isopropyl ether, which can be recycled, and propene polymers.

It is not essential that the propene used be pure, and a propane/propene mixture is often used, with a resulting saving in separation costs.

Manufacture of isopropanol by the sulphuric acid process is carried out in the UK by BP, Courtauld, and Shell, at scales of 30 000, 20 000 and 130 000 tons per annum respectively.

CATALYTIC HYDRATION

The thermodynamics of the gas-phase hydration of propene are even less favourable than those of ethylene hydration, so that under equivalent conditions lower conversions are obtained than with ethylene. However, since propene is more reactive than ethylene, lower temperatures can be used, and, for example, hydration can be carried out at 500 to 525 K and 40 atm over a solid phosphoric acid catalyst to give a 3·8 % conversion.

The first commercial operation of a heterogeneous-catalysed direct hydration process was by ICI in 1951. In this process propene is reacted with water over a tungsten oxide catalyst at 520 to 570 K and 200 to 300 atm, a molar ratio of water to propene of about 2·5:1 being used. Under these conditions water is present in both the gas and the liquid phase. The presence of water in the liquid phase markedly affects the position of the equilibrium since iso-propanol is much more soluble in water than is propene, and is preferentially removed from the gas phase as it is formed. Thus, higher conversions are attainable than when operating solely in the gas phase. In order to operate in this way it is essential that the catalyst is not inactivated by liquid water; a solid phosphoric acid catalyst could not be used under these conditions since the phosphoric acid would be leached out.

ICI operate this process in the UK at a scale of 45 000 tons per annum.

Hibernia Chemie started operation of a process using a solid phosphoric acid catalyst in West Germany in 1966. It is reported that this process will be used by Shell in their new plant under construction at Pernis in Holland, and by BP in a plant under construction at Baglan Bay in the UK.

Hydration of Butenes

Hydration of but-1-ene and but-2-ene gives s-butanol, and hydration of isobutene gives t-butanol:

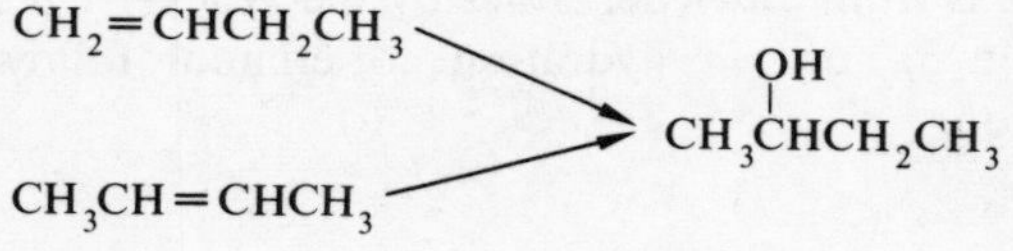

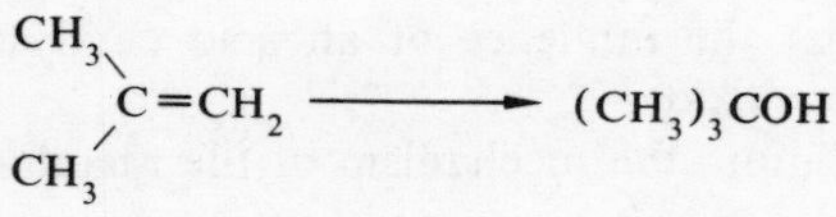

$$CH_2=CHCH_2CH_3$$
$$CH_3CH=CHCH_3$$
$$\longrightarrow CH_3\overset{\overset{\displaystyle OH}{|}}{C}HCH_2CH_3$$

$$\begin{matrix} CH_3 \\ \diagdown \\ C=CH_2 \\ \diagup \\ CH_3 \end{matrix} \longrightarrow (CH_3)_3COH$$

s-Butanol is of importance as an intermediate in the manufacture of methyl ethyl ketone (see page 219), and of s-butyl acetate; manufacture of t-butanol by hydration of isobutene is of little commercial importance.

Hydration of the n-butenes is carried out by the sulphuric acid process. The conditions required for absorption in sulphuric acid are milder than those required for propene and, for example, 70% sulphuric acid at 310 K can be used. Direct hydration over a solid catalyst does not appear to be used.

Hydration of Acetylene

As the reader will probably know, acetylene may be hydrated by treatment with dilute sulphuric acid containing mercury salts:

$$CH \equiv CH + H_2O \longrightarrow CH_3CHO$$

This reaction provides a route to acetaldehyde which has been of commercial importance, particularly in Germany. However, the importance of this process is exaggerated by a number of textbooks of organic chemistry, and it is worth noting that it has never been used on the large scale in the UK, and has been of only minor importance in the USA.

The hydration is carried out by passing acetylene into sulphuric acid of about 20% concentration, containing mercury and iron salts, at about 340 K. Acetaldehyde in about 95% yield is recovered from the aqueous solution by distillation. The mechanism of the reaction is not well established.

As we have seen, acetylene is an expensive two-carbon raw material, and this process is now obsolescent. Most acetaldehyde manufacture is from ethylene, either by the Wacker Chemie process (see Chapter 3), or *via* hydration to ethanol followed by dehydrogenation.

Hydration of Ethylene Oxide

Hydration to ethylene glycol, which may be carried out in neutral solution or under the influence of an acid catalyst, is the major outlet for ethylene oxide.

In neutral solution, the mechanism of the reaction is thought to be as follows:

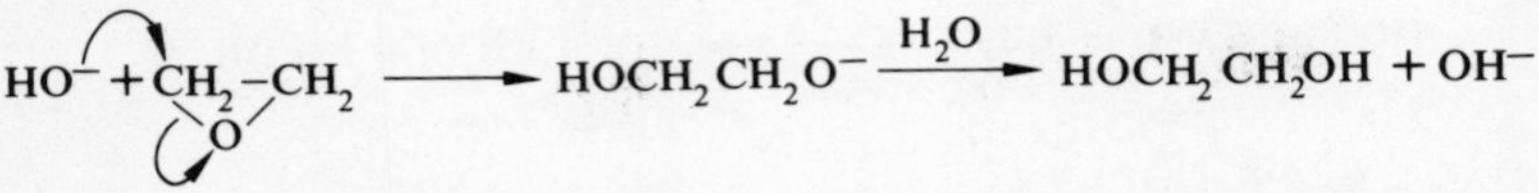

$$HO^- + CH_2-CH_2 \longrightarrow HOCH_2CH_2O^- \xrightarrow{\ H_2O\ } HOCH_2CH_2OH + OH^-$$

Under these conditions a fairly high temperature (about 470 K) is necessary to achieve a reasonable rate of reaction, and to keep the reactants in the liquid phase the reaction has to be done under about 14 atm pressure.

Acid catalysts facilitate the reaction by protonating the ethylene oxide to give an oxonium complex, which then undergoes attack by water by an S_N1 or S_N2 mechanism:

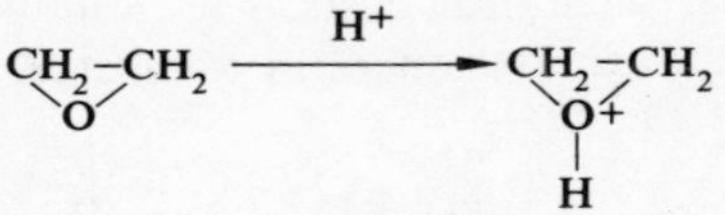

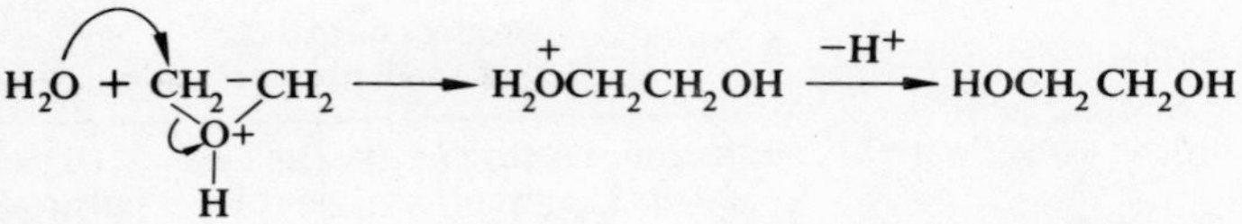

$$CH_2-CH_2 \xrightarrow{\ H^+\ } CH_2-CH_2$$

$$H_2O + CH_2-CH_2 \longrightarrow H_2OCH_2CH_2OH \xrightarrow{\ -H^+\ } HOCH_2CH_2OH$$

or

$$CH_2-CH_2 \longrightarrow {}^+CH_2CH_2OH \xrightarrow{\ H_2O\ } H_2OCH_2CH_2OH \xrightarrow{\ -H^+\ } HOCH_2CH_2OH$$

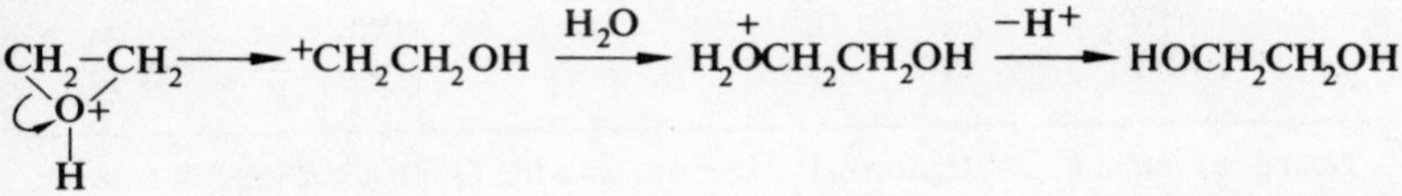

Much less vigorous conditions are required in this case. Thus, in 0·5% sulphuric acid the reaction may be carried out at 320 to 340 K.

The acid-catalysed process has the advantage of not requiring pressure equipment, and of employing lower temperatures, but on the other hand it requires corrosion-resistant materials of construction, and separation of sulphuric acid from the product is quite troublesome. The uncatalysed process appears to be the more widely used.

In both processes the glycol produced can itself carry out a nucleophilic attack on ethylene oxide to give diethylene glycol, which itself can react further to give triethylene glycol and so on:

$$HOCH_2CH_2OH + CH_2{-}CH_2 \longrightarrow HOCH_2CH_2OCH_2CH_2OH$$

diethylene glycol

$$HOCH_2CH_2OCH_2CH_2OH + CH_2{-}CH_2 \longrightarrow HOCH_2CH_2OCH_2CH_2OCH_2CH_2OH$$

triethylene glycol

The extent to which polyethylene glycols are formed is markedly affected by the ratio of water to ethylene oxide used (see Table 6-1). In practice, the economic advantage of obtaining a high yield by using a high water/ethylene oxide ratio has to be balanced against the disadvantages of the larger reactor volume requirements at high dilution, and the higher separation costs. The ratio used depends on the particular economic circumstances in each case; it is generally between 5:1 and 8:1.

TABLE 6-1

ETHYLENE OXIDE HYDRATION

Molar ratio (ethylene oxide : water)	Product distribution (% of total ethylene oxide)			
	Ethylene glycol	Diethylene glycol	Triethylene glycol	Higher polyglycols
1 : 10·5	82·3	12·7	—	—
1 : 6·7	76·5	18·5	—	—
1 : 4·2	65·7	27·0	2·3	—
1 : 2·1	47·2	34·5	13·0	0·3
1 : 0·61	15·7	26·0	19·8	33·5

Source of data: C. Matignon, H. Moureu, and M. Dode, *Bull. Soc. Chim. Fr.*, **1**, (5), 1308 (1934). By permission of the Société Chimique de France.

Bases also catalyse the hydration. However, they cause a greater increase in the rate of glycolysis than of hydrolysis and promote the formation of polyethylene glycols, and for this reason base catalysis is not used in commercial processes.

Major uses of ethylene glycol are as antifreeze for car engines, and in the manufacture of poly(ethylene terephthalate) (see Chapter 7).

ESTERIFICATION

Esterification is applied in the organic chemical industry over a wide range of scales. A small number of esters are made on a large scale, e.g. various phthalates, ethyl and butyl acetates, cellulose acetate, and a large number of other esters are made on a much smaller scale, e.g. for use in flavouring and perfumery materials. Esterification is also important in the manufacture of a number of polymers (see Chapter 7).

Esterification of alcohols by carboxylic acids is normally carried out in the presence of acidic catalysts, which function by protonating the carbonyl oxygen of the carboxyl group, and thus increasing the susceptibility of the attached carbon atom to nucleophilic attack:

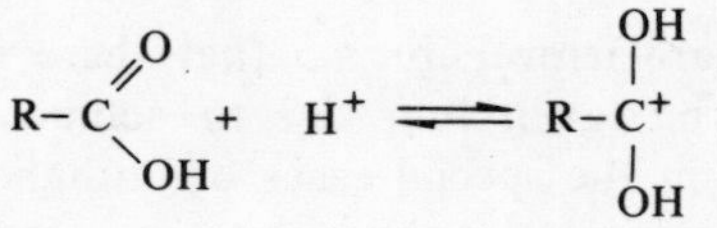

$$R-C\overset{O}{\underset{OH}{}} + H^+ \rightleftharpoons R-\overset{OH}{\underset{OH}{C^+}}$$

Attack of the carbonium ion by the alcohol, followed by the loss of water and a proton, leads to the formation of the ester:

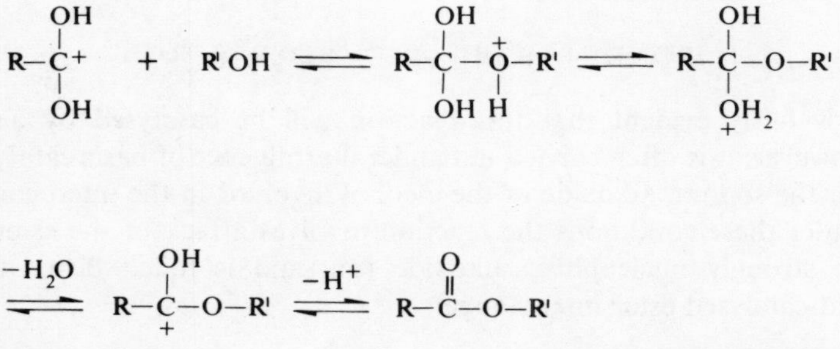

$$R-\overset{OH}{\underset{OH}{C^+}} + R'OH \rightleftharpoons R-\overset{OH}{\underset{OH}{C}}-\overset{+}{\underset{H}{O}}-R' \rightleftharpoons R-\overset{OH}{\underset{\overset{OH_2}{+}}{C}}-O-R'$$

$$\overset{-H_2O}{\rightleftharpoons} R-\overset{OH}{\underset{+}{C}}-O-R' \overset{-H^+}{\rightleftharpoons} R-\overset{O}{\overset{\|}{C}}-O-R'$$

Overall, the reaction is reversible, the reverse reaction being, of course, the hydrolysis of an ester to its constituent acid and alcohol:

$$RCO_2H + R'OH \rightleftharpoons RCO_2R' + H_2O$$

One of the technological problems in carrying out esterifications is to find a method of displacing this equilibrium so as to obtain high conversions to ester. This is often achieved by removing water and/or ester by distillation as the reaction proceeds.

Acid chlorides and acid anhydrides are much more readily attacked by nucleophiles than carboxylic acids, and will bring about esterification under much milder conditions:

$$\underset{\substack{\| \\ O}}{RCCl} + R'OH \longrightarrow \underset{\substack{\| \\ O}}{RCOR'} + HCl$$

$$\begin{array}{c} RC{\overset{O}{\diagup}} \\ \diagdown O \\ RC{\diagdown}_{O} \end{array} + R'OH \longrightarrow \underset{\substack{\| \\ O}}{RCOR'} + RCO_2H$$

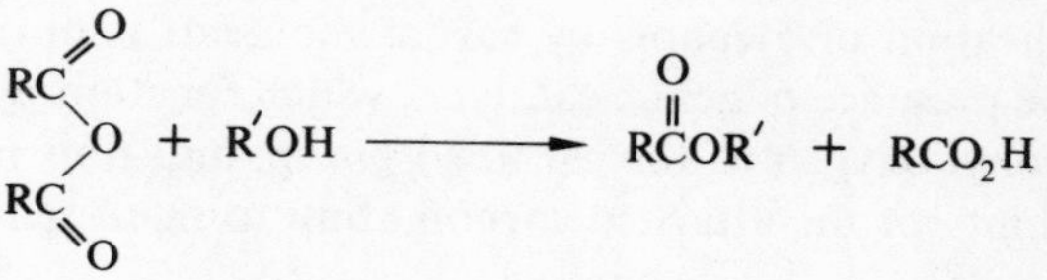

These reactions are irreversible so that there is no necessity for special steps to be taken in order to achieve high conversions. However, except in the special cases of phthalic and maleic anhydrides, acid chlorides and anhydrides are more expensive sources of an acyl group than the corresponding acid, and they are only used where they have particular advantages.

Esters may also be made from other esters by *transesterification* or *ester interchange*:

$$RCO_2R' + R''OH \rightleftharpoons RCO_2R'' + R'OH$$

It is fairly evident that this reaction will be catalysed by acids. However, it is often carried out under the influence of basic catalysts, e.g. the sodium alkoxide of the alcohol involved in the interchange. Under these conditions the reaction involves attack of the ester by the strongly nucleophilic alkoxide ion, and is much faster than acid-catalysed ester interchange:

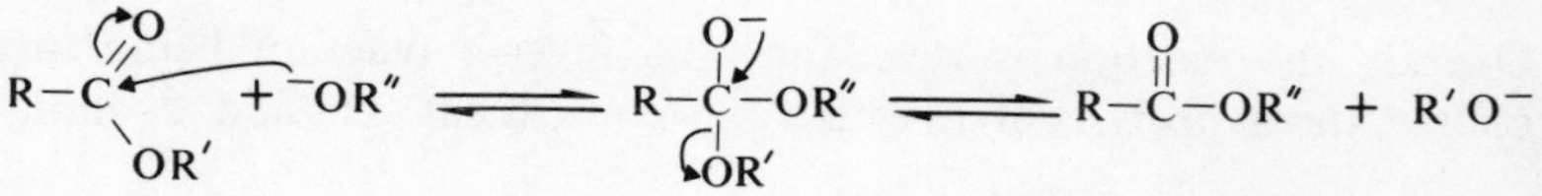

$$R-C{\overset{O}{\diagdown}}{\diagdown}_{OR'} + {}^-OR'' \rightleftharpoons R-\underset{\substack{| \\ OR'}}{\overset{O^-}{C}}-OR'' \rightleftharpoons R-\underset{\substack{\| \\ O}}{C}-OR'' + R'O^-$$

$$R'O^- + R''OH \rightleftharpoons R'OH + R''O^-$$

The most important commercial application of ester interchange is in the manufacture of poly(ethylene terephthalate), discussed in Chapter 7.

Ethyl Acetate

Ethyl acetate is made on a moderately large scale (US production

1967, *ca.* 52 000 tons). By far its major use is as a solvent, e.g. in lacquers.

In manufacture by batchwise operation, acetic acid and an excess of ethanol, together with about 1% of sulphuric acid as catalyst, are reacted at the boiling point of the mixture in a vessel fitted with a fractionating column. The distillate from the top of the column consists of a constant boiling mixture of 82·6% ethyl acetate, 8·4% ethanol, and 9% water:

$$C_2H_5OH + CH_3CO_2H \rightleftharpoons CH_3CO_2C_2H_5 + H_2O$$

Further ethanol and acetic acid can be added to the reactor as the reaction proceeds, but as the amount of water removed in the distillate is less than the amount being formed in the reaction, water accumulates in the reactor. The dilute aqueous solution in the reactor has to be periodically discarded. The yield is about 95%.

The reaction can be carried out continuously by feeding the reactants and catalyst into a distillation column in which both reaction and separation occur, a mixture of ester, water and alcohol being taken from the top of the column, and water from the bottom of the column. The yield in this case is about 99%.

Butyl Acetates

n-Butyl acetate may be made by a process similar to that for the batchwise manufacture of ethyl acetate described above:

$$C_4H_9OH + CH_3CO_2H \rightleftharpoons CH_3CO_2C_4H_9 + H_2O$$

In this case, water does not accumulate in the reactor, but is all removed in the distillate as the reaction proceeds, in a ternary azeotrope of 35·3% n-butyl acetate, 27·4% n-butanol and 37·3% water. On completion of esterification, excess n-butanol is removed in a fraction consisting of 53% n-butyl acetate and 47% n-butanol, and finally n-butyl acetate is distilled. The first two fractions are treated for the recovery of ester and n-butanol. The overall yield of n-butyl acetate is about 95%.

Isobutyl acetate and s-butyl acetate are made by similar processes. t-Butanol cannot be satisfactorily esterified in this way, since the reaction proceeds only very slowly.

Practically the entire use of butyl acetates is as solvents in surface coatings. Combined production of butyl acetates is about the same as the production of ethyl acetate.

Cellulose Acetate

Cellulose is a naturally occurring high molecular weight polymer made up of anhydroglucose units:

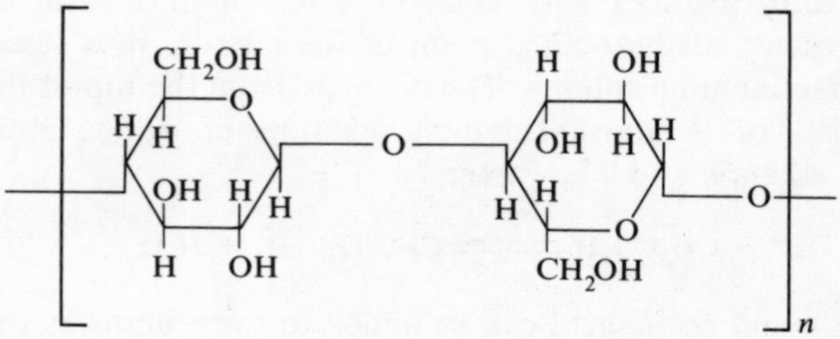

cellulose

It occurs widely, in the cell walls of plants. The main commercial sources are cotton and softwoods which contain about 90% and 50% of cellulose respectively. Treatment of cotton with hot dilute aqueous sodium hydroxide followed by bleaching, gives cellulose of a purity of 98% or more, and a product of similar purity can be obtained by mechanical and chemical treatment of wood.

Naturally occurring cellulose fibres, e.g. cotton and linen, have been used in textile applications for thousands of years, and paper has been made from cellulosic materials for about 1 800 years. In the last 100 years, starting with the development of celluloid and other cellulose nitrate plastics in 1870, the use of derivatives of cellulose in other types of application has developed. In such applications, the ready-made polymeric structure of cellulose is utilised to give products which can be used as plastics, surface coatings and man-made fibres. In this way it was possible to make a number of technologically valuable products long before the production of synthetic high-polymeric materials of equivalent properties was possible. The development of synthetic materials has now diminished the relative importance of these products, but a number of them are still used in substantial amounts, among them being cellulose acetate.

Cellulose acetate first became industrially important during World War I when it was introduced as a substitute for cellulose nitrate in dopes* for aircraft. Use of the highly inflammable cellulose nitrate dopes made aircraft extremely vulnerable to hits by tracer and incendiary bullets, and cellulose acetate was a great improve-

* Dopes are lacquers used for stiffening and waterproofing the fabrics which made up the outer covering of aircraft at that time.

ment in this respect. At the end of the war, demand for material for this application diminished markedly, and the availability of supplies of cellulose acetate led to the development of acetate rayon, and later to the use of cellulose acetate in plastics applications.

The technological importance of cellulose acetate stems from the fact that it is soluble in organic solvents and is fusible, and so can, for example, be spun into fibres, cast into film, and moulded, whereas cellulose itself is infusible and insoluble. The change in properties on esterification is due to the elimination of the strong hydrogen bonding which exists between the hydroxyl groups of the cellulose molecules, and which greatly restrict molecular mobility in cellulose.

MANUFACTURE

Esterification of cellulose with acetic acid is not practicable, since under the conditions required to achieve an acceptable rate of reaction, i.e., high temperatures and acid catalysis, very severe degradation of the polymer chain occurs. Acetylation is therefore carried out with acetic anhydride. Acetic acid is generally used as solvent for the reaction, though some processes use methylene chloride, and sulphuric acid in amounts up to 15% of the weight of the cellulose is used as catalyst; the reaction temperature is in the range 280 to 320 K. During acetylation, some cleavage of polymer chains occurs, and the extent of this reaction has to be carefully controlled in order to obtain a product with the correct properties:

$$[C_6H_7O_2(OH)_3]_n + 3n(CH_3CO)_2O$$

$$\longrightarrow [C_6H_7O_2(O_2CCH_3)_3]_n + 3nCH_3CO_2H$$

The product obtained is the triacetate, i.e. the completely acetylated product, whereas for many applications a lower degree of substitution is required. Thus, for acetate rayon manufacture a product with an average of 2·3–2·5 acetyl groups per glucose unit is required, since such a product is soluble in acetone and can be spun from solution in this solvent. (See Chapter 8 for methods of spinning fibres.) Cellulose triacetate is not soluble in acetone, or in most other common solvents, and for many years was for this reason not used in fibre manufacture. Cellulose triacetate fibres, e.g. 'Tricel', are now made by spinning from a solution in, e.g., methylene chloride.

Direct production of a partially acetylated cellulose of acceptable quality is not feasible, so that to obtain material of the desired degree of substitution the acetylation reaction mixture, containing triacetate, is diluted with water and 'aged' at slightly above room temperature until an appropriate amount of hydrolysis has taken place. The reaction mixture is then diluted further with water and the product separates out and is filtered off and dried.

The acetic acid by-product is recovered in approximately 30–35% solution together with the acetic acid used as solvent. This solution is concentrated by a combination of liquid–liquid extraction and extractive and azeotropic distillation, and part of the acid is then re-converted into anhydride by pyrolysis to ketene followed by reaction of this with more acid (see page 90). Efficient recovery of acetic acid is of prime importance in the economics of the process.

Cellulose acetate is manufactured in quite substantial quantities (US consumption 1965, about 300 000 tons). Its major use is in fibres, but it still finds some plastics applications.

Dialkyl Phthalates

Dialkyl phthalates with up to 13 carbon atoms in the alkyl groups are of very great importance as plasticisers for poly(vinyl chloride) and other polymers (see Chapter 8). They are made by reaction of the appropriate alcohol with phthalic anhydride, e.g.:

Phthalic anhydride, obtained by oxidation of o-xylene or naphthalene (see Chapter 3), is cheaper than phthalic acid.

The reaction is carried out in one stage under conditions similar to those used in the manufacture of ethyl acetate and butyl acetates. Sulphuric acid is used as catalyst, and water is removed by distillation as the reaction proceeds; since the products have high boiling points, no co-distillation of product with the water occurs. At the completion of reaction the product is purified by vacuum distillation.

A range of products from dimethyl- to ditridecyl phthalates is made, but the most important are those of various C_8 alcohols, these being the most generally used plasticisers for poly(vinyl chloride). Total manufacture of dialkyl phthalates in the USA in 1966 was about 340 000 tons; phthalates of C_8 alcohols accounted for about half of this tonnage. In the UK in 1970, production of dialkyl phthalates was 99 000 tons.

HYDROGENATION AND HYDROGENOLYSIS

Hydrogenation, that is, reduction of compounds by reaction with molecular hydrogen, finds a number of applications in chemical manufacture. As the reader will know, methods of reduction other than hydrogenation are available, but these generally involve higher raw material costs than hydrogenation and consequently tend to be used only in small-scale operations.

Hydrogen

The cost of hydrogen is an important factor in the economics of hydrogenation processes, so that it is appropriate to start with a discussion of this raw material.

A dominant factor in hydrogen economics is that as a consequence of its low density, low critical temperature (33·3 K) and high critical pressure (12·8 atm), it is an extremely expensive material to transport. Transportation of hydrogen as a liquid in road or rail tankers, a method of transport used for a number of gases, e.g. ethylene, vinyl chloride, oxygen, requires the construction of tankers which will contain the liquid at very low temperatures and under pressure. The engineering involved is expensive, as is the liquefaction process itself. Further, the low density of liquid hydrogen (70 kg m^{-3}) increases the cost in terms of weight of material transported. Bulk transport as a liquid is possible, and has been carried out in the USA in connection with the space programme, but transport in this form for commercial chemical operations is not economically feasible. Transport by pipe-line over long distances is economically unattractive because of the low density of the gas. The only method of transport used in the UK is as compressed gas in outsize versions of laboratory gas cylinders; as can be imagined, the costs are high.

Two important consequences follow from the factors discussed above. Firstly, if hydrogen is required as a feedstock for a chemical process, it is necessary to generate it near the consuming plant or for there to be by-product hydrogen available, except where the quantities required are so small that the use of cylinder hydrogen is feasible. Secondly, if hydrogen is produced as a by-product of a process, its value depends on the circumstances at the site concerned. If it can be used as a raw material in another process, then its value can be taken to be what it would cost to make hydrogen in the quantities required at that site. If there is no such outlet, and it is used as a fuel in the factory, then it will have only 'fuel value'—about £17/ton in the UK (1970). This is substantially less than its value as a chemical raw material, and there is consequently considerable advantage in locating plants with heavy hydrogen consumptions near plants which give hydrogen as a by-product, e.g. near a catalytic reformer (see Chapter 2).

Where by-product hydrogen cannot be used, the most common method of generation is steam reforming of hydrocarbons (see Chapter 2). If desired, ready-made hydrogen plants with capacities ranging from about 60 kg/day upwards can be purchased. Naturally, the smaller the plant, the higher the cost of the hydrogen generated.

Catalysts

At the time of writing, all commercially important hydrogenations, if we exclude hydroformylation processes discussed later this chapter, are carried out in the presence of solid catalysts. As has already been pointed out, investigation of the mechanisms of heterogeneous-catalysed reactions is difficult, and present understanding of such hydrogenations is far from complete. This being so, we shall not discuss these mechanisms in detail.

A variety of catalysts are used for commercial hydrogenations, amongst the most important being nickel, palladium, platinum, copper oxide–chromium oxide and molybdenum and tungsten sulphides. With metal hydrogenation catalysts a large surface area is achieved in one of three ways: (i) by use of the metal in a colloidal dispersion, (ii) by use of supported catalysts in which the metal is deposited on an inert supporting material with a high surface area, and (iii) by use of Raney-type catalysts, where one component of an alloy, usually aluminium, has been dissolved away to leave the active metal as a high surface area skeleton.

Platinum and palladium catalysts are extremely active, but are expensive. The other catalysts are cheaper, but somewhat less active, so that more vigorous conditions, i.e., higher temperatures and pressures may be required when using them. The sulphide catalysts are useful for feeds containing sulphur compounds, which would poison the metallic catalysts.

Commercial hydrogenations are carried out in both the gas and liquid phase. In the gas phase, both fixed- and fluidised-bed reactors are used. In liquid-phase operation, the catalyst is often suspended in the reaction mixture; less often a fixed bed of catalyst is used.

Hydrogenation of Benzene

The hydrogenation of benzene to cyclohexane is carried out on a large scale, the cyclohexane being required mainly for the production of monomers for nylon 66 and nylon 6. UK capacity for cyclohexane in 1969 was about 330 000 tons, and has since been substantially increased.

TABLE 6-2

PETROLEUM CONSTITUENTS WITH BOILING POINTS ABOUT 354 K

Compound	b.p. (K)
cyclohexane	354·0
benzene	353·3
2,2-dimethylpentane	352·5
2,4-dimethylpentane	354·0
2,2,3-trimethylbutane	354·1

The first question that arises in connection with this process is why it is necessary to operate it at all. Cyclohexane occurs in petroleum and it might be supposed that direct isolation from naphtha would provide an adequate source. However, such isolation is difficult and expensive because of the occurrence in petroleum of a number of other hydrocarbons with boiling points the same as or very close to that of cyclohexane (see Table 6-2). Separation of cyclohexane by fractional distillation is thus impossible, and the production of high-purity material requires a complex system of processing. Consequently, most cyclohexane is made by benzene hydrogenation.

The hydrogenation of benzene is exothermic and reversible:

$$C_6H_6 + 3H_2 \rightleftharpoons C_6H_{12} \qquad \Delta H = -218 \text{ kJ/mol}$$

Conversion to cyclohexane at equilibrium will evidently be favoured by low temperatures and high pressures. Adequate rates of reaction can be achieved at about 470 K, and at this temperature only moderate pressures are required to give practically complete conversion at equilibrium. Further, by correct choice of catalyst and conditions practically quantitative yields can be obtained, and it is thus possible to design processes so that the reactor effluent requires only treatment to remove volatile impurities introduced with the hydrogen to produce material of saleable quality. Side reactions which can occur, and are guarded against in process design, are isomerisation to methylcyclopentane, and cracking reactions. Cracking is favoured by high temperatures and high hydrogen pressures. The extent to which isomerisation occurs depends on the catalyst used. As has been indicated, in a well-designed process, these reactions occur to a negligible extent.

A number of versions of the process are operated; both gas- and liquid-phase operation are used, and both nickel and platinum are used as catalysts. In the Institute Français du Petrole (IFP) process, hydrogenation is carried out in the liquid phase, with a Raney nickel catalyst suspended in the reaction mixture, at a temperature in the range 470 to 510 K, and about 40 atm pressure. The product is removed from the reactor in the vapour phase. In order that the proportion of benzene in the product stream from the reactor be low, the concentration of benzene in the reaction mixture must be low, i.e. the reaction must be taken to high conversion. This can be achieved without the use of an excessively large reactor since the reaction is zero order in benzene, benzene being very strongly adsorbed on the catalyst, so that the rate of reaction is not affected by benzene concentration.

The gas stream from the main reactor, which still contains a small amount of benzene ($<5\%$), is passed through a second reactor where the hydrogenation is completed in the vapour phase over a supported nickel catalyst. The product is then passed through a separation system to remove hydrogen and any low boiling constituents introduced with the hydrogen, e.g. methane, and cyclohexane of about $99 \cdot 95\%$ purity is obtained.

Hydrogenation of Fats and Oils

Hydrogenation of animal and vegetable fats and oils is a process of considerable commercial importance. It finds its major applications in the manufacture of margarine and other edible fats, and cooking oils. It is also used in the modification of fats and oils for use in other applications, e.g. for soap manufacture and for industrial greases.

The composition of fats and oils has been briefly discussed in Chapter 1 where it was seen that they are esters of glycerol and fatty acids. Many of them, particularly those of vegetable and marine origin, involve unsaturated fatty acids, and these have lower melting points, and are more susceptible to oxidation than their saturated analogues. This is illustrated by a comparison of trilinolein and tristearin:

$$CH_2O_2C(CH_2)_7CH=CHCH_2CH=CH(CH_2)_4CH_3$$
$$CHO_2C(CH_2)_7CH=CHCH_2CH=CH(CH_2)_4CH_3$$
$$CH_2O_2C(CH_2)_7CH=CHCH_2CH=CH(CH_2)_4CH_3$$

trilinolein
(triglyceride of linoleic acid)
m.p. 260 K
readily oxidised by air

$$CH_2O_2C(CH_2)_{16}CH_3$$
$$CHO_2C(CH_2)_{16}CH_3$$
$$CH_2O_2C(CH_2)_{16}CH_3$$

tristearin
(triglyceride of stearic acid)
m.p. 346 K
fairly stable to oxidation

Hydrogenation is carried out with the purpose of reducing such unsaturation, and hence increasing the melting point and/or the resistance to oxidative degradation of the material. Thus, hydrogenation of vegetable oils, which are liquid at room temperature, gives solid fats which can be used in margarine manufacture.

Hydrogenation is carried out in the liquid phase with a supported nickel catalyst in suspension in the oil, temperatures from 400 to 470 K and pressures up to about 4 atm being used. Batchwise operation appears to be most common, batch sizes of up to about 22 tons being made. In these operations the object is not to produce a single chemical compound, but to obtain a product with a particular set of technological properties, and processes are designed to this end. In margarine manufacture for instance, the spreading properties which are determined by, amongst other things, the melting range of the product, are of great importance, and there is a great deal of know-how involved in achieving the desired properties.

Hydrogenation of Adiponitrile

Hydrogenation of adiponitrile provides the major route to hexa-methylenediamine, one of the monomers for nylon 66 manufacture:

$$NC\,(CH_2)_4\,CN \;+\; 4H_2 \longrightarrow H_2NCH_2\,(CH_2)_4\,CH_2NH_2$$

The reaction can be carried out in the presence of a supported nickel or cobalt catalyst at 390 to 440 K and 270 to 400 atm. Side reactions can occur leading to the formation of hexamethyleneimine and polyamines:

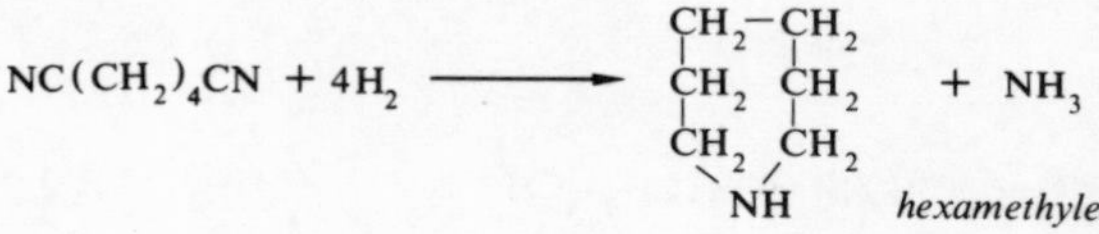

$$NC(CH_2)_4CN + 4H_2 \longrightarrow
\begin{array}{c}
CH_2\!-\!CH_2 \\
CH_2 \;\; CH_2 \\
CH_2 \;\; CH_2 \\
\diagdown\!\diagup \\
NH
\end{array}
\;+\; NH_3$$

$\qquad\qquad\qquad\qquad\qquad\qquad\qquad$ *hexamethyleneimine*

$$2NC(CH_2)_4CN + 8H_2 \longrightarrow H_2N(CH_2)_6NH(CH_2)_6NH_2 + NH_3$$

It is assumed that these side reactions involve nucleophilic attack by amines on imines formed as intermediates in the hydrogenation:

$$RC\equiv N \xrightarrow{\;H_2\;} RCH=NH \xrightarrow{\;H_2\;} RCH_2NH_2$$

$$RCH=NH + RCH_2NH_2 \rightleftharpoons R-\overset{\displaystyle NH_2}{\underset{\displaystyle H}{C}}-NHCH_2R \rightleftharpoons RCH=NCH_2R + NH_3$$

$$RCH=NCH_2R \xrightarrow{\;H_2\;} RCH_2NHCH_2R$$

The formation of by-products is largely suppressed by carrying out the hydrogenation in the presence of ammonia, which presumably has the effect of displacing the equilibria shown above to the left. Yields of over 95% are claimed.

Adiponitrile is made by a number of processes. One of the most widely used involves dehydration of the ammonium salt of adipic acid:

$$NH_4^+ \;{}^-O_2C\,(CH_2)_4\,CO_2^-\; NH_4^+ \longrightarrow NC(CH_2)_4CN + 4H_2O$$

This reaction may be brought about by treating adipic acid with an excess of ammonia at 570 to 720 K in the presence of a phosphoric acid or a phosphate catalyst.

Hydrogenation of Nitroaromatics

We have already seen (see Chapter 5) that hydrogenation of nitrobenzene and nitrotoluenes provides the major method of manufacture of aniline, toluidines and tolylenediamines.

Hydrogenation of nitrobenzene is generally carried out in the vapour phase, and both fixed- and fluidised-bed reactors are used. In the process used by American Cyanamid the reaction is carried out in a fluidised-bed reactor over a copper-on-silica catalyst at about 540 K and just over atmospheric pressure. The heat of reaction is removed by cooling coils immersed in the fluidised bed. A large excess of hydrogen is used (molar ratio hydrogen:nitrobenzene about 9:1). The yield is at least 98 %. Liquid-phase hydrogenation of nitrobenzene appears to give slightly lower yields.

Hydrogenation of 2,4- and 2,6-dinitrotoluenes is carried out in the liquid phase, because with these compounds there is a considerable explosion hazard at the temperatures necessary for gas-phase operation. Nickel, platinum or palladium catalysts are used, and a variety of conditions are employed. In one process a solution of the dinitro compounds in methanol, with a Raney nickel catalyst in suspension, is treated with hydrogen at about 360 K and 50 atm.

Hydrogenolysis of Fatty Acid Esters

Hydrogenolysis, by treatment with hydrogen under forcing conditions, is an important method of preparation of straight-chain alcohols from fats and oils. The reaction can be carried out on triglycerides themselves, but the glycerol liberated causes difficulties in the isolation of the alcohols, and it is more usual to use methyl esters obtained by ester interchange of the triglycerides with methanol in the presence of sodium methoxide:

$$3MeOH + \begin{array}{c} CH_2O_2CR \\ | \\ CHO_2CR \\ | \\ CH_2O_2CR \end{array} \xrightarrow{\text{NaOMe}} \begin{array}{c} CH_2OH \\ | \\ CHOH \\ | \\ CH_2OH \end{array} + 3RCO_2Me$$

The hydrogenolysis is carried out at about 560 to 640 K and 200 atm, with the catalyst, a complex mixture of cupric oxide and chromic oxide, in suspension in the reaction mixture. Continuous processing is normally used:

$$RCO_2Me + 2H_2 \longrightarrow RCH_2OH + MeOH$$

Reduction of fatty acid esters with sodium and alcohol (the Bouveault–Blanc reaction) can also be used, but the raw material costs of this process are substantially higher than those in hydrogenolysis.

Since the fats and oils used as raw materials consist of mixed triglycerides of a number of fatty acids, mixtures of alcohols are produced by hydrogenolysis. These are fractionated into cuts to suit market requirements; pure individual alcohols are not normally isolated. Thus, commercial grades of lauryl alcohol (dodecan-1-ol) may contain varying amounts of C_{10}, C_{14} and C_{16} alcohols. Such mixtures are in most cases perfectly adequate for the applications for which the alcohols are required.

Fatty alcohols can be made synthetically (see page 245), but hydrogenolysis is still by far the most important source. There is no synthetic manufacture in the UK.

The fatty alcohols have a wide variety of applications, the most important being in the manufacture of surface-active agents. An important example of this type of application is the manufacture of sodium lauryl sulphate:

$$C_{12}H_{25}OH \xrightarrow{H_2SO_4} C_{12}H_{25}OSO_3H \xrightarrow{NaOH} C_{12}H_{25}OSO_3Na$$

Sodium lauryl sulphate is widely used in toothpastes, shampoos, and light duty domestic detergents.

Hydrodealkylation

As was indicated in Chapter 2, the proportions of benzene, toluene and xylenes produced by catalytic reforming are not in line with requirements for these materials in chemical applications, the amounts of toluene and xylenes produced being far higher than is required. Hydrodealkylation processes provide a means of redressing this imbalance by allowing benzene to be prepared from toluene and C_8 aromatics, e.g.:

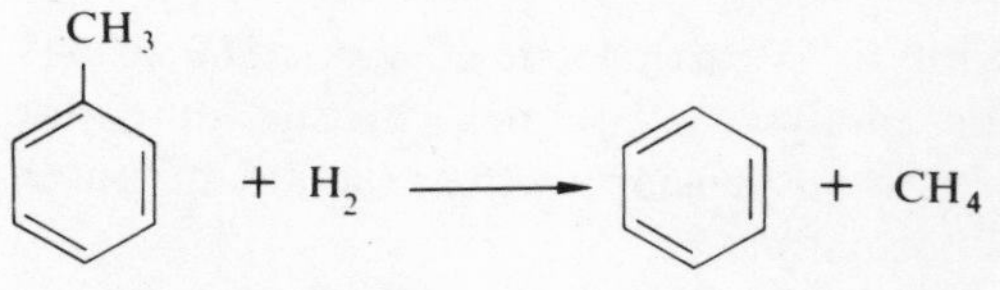

Hydrodealkylation may be carried out catalytically or purely thermally, and at present both types of process are used to approximately the same extent. Toluene is the feedstock generally used for benzene manufacture.

In catalytic hydrodealkylation, hydrogen and the aromatic feedstock in a molar ratio of 5–10:1 are reacted at 810 to 920 K and 33 to 53 atm over, e.g., a chromia–alumina catalyst. In thermal hydrodealkylation the temperatures used tend to be somewhat higher (920 to 1 030 K). In both cases, yields of benzene of up to 99% are claimed. Some biphenyl is formed, but can be recycled to the reactor.

In the USA hydrodealkylation is also used for the production of naphthalene from petroleum-derived methylnaphthalenes.

DEHYDROGENATION

Dehydrogenations form one particular class of oxidation reactions, and are, in fact, sometimes carried out in the presence of oxygen. However, it is convenient to consider them in a separate category to oxidation.

In the absence of oxygen, dehydrogenations are reversible. They are endothermic, so that high conversion to products at equilibrium is favoured by high temperatures; since an increase in the number of molecules occurs during reaction, it is also favoured by low pressures. This situation is, naturally enough, the exact reverse of that obtaining in hydrogenation. Dehydrogenations are generally carried out in the vapour phase in the presence of solid catalysts.

Dehydrogenation of n-Butane and n-Butenes

Butadiene is required in substantial quantities for the manufacture of synthetic rubbers and other polymers (UK production 1970, 170 000 tons). As we saw in Chapter 2, butadiene is a by-product in the production of ethylene by naphtha cracking, and in countries where this is the main method of ethylene manufacture, e.g. Europe and Japan, most butadiene is obtained from this source. In the USA, however, a high proportion of ethylene is made by cracking ethane and propane, and very little butadiene is produced in these reactions. The amount of by-product butadiene available is thus inadequate to meet the demand, and it is necessary to operate processes specifically for butadiene production. The processes used involve dehydrogenation of n-butane or n-butenes:

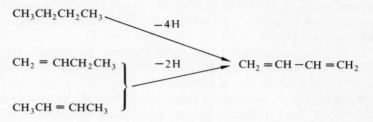

It has been known for some time that butadiene can be made by dehydrogenating n-butenes, and processes involving this reaction were patented as early as 1910. Large-scale application of this type of process did not, however, occur until World War II, when the loss of natural rubber supplies led to the institution of a crash synthetic rubber programme in the USA. Since the major rubber chosen for development was SBR, a copolymer of butadiene and styrene, large quantities of butadiene (and styrene) were required. Butane and butene dehydrogenation processes were rapidly developed and put into production, and by 1944 the installed capacity for butadiene in the USA was 275 000 tons per annum based on butenes and 75 000 tons per annum based on butane. There was also a substantial amount of capacity by a process based on ethanol (now obsolete), and some manufacture by thermal cracking.

n-BUTENES

Dehydrogenation of n-butenes can be carried out over a variety of catalysts. Shell '205' catalyst, for instance, consists of iron oxide containing smaller amounts of chromic oxide and potassium carbonate; Dow 'B' catalyst is a calcium nickel phosphate.

Operating temperatures are usually in the range 870 to 960 K. Although the use of higher temperatures increases the rate of reaction and favourably affects the position of the equilibrium, it also tends to cause thermal cracking of the feed, and polymerisation of the butadiene. Steam is used as a diluent to reduce the partial pressure of hydrocarbons and thus to increase the conversion achievable. It also reduces the rate of deposition of carbon on the catalyst by converting it to carbon monoxide by the water gas reaction (see page 14). Molar ratios of steam/butene of 8:1 and 20:1 are used with the Shell and Dow catalyst respectively. The reaction is carried out in fixed-bed adiabatic reactors, the heat of reaction being supplied by superheating the steam before it is fed to the reactor.

Although the use of steam reduces carbon deposition, it does not completely prevent it, and the catalyst loses activity during use. It thus has periodically to be reactivated. In the case of the Shell catalyst, regeneration is required for about 1 hour in every 24 hours, and is brought about by passing steam through the catalyst bed. With the Dow catalyst it is required for 15 minutes in every 30 minutes, and is carried out with a mixture of steam and air. The disadvantages of higher steam ratio and more frequent regeneration when using the Dow catalyst are offset by the fact that it gives higher yields at higher conversions than the Shell catalyst (about 90% at 45% conversion compared with about 74% at 27% conversion).

n-BUTANE

Manufacture of butadiene from butane poses greater problems than manufacture from butenes. For one thing, under similar conditions of temperature and pressure the equilibrium conversion to butadiene is considerably less than with the butenes. Thus, at 1 020 K and atmospheric pressure, the maximum conversion to butadiene attainable is 14% compared with about 50% from n-butenes. The equilibrium is, however, very sensitive to pressure and at the same temperature and 0·167 atm the equilibrium conversion to butadiene is 49%.

The situation is further complicated by the fact that steam dilution cannot be used to achieve low partial pressures because the catalysts used (alumina–chromia) are deactivated by steam. It is thus necessary to operate at reduced absolute pressures, which introduces various problems, e.g. the necessity to guard against the entry of air into the system, so as to avoid the possibility of explosions. Also, since the beneficial effect of steam on catalyst fouling is not obtained, frequent catalyst regeneration is necessary.

The first commercially successful process for the one-stage conversion of butane to butadiene was that developed by the Houdry Process Corporation in the USA during World War II. This is a cyclic process in which the catalyst is periodically regenerated by burning off the carbon deposited during reaction, the heat generated being used to supply the heat of reaction for the dehydrogenation. The butane is heated to about 890 K and passed at about 0·2 atm into a reactor containing a bed of catalyst at about the same temperature. The endothermic reaction causes the temperature of the catalyst bed to drop, and when it reaches about 870 K the butane feed is cut off, the reactor is vacuum-purged to remove hydrocarbons, and then preheated air is passed to burn off the carbon which has been deposited. The heat of combustion of the carbon raises the temperature of the catalyst bed to about 890 K. The reactor is purged to remove air, and the next dehydrogenation cycle is begun. The whole cycle of operations takes about 15 to 30 minutes; a plant normally has at least five reactors so that a continuous output of product may be maintained. A yield of about 60% at a conversion of about 30% is obtained.

OXIDATIVE DEHYDROGENATION

In recent years a large amount of development work has been carried out on processes in which dehydrogenation of n-butane and/or n-butenes is brought about under oxidising conditions. These processes use lower reaction temperatures than conventional dehydrogenation processes, and higher yields and conversions are claimed. They fall into two main classes.

In the direct oxidative processes, hydrocarbon and air are reacted over a catalyst, typically at temperatures in the range 720 to 820 K. A wide variety of catalysts have been suggested, but mixtures of bismuth and molybdenum oxides and of tin and antimony oxides appear to be favoured. Yields of about 90% at high conversions have been claimed for the dehydrogenation of n-butenes; this type of process does not appear to function well with butane feed.

The other major class of oxidative dehydrogenation involves dehydrogenation by reaction with iodine at about 770 to 820 K. In this case n-butane as well as n-butenes can be used as feed:

$$CH_3CH_2CH_2CH_3 \ + \ 2I_2 \ \longrightarrow \ CH_2{=}CH{-}CH{=}CH_2 \ + 4HI$$

The hydrogen iodide formed is converted back to iodine by direct or indirect oxidation with air:

$$4HI \ + \ O_2 \ \longrightarrow \ 2I_2 \ + \ 2H_2O$$

It is suggested that iodine brings about dehydrogenation by reactions of the type shown in the scheme below:

$$I_2 \ \longrightarrow \ 2I\cdot$$

$$CH_3-CH_2-CH_2-CH_3 \;+\; I\cdot \longrightarrow CH_3-CH_2-\overset{\cdot}{C}H-CH_3 \;+\; HI$$

$$CH_3-CH_2-\overset{\cdot}{C}H-CH_3 \;+\; I_2 \longrightarrow CH_3-CH_2-\underset{\underset{I}{|}}{C}H-CH_3 \;+\; I\cdot$$

$$CH_3-CH_2-\underset{\underset{I}{|}}{C}H-CH_3 \;+\; I\cdot \longrightarrow CH_3-\overset{\cdot}{C}H-\underset{\underset{I}{|}}{C}H-CH_3 \;+\; HI$$

$$CH_3-\overset{\cdot}{C}H-\underset{\underset{I}{|}}{C}H-CH_3 \longrightarrow CH_3-CH=CH-CH_3 \;+\; I\cdot$$

A number of versions of this type of process have been suggested in the patent literature. It appears that a favoured method is to carry out the dehydrogenation in the presence of a hydrogen iodide acceptor, and then to regenerate hydrogen iodide from the acceptor by oxidation with air. A variety of metal oxides and hydroxides can be used as acceptors.

It is reported that iodine oxy-dehydrogenation processes are operated in two plants, by Petro-Tex in the USA, and by Shell in France.

Dehydrogenation of Ethylbenzene

$$PhCH_2CH_3 \longrightarrow PhCH=CH_2 \;+\; H_2$$

At present practically all styrene is made by the dehydrogenation of ethylbenzene, although as was indicated in Chapter 3, styrene is a co-product in a recently developed propylene oxide process, and this may become an important source in the future. Virtually all the styrene made is used in the manufacture of polymers, notably polystyrene and SBR rubber which account for about 50 and 25 % of consumption in the UK respectively. UK capacity for styrene was 275 000 tons per annum in 1970 and there are plans in hand to more than double this in the next few years.

The catalysts used in the dehydrogenation of ethylbenzene are of the same type as those used in the dehydrogenation of butenes; iron oxide promoted with chromic oxide and potassium carbonate is widely used. There are two main versions of the dehydrogenation process, which vary in the type of reactor used. In the version originally developed in the USA, the catalyst is held as a single bed in an adiabatic reactor, and the heat of reaction is supplied by mixing superheated steam with the ethylbenzene feed to the reactor. In the version developed in Germany, a tubular catalytic reactor is used, and the heat is supplied by circulating hot gases around the outside of the tubes.

When an adiabatic reactor is used, the endothermic reaction causes the temperature to drop as the gases pass through the catalyst bed, and in order that the temperature at the exit from the bed be sufficiently high to allow an acceptable conversion to be achieved, the gases entering the bed have to be at a relatively high temperature. Typically the reactor is fed with ethylbenzene at 790 to 820 K and steam at 990 K in the proportions of 2·6 kg steam/kg ethylbenzene, the temperature of the resulting mixture being about 920 K. The effluent from the bed is at about 860 K. The high temperatures in the first part of the bed cause a significant amount of cracking of ethylbenzene to benzene and toluene to occur, these being formed in yields of about 3 and 5% respectively. Yields of styrene of 88 to 91% and conversions of about 40% are obtained.

In a tubular reactor much closer control of the temperature profile along the bed is possible, and cracking of ethylbenzene can be minimised. Yields of about 93% are obtained at 40% conversion. However, this yield advantage is counterbalanced by the fact that a tubular reactor has a substantially higher capital cost than an adiabatic reactor. Adiabatic reactors are most often used.

Since styrene readily polymerises at elevated temperatures, it is isolated from the reaction products by fractional distillation under reduced pressure, so as to minimise processing temperatures.

Dehydrogenation of Alcohols
The dehydrogenation of alcohols is a reaction of considerable commercial importance; it provides one of the major methods of manufacture of aldehydes and ketones.

METHANOL
Formaldehyde has been made commercially by the dehydrogenation of methanol since 1888, and this is still very much the most important method of manufacture. In the UK all formaldehyde is made in this way; in the USA about 85% of manufacture is by this route, the other 15% being made by free-radical gas-phase oxidation of lower alkanes.

There are two major types of process used for the dehydrogenation. In what appears to be the more important type, methanol and a limited amount of air are allowed to react over a silver or copper catalyst at temperatures ranging from 720 to 870 K. Under these conditions, two overall reactions occur, oxidation, which is exothermic, and dehydrogenation, which is endothermic:

$$CH_3OH + \tfrac{1}{2}O_2 \longrightarrow CH_2=O + H_2O \qquad \Delta H = -154 \ \text{kJ/mol}$$

$$CH_3OH \longrightarrow CH_2=O + H_2 \qquad \Delta H = +84 \ \text{kJ/mol}$$

The methanol/air ratio is adjusted so that the heats of reaction balance out to give the reaction temperature required. Thus, no heating or cooling is necessary and a simple adiabatic reactor can be used. The methanol/air mixture fed normally contains 30 to 50% methanol by volume; yields of 85 to 95% at conversions of about 65% are obtained.

In the other major process the reaction is carried out with an excess of air over an oxide catalyst, typically molybdenum oxide promoted with iron oxide, at 620 to 720 K. A more complex reactor is required in this case since it is necessary to provide for cooling, but it is claimed that practically complete conversion of methanol is achievable. Yields of over 90% are claimed.

Formaldehyde is a gas (b.p. 254 K). It cannot be stored or transported in the free state, either as a gas or liquid, since it very readily polymerises, and consequently it is normally supplied in aqueous solution, typically of 37% concentration. Even in solution polymerisation can occur, and formaldehyde solution usually contains a small percentage of methanol, which acts as a stabiliser.

Since about 1·7 kg of water has to be transported with every kg formaldehyde, transport costs in terms of 100% formaldehyde are high. In consequence, long-distance transport of large quantities of formaldehyde tends to be avoided, and there has been less tendency than with most other large-tonnage organic chemicals to concentrate manufacture in a small number of large plants. It is quite common for large-tonnage users to make their own formaldehyde from purchased methanol.

The major use of formaldehyde is in the manufacture of phenol-formaldehyde-, melamine-formaldehyde-, and urea-formaldehyde resins (see Chapter 7). UK production in 1970 was 120 000 tons.

ETHANOL

Dehydrogenation of ethanol, initially made by fermentation but now usually prepared by hydration of ethylene, has for many years provided a major route to acetaldehyde:

$$CH_3CH_2OH \xrightarrow{\quad -2H \quad} CH_3CHO$$

The reaction may be carried out over a silver catalyst by a combined oxidation–dehydrogenation process of the type used in formaldehyde manufacture, yields of about 90% at about 50% conversion being obtained. Dehydrogenation over a chromium oxide-promoted copper catalyst at about 530 to 560 K is also used, and gives about the same yields and conversions. In this case it is necessary to supply heat to the reactor.

The Wacker Chemie process (see Chapter 3) now provides a one-step route from ethylene to acetaldehyde, and it seems likely that manufacture by dehydrogenation of ethanol will eventually fall into disuse.

The largest use of acetaldehyde is in the manufacture of acetic acid and acetic anhydride. It has a number of other uses as an intermediate, of which substantially the most important is in the manufacture of n-butanol *via* the aldol condensation (see Fig. 1-1). It should be noted that both these outlets are subject to competition from other routes, notably direct oxidation of lower alkanes in the case of acetic acid, and the Oxo process in the case of n-butanol (see page 226).

ISOPROPANOL

Acetone is made by the dehydrogenation of isopropanol:

$$\underset{\underset{OH}{|}}{CH_3CHCH_3} \xrightarrow{\quad -2H \quad} \underset{\underset{O}{\|}}{CH_3CCH_3}$$

The dehydrogenation is most commonly carried out over a copper or brass catalyst at about 670 to 750 K, yields of over 90% at conversions of 75 to 78% being obtained. Reaction over a zinc oxide-on-pumice catalyst at about 650 K can also be used, as can balanced oxidation–dehydrogenation reactions.

In recent years, substantial quantities of acetone have become available as a by-product of phenol manufacture by the cumene process, and to a lesser extent, of acetic acid manufacture by the oxidation of light naphtha, and by-product acetone now makes up a significant proportion of total acetone production. Total acetone production in the UK in 1970 was 142 000 tons.

Acetone is used as a solvent, and as an intermediate in the manufacture of a number of chemicals, e.g., bisphenol A (see page 174), methyl isobutyl ketone, and methyl methacrylate:

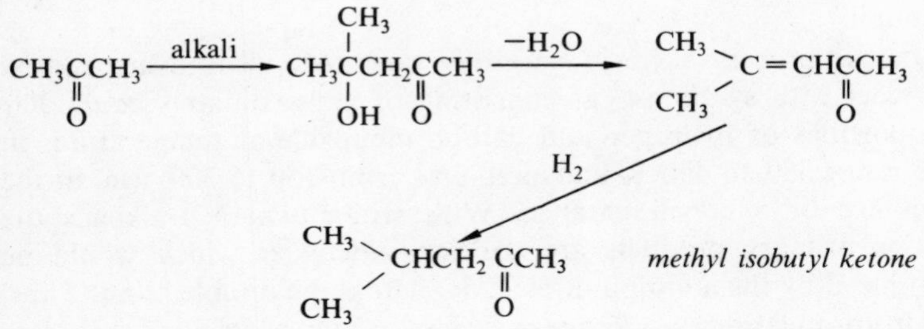

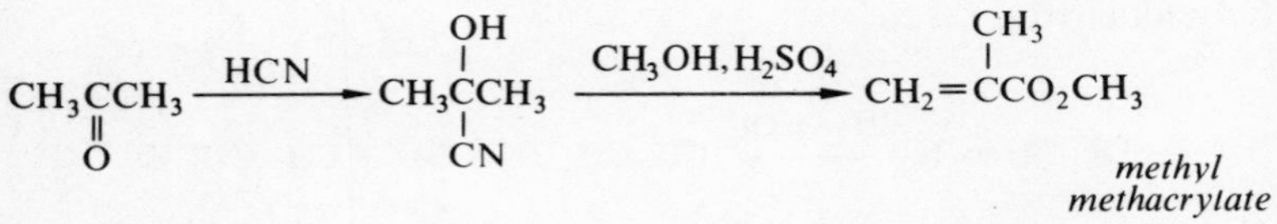

$$CH_3CCH_3 \xrightarrow{\text{alkali}} CH_3\underset{OH}{C}CH_2\underset{O}{C}CH_3 \xrightarrow{-H_2O} (CH_3)_2C=CHCCH_3$$

$$\xrightarrow{H_2} (CH_3)_2CHCH_2CCH_3 \quad \textit{methyl isobutyl ketone}$$

$$CH_3CCH_3 \xrightarrow{HCN} CH_3\underset{CN}{\overset{OH}{C}}CH_3 \xrightarrow{CH_3OH, H_2SO_4} CH_2=\underset{CH_3}{C}CO_2CH_3$$

methyl methacrylate

S-BUTANOL

Methyl ethyl ketone is made from s-butanol by dehydrogenation over a copper catalyst under conditions similar to those used in the manufacture of acetone from isopropanol:

$$CH_3\underset{OH}{C}HCH_2CH_3 \longrightarrow CH_3\underset{O}{C}CH_2CH_3 + H_2$$

Its major use is in solvent applications.

HYDROFORMYLATION

Hydroformylation, or as it is generally called, the Oxo* reaction, provides a very valuable method of manufacturing primary alcohols from propanol upwards. The reaction, which involves the treatment of alkenes with carbon monoxide and hydrogen in the presence of a cobalt catalyst, was discovered in Germany in 1938, and a plant to

* The name 'Oxo' was given to the reaction by the German workers because the initial products are carbonyl, or in German, 'oxo', compounds.

operate the Oxo process was built in Germany during World War II. At the end of the war, British and American investigation teams obtained details of the process technology, and research and development work was started on the reaction by a number of companies. The Enjay Chemical Company started pilot plant operation in the USA in 1948, and in the UK ICI started full-scale production in 1951. World Oxo capacity in 1966 was about 850 000 tons per annum.

In the process as most generally operated, alkenes are caused to react with synthesis gas consisting of approximately equimolar proportions of hydrogen and carbon monoxide at temperatures in the range 380 to 450 K and pressures from 150 to 300 atm, in the presence of a cobalt catalyst. With straight-chain 1-alkenes the major primary products are the two aldehydes which would be produced by the addition of H–CHO across the double bond. Thus, hydroformylation of propene gives n-butyraldehyde and iso-butyraldehyde:

$$CH_3 CH{=}CH_2 \xrightarrow{H_2+CO} CH_3CH_2CH_2CHO + CH_3 \underset{\underset{CHO}{|}}{C}HCH_3$$

The aldehydes are hydrogenated to alcohols in a separate step.

With linear alkenes above propene, migration of the double bond can occur during the reaction, and more than two aldehydes may be formed, e.g.:

$$CH_3CH_2CH_2CH{=}CH_2$$

$$\downarrow H_2 + CO$$

$$CH_3CH_2CH_2CH_2CH_2CHO + CH_3CH_2CH_2\underset{\underset{CHO}{|}}{C}HCH_3 + CH_3CH_2\underset{\underset{CHO}{|}}{C}HCH_2CH_3$$

major products   *minor product*

As a consequence of this migration, the products obtained from linear alkenes tend to be the same irrespective of the position of the double bond in the starting material.

Branching of the chain at or near the double bond hinders addition of a formyl group, and tends to cause it to enter at the position most remote from the branching. For example, hydro-formylation of isobutene leads almost entirely to the formation of 3-methylbutyraldehyde:

$$\begin{array}{c} CH_3 \\ \diagdown \\ \diagup \quad C{=}CH_2 \\ CH_3 \end{array} \xrightarrow{\quad H_2{+}CO \quad} \begin{array}{c} CH_3 \\ | \\ CH_3CHCH_2CHO \end{array} \quad + \quad \begin{array}{c} CH_3 \\ | \\ CH_3CCH_3 \\ | \\ CHO \end{array}$$

$$\qquad\qquad\qquad\qquad\qquad\qquad\qquad \textit{major product} \qquad\quad \textit{ca. 5\%}$$

A branched-chain alkene with a non-terminal double bond gives essentially the same products as the corresponding 1-alkene, because of double bond migration during the reaction.

Mechanism

The chemistry of the Oxo process is very complex, and there are a number of aspects of the mechanism of the reaction which are still uncertain.

There is general agreement that catalysis of the reaction is by cobalt carbonyl compounds in solution in the reaction mixture. Cobalt carbonyls may be added as such or may be formed *in situ* by the reaction of other cobalt compounds, e.g. salts, with synthesis gas. An example of the type of reaction involved in their formation is:

$$2Co(A)_2 + 8CO + 2H_2 \longrightarrow Co_2(CO)_8 + 4HA$$

$A = organic\ anion$

The composition of the cobalt-containing species in the reaction mixture is governed by the operating conditions, and in particular is heavily dependent on the partial pressure of carbon monoxide. Cobalt carbonyls are unstable at elevated temperatures and a high partial pressure of carbon monoxide is necessary in the process to prevent decomposition of the catalyst system to metallic cobalt at the operating temperature, by reactions of the following type:

$$Co_2(CO)_8 \Longrightarrow 2Co + 8CO$$

The Oxo reaction probably involves hydridotetracarbonylcobalt, $HCo(CO)_4$, and/or hydridotricarbonylcobalt, $HCo(CO)_3$, as the active catalytic species. Hydridotetracarbonylcobalt is formed by reaction of cobalt carbonyls with hydrogen, e.g.:

$$Co_2(CO)_8 \ + \ H_2 \ \rightleftharpoons \ 2HCo(CO)_4$$

It is suggested that it dissociates to carbon monoxide and hydridotricarbonylcobalt, though there is little direct evidence for the existence of this compound:

$$HCo(CO)_4 \ \rightleftharpoons \ HCo(CO)_3 \ + \ CO$$

The formation of aldehydes is thought to involve a series of reactions of the following type:

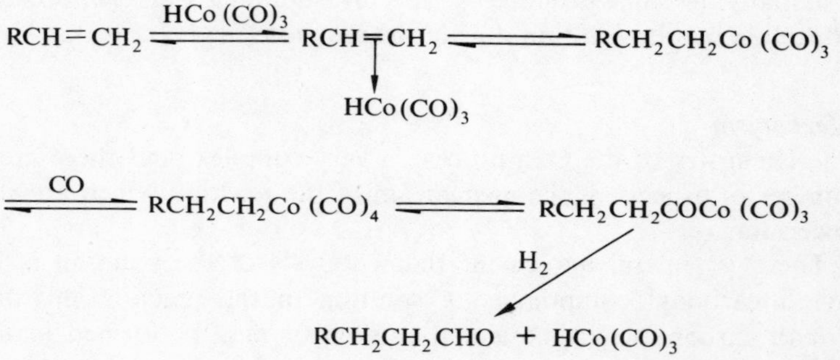

$$RCH = CH_2 \xrightarrow{\text{HCo(CO)}_3} RCH \overset{\displaystyle -}{\underset{\downarrow}{\text{-}}} CH_2 \rightleftharpoons RCH_2CH_2Co(CO)_3$$
$$HCo(CO)_3$$

$$\xrightarrow{\text{CO}} RCH_2CH_2Co(CO)_4 \rightleftharpoons RCH_2CH_2COCo(CO)_3$$
$$\xrightarrow{H_2}$$
$$RCH_2CH_2CHO \ + \ HCo(CO)_3$$

The alternative mode of rearrangement of the π-complex to that shown results in the formation of branched-chain aldehyde:

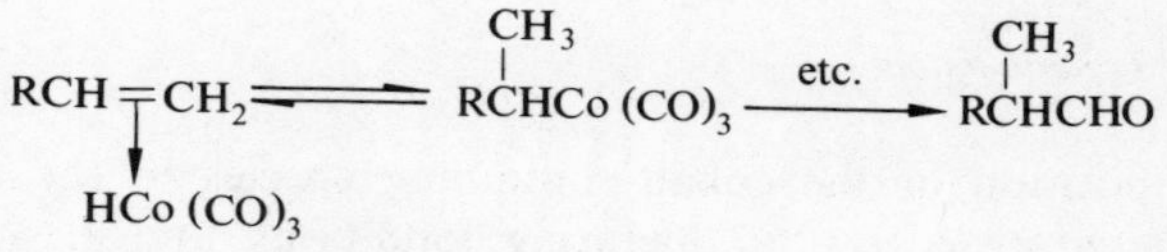

$$RCH \overset{\displaystyle -}{\underset{\downarrow}{\text{-}}} CH_2 \rightleftharpoons R\overset{CH_3}{\underset{|}{C}}HCo(CO)_3 \xrightarrow{\text{etc.}} R\overset{CH_3}{\underset{|}{C}}HCHO$$
$$HCo(CO)_3$$

Similar schemes involving the hydridotetracarbonylcobalt can be postulated. In this case, formation of a π-complex from the coordinatively saturated hydridotetracarbonylcobalt occurs with the expulsion of a molecule of carbon monoxide:

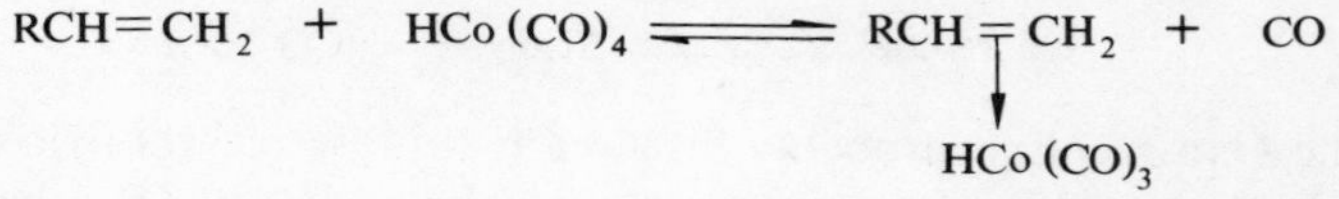

$$RCH = CH_2 \ + \ HCo(CO)_4 \rightleftharpoons RCH \overset{\displaystyle -}{\underset{\downarrow}{\text{-}}} CH_2 \ + \ CO$$
$$HCo(CO)_3$$

The isomerisation of alkenes in the Oxo reaction can be accounted for by reactions of the type discussed above:

$$RCH_2CH=CH_2 + HCo(CO)_3 \rightleftharpoons RCH_2CH\overset{}{\underset{\underset{HCo(CO)_3}{\downarrow}}{\text{—}}}CH_2 \rightleftharpoons RCH_2\overset{CH_3}{\underset{}{\overset{|}{C}}}HCo(CO)_3$$

$$\rightleftharpoons RCH\underset{\underset{HCo(CO)_3}{\downarrow}}{\text{—}}CHCH_3 \rightleftharpoons RCH=CHCH_3 + HCo(CO)_3$$

Process Operation

There has been very little published, other than in patents, on methods of operation of the Oxo process, so that firm details of commercial practice are unusually sparse. One might deduce from this that Oxo technology is highly complex, and that even general information about process operation is considered to be of potential value to competitors.

Production of alcohols by the Oxo process is most often carried out as a two-stage operation in which aldehydes are produced in the hydroformylation stage, and then reduced to alcohols in a separate stage. However, in some recently developed modifications of the process (see below), one-stage operation is possible.

In the conventional Oxo process, the hydroformylation reaction is carried out at temperatures in the range 380 to 450 K, and pressures from 150 to 300 atm. The cobalt can be introduced into the system in a number of ways, and this is one of the major sources of variation in hydroformylation process design. When the alkene being used is a liquid, the cobalt may be conveniently introduced in solution in the feed as, e.g., cobalt naphthenate, or as preformed cobalt carbonyl. With gaseous alkenes a solvent is often used, and the cobalt naphthenate or carbonyl can be dissolved in this. Alternatively, the cobalt may be introduced as a slurry of oxide or salt in liquid alkene or solvent

The product stream leaving the hydroformylation reactor contains cobalt in solution, and this has to be removed and recovered before the material is passed to the next stage of the process. There is a variety of ways in which this can be done, the choice of method of cobalt removal being interlinked with the choice of method of adding the catalyst to the reactor. It has already been indicated that cobalt carbonyls are unstable at elevated temperatures and low partial pressures of carbon monoxide, and one method of cobalt removal consists essentially of releasing the pressure from the hot reaction products and allowing decomposition to occur, after which the metallic cobalt can be separated from the organic reaction products and converted into a form suitable for recycling to the reactor. Decomposition can also be carried out with water or aqueous acid, in which case the cobalt is obtained as a mixture of metallic cobalt, cobalt oxides and cobalt salts.

These methods have a number of disadvantages, amongst them being the fact that the production of solids during catalyst removal can lead to handling difficulties. More recently developed methods involve extraction of hydridotetracarbonylcobalt from the product stream. Thus, if the product stream is contacted with an aqueous solution of magnesium bicarbonate the acidic hydridotetracarbonylcobalt is extracted as a water-soluble salt. Acidification of the extract liberates the hydridotetracarbonylcobalt, which can be transferred to fresh alkene feed by liquid–liquid extraction, or can be carried into the reactor in a stream of synthesis gas.

The decobalted reaction product is generally hydrogenated without further purification, either in the vapour phase at 320 to 520 K and approximately atmospheric pressure, or in the liquid phase at similar temperatures and under 10 to 240 atm pressure. Catalysts normally used are molybdenum sulphide, nickel or cobalt.

If, as is sometimes the case, aldehydes are required as products, they are isolated from the decobalted product stream by distillation.

PRODUCT DISTRIBUTION

Straight-chain alcohols are generally more valuable than branched-chain alcohols, and a considerable amount of effort has been expended in attempts to increase the proportion of straight-chain products. The reaction temperature and partial pressure of carbon monoxide have a substantial effect on the product distribution, the formation of straight-chain products being favoured by low temperatures and high carbon monoxide pressure. Thus, hydroformylation of propene at 373 K and 250 atm gives about 81% n-butyraldehyde, compared with about 54% at 453 K and 250 atm; at 453 K and 1 000 atm about 74% n-butyraldehyde is obtained. One suggestion that has been made to account for this effect is that both hydridotetracarbonylcobalt and hydridotricarbonylcobalt are catalytically active, and that the reaction involving the more heavily co-ordinated tetracarbonyl gives a higher ratio of straight- to branched-chain products, possibly for steric reasons. The extent of dissociation of hydridotetracarbonylcobalt to hydridotricarbonylcobalt would be expected to be reduced by low temperatures and high partial pressures of carbon monoxide.

The extent to which advantage can be taken of the effects on the product distribution of temperature and pressure is limited by two factors. Firstly, although low temperatures give higher yields of the most valuable products, they also reduce the rate of reaction, so that for a given rate of output a larger reactor volume is required than at higher temperatures; this leads to an increase in capital costs. Secondly, high operating pressures also increase the capital costs and, in addition, the energy costs for gas compression; 300 atm appears to be the highest pressure at which the process is operated commercially.

A significant development in Oxo technology was made by Shell during the 1960s. In the Shell process, the catalyst is modified by the addition of, e.g., tributylphosphine and the active species is probably a complex of the type $HCo(CO)_3(PBu_3)$, in which one of the carbon monoxide ligands has been replaced by the phosphine. Such catalyst systems have substantially different properties from the original Oxo catalysts. For instance, the phosphine ligand stabilises the complex to such an extent that it can exist under much lower partial pressures of carbon monoxide. It is reported that the operating pressure in the Shell process is about 30 atm, so that substantial savings in capital and energy costs can be made. A second important feature of this type of catalyst system is that it leads to the production of a higher ratio of straight-chain to branched-chain product. This has been attributed to a combination of steric and electronic effects of the phosphine ligand in the formation of alkylcarbonyl from the π-complex. A further factor is that the phosphine-modified catalyst has hydrogenation activity in addition to hydroformylation activity, so that alcohols can be made by a one-stage process.

There are, however, some drawbacks to this process, the most important being that reaction is slower than when the conventional catalyst is used, and consequently larger amounts of the modified catalyst are required. The economics of the process are heavily dependent on very efficient recovery and re-use of the catalyst.

Another development, not yet in commercial use, involves the use of rhodium-containing catalysts, e.g. $RhH(CO)(Ph_3P)_3$. These also allow the reaction to be carried out at relatively low pressures, but are much more active than cobalt-containing catalysts so that much smaller amounts are required. They have the further advantage of not catalysing isomerisation, so that more control over the products obtained is possible than when cobalt catalysts are used.

OXO PRODUCTS

A wide range of products from propionaldehyde up are made by the Oxo process. The most important are the C_8–C_{13} alcohols used in the manufacture of plasticisers, and n- and iso-butanols and n-butyraldehyde.

Propionaldehyde is made, by hydroformylation of ethylene, on a relatively small scale (US production 1966, *ca*. 13 000 tons). It is not made in the UK. The major use of propionaldehyde is oxidation to propionic acid, but some is hydrogenated to n-propanol, which has some small-tonnage uses, and there are some other minor uses. Since propionic acid is produced in substantial quantities as a by-product in the manufacture of acetic acid by oxidation of light naphtha its manufacture from propionaldehyde is not commercially attractive in areas where such by-product material is available in large quantities. In the UK, for instance, by-product propionic acid is available in quantities substantially greater than the demand.

Most Oxo operators make C_4 products. The major outlet for n-butanol is in lacquer manufacture, where it and its esters are used as solvents; isobutanol is also used in these applications. n-Butyraldehyde is used extensively for the manufacture of 2-ethylhexan-1-ol *via* an aldol condensation:

$$2C_3H_7CHO \xrightarrow{\text{alkali}} \overset{\overset{\text{OH}}{|}\;\overset{C_2H_5}{|}}{C_3H_7CHCHCHO} \xrightarrow{-H_2O} \overset{\overset{C_2H_5}{|}}{C_3H_7CH=CCHO}$$

$$\xrightarrow{H_2} \overset{\overset{C_2H_5}{|}}{C_3H_7CH_2CHCH_2OH} \qquad \textit{2-ethylhexan-1-ol}$$

2-Ethylhexanol has been well established in plasticiser manufacture for many years and, despite the rather cumbersome process involved in its manufacture, has managed so far to stave off competition from the C_8 alcohols obtained by hydroformylation of C_7 alkenes.

C_5 alcohols have a number of applications, for example as intermediates in the manufacture of additives for diesel fuels and lubricating oils. C_6 Oxo alcohols have similar applications, and are also used to a fairly minor extent in plasticiser manufacture.

A number of products in the C_8–C_{13} range are made for use in plasticiser manufacture. Thus, a mixture of alcohols known as 'iso-octanol'* is made by hydroformylation of mixed C_7 alkenes obtained by reaction of propene and n-butenes under the influence of an acid catalyst. It consists mainly of dimethylhexanols and methylheptanols. Hydroformylation of a mixture of C_6–C_8 terminal straight-chain alkenes obtained by wax cracking gives a mixture of normal- and 2-methyl-alcohols which has some technical advantages over iso-octanol. 'Nonanol', consisting mainly of 2,5,5-trimethylhexan-1-ol, is made from di-isobutene, and 'isodecanol' a C_{10} mixture, is made from propylene trimer. 'Tridecanol', a mixture of C_{13} alcohols, is made from propylene tetramer.

The Oxo process is operated in the UK by ICI and Shell, the reported capacities of the plants being about 200 000 and 100 000 tons per annum of mixed Oxo products respectively. The Shell plant utilises the phosphine-modified process.

* The term 'iso-octanol' is a commercial term; it does not have any relationship to the use of the term 'iso' in systematic nomenclature.

Chapter 7

Polymerisation

POLYMERS form an extremely important group of end-use products of the chemical industry. They owe their importance to their mechanical properties, which allow them to compete with traditional materials such as wood, metals, natural fibres and natural rubber, and which have also led to the development of entirely new types of uses and applications. Total production of synthetic polymers in the UK in 1970 was about 2 million tons.

Except in the case of naturally occurring polymers such as cellulose, polymers are produced from low molecular weight compounds, *monomers*, by the process of *polymerisation*. Polymerisation reactions can be classified in a number of ways. In 1929, Carothers, the inventor of nylon, suggested the division into *addition* and *condensation polymerisation*. In addition polymerisation the molecular formula of the repeating unit in the polymer is identical with that of the monomer from which the polymer is derived. Thus, polymerisation of vinyl chloride, which has a molecular formula C_2H_3Cl, gives poly(vinyl chloride), with a repeating unit with the same formula:

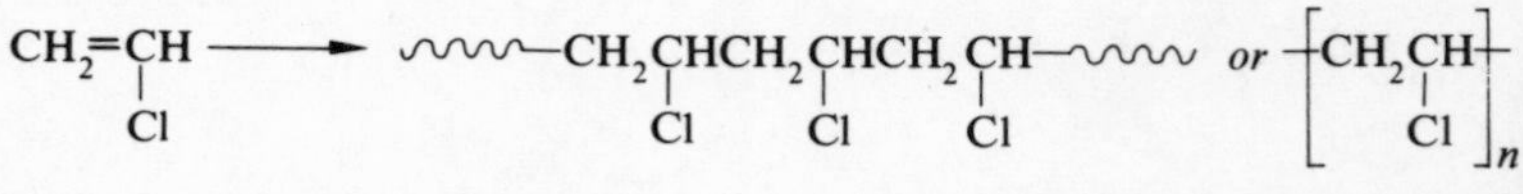

$$CH_2{=}\underset{\underset{Cl}{|}}{CH} \longrightarrow \;\text{\textasciitilde\textasciitilde\textasciitilde}{-}CH_2\underset{\underset{Cl}{|}}{CH}CH_2\underset{\underset{Cl}{|}}{CH}CH_2\underset{\underset{Cl}{|}}{CH}{-}\text{\textasciitilde\textasciitilde\textasciitilde}\; or \left[CH_2\underset{\underset{Cl}{|}}{CH}\right]_n$$

poly(vinyl chloride)

In condensation polymerisation, loss of a simple molecule occurs as each structural unit is introduced into the polymer chain, so that the repeating unit in the polymer contains fewer atoms than the monomer or monomers used. Thus, reaction of terephthalic acid with ethylene glycol gives poly(ethylene terephthalate) and water:

$$n\,HOCH_2CH_2OH + n\,HO_2CC_6H_4CO_2H \longrightarrow \left[OCC_6H_4CO_2CH_2CH_2O\right]_n + 2nH_2O$$

poly(ethylene terephthalate)

228

Although this classification is still widely used, a more useful division from the point of view of reaction mechanism is into *chain-growth polymerisation* and *step-growth polymerisation*. In chain-growth polymerisation a particular polymer chain is formed in a single chain reaction, whereas in step-growth polymerisation it is formed by the occurrence of a succession of separate reactions between functional groups. Most addition polymerisations are chain-growth polymerisations, and most condensation polymerisations are step-growth polymerisations, but there are exceptions.

CHAIN-GROWTH POLYMERISATION

The most important group of commercial polymers prepared by chain-growth polymerisation are those obtained by the polymerisation of monomers containing carbon–carbon double bonds, often called *vinyl polymerisation*. This group includes, for example, polyethylene, poly(vinyl chloride), poly(vinyl acetate), polystyrene, poly(methyl methacrylate) and most of the synthetic rubbers. Vinyl monomers can be polymerised in a number of ways, of which the most important method at present is free-radical vinyl polymerisation. Chain-growth polymerisations of some other classes of monomers, notably carbonyl compounds and 1,2-epoxides, are also of some commercial importance.

Free-Radical Vinyl Polymerisation
As the name implies, this type of polymerisation involves free-radical intermediates.

Monomers used in free-radical vinyl polymerisation, and for that matter in other types of vinyl polymerisation, are most commonly monosubstituted ethylenes. 1,1-Disubstituted ethylenes are used to some extent, and 1,2-disubstituted and polysubstituted ethylenes are used in a limited number of cases. By no means all vinyl compounds can be polymerised by a free-radical mechanism.

MECHANISM
In this general discussion we shall represent the monomer as $CH_2{=}CHX$.

In common with other chain reactions, free-radical vinyl polymerisations involve initiation, propagation, and termination steps. Initiation of a kinetic chain involves the production of a free radical from an initiator, and addition of this radical to a molecule of monomer:

$$I \text{ (initiator)} \longrightarrow 2R\cdot$$

$$R\cdot \ + \ CH_2{=}\underset{\underset{X}{|}}{CH} \longrightarrow RCH_2\underset{\underset{X}{|}}{CH}\cdot$$

The new radical so formed can then add to a further molecule of monomer:

$$RCH_2\underset{\underset{X}{|}}{CH}\cdot \ + \ CH_2{=}\underset{\underset{X}{|}}{CH} \longrightarrow RCH_2\underset{\underset{X}{|}}{CH}CH_2\underset{\underset{X}{|}}{CH}\cdot$$

This type of reaction, addition of monomer to a growing radical, is the propagation step of the kinetic chain, and its repetition results in the formation of long-chain radicals:

$$RCH_2\underset{\underset{X}{|}}{CH}CH_2\underset{\underset{X}{|}}{CH}\cdot \xrightarrow{\;n\text{ propagation steps}\;} R{\left[CH_2\underset{\underset{X}{|}}{CH}\right]}_{n+1}CH_2\underset{\underset{X}{|}}{CH}\cdot$$

Termination of kinetic chains can occur by coupling of radicals, or by disproportionation:

$$R{\left[CH_2\underset{\underset{X}{|}}{CH}\right]}_{n'}CH_2\underset{\underset{X}{|}}{CH}\cdot \ + \ \cdot\underset{\underset{X}{|}}{CH}CH_2{\left[\underset{\underset{X}{|}}{CH}CH_2\right]}_{n''}R \longrightarrow R{\left[CH_2\underset{\underset{X}{|}}{CH}\right]}_{n'}CH_2\underset{\underset{X}{|}}{CH}-\underset{\underset{X}{|}}{CH}CH_2{\left[\underset{\underset{X}{|}}{CH}CH_2\right]}_{n''}R$$

$$R{\left[CH_2\underset{\underset{X}{|}}{CH}\right]}_{n'}CH_2\underset{\underset{X}{|}}{CH}\cdot \ + \ \cdot\underset{\underset{X}{|}}{CH}CH_2{\left[\underset{\underset{X}{|}}{CH}CH_2\right]}_{n''}R \longrightarrow R{\left[CH_2\underset{\underset{X}{|}}{CH}\right]}_{n'}CH{=}\underset{\underset{X}{|}}{CH} \ + \ CH_2CH_2{\left[\underset{\underset{X}{|}}{CH}CH_2\right]}_{n''}R$$

Once a chain is initiated, propagation and termination occur over a very short period, typically less than 1 second.

A variety of compounds can be used as initiators. For example, organic peroxides, which readily undergo thermal cleavage across the weak oxygen–oxygen bond, are widely used, e.g.,

$$\underset{\text{\textit{benzoyl peroxide}}}{Ph\overset{\overset{O}{\|}}{C}O-O\overset{\overset{O}{\|}}{C}Ph} \longrightarrow 2Ph\overset{\overset{O}{\|}}{C}O\cdot$$

Some other examples of initiating systems will be encountered later in this chapter.

The reactions discussed above would lead us to expect that initiator residues will be incorporated into the polymer, at one or both ends of the chain depending on which termination step is operating, and in a number of cases the presence of these residues at chain ends has been demonstrated. Since they make up only a small proportion of the total polymer molecule, in most polymers they have no appreciable effect on properties.

In the propagation reaction, radicals have been shown attacking the monomer at the methylene group, giving what is called a 'head-to-tail' structure. This is the predominant structure which is in fact found in vinyl polymers produced by free-radical polymerisation; poly(vinyl chloride) and polystyrene, for example, have been shown to be practically completely made up of head-to-tail structures. Addition to the methylene group is favoured, (i) because it is less sterically hindered than addition at the substituted carbon atom, and (ii) in some cases, because it gives a radical in which resonance stabilisation is possible, e.g.:

$$R\cdot \ + \ CH_2 = CH \qquad\qquad RCH_2 - CH\cdot \qquad\qquad RCH_2 - CH$$

etc.

When termination by coupling occurs, there will be a head-to-head structure formed; the presence of such structures has been demonstrated in a number of polymers.

Molecular Weight

The molecular weight of polymer produced depends on the relative rates of the propagation and termination reactions, production of high molecular weight polymer requiring that the propagation reactions be much faster than the termination reactions. The ratio of the rates of these reactions is markedly affected by the concentration of free radicals in the system, since the termination reactions, which involve encounters between two radicals, are affected more by an increase in concentration than the propagation reactions, which involve encounters between a radical and a monomer molecule. Thus, for a given system, the higher the rate of initiation, the lower the molecular weight of polymer obtained.

Since individual polymer chains grow for different times before undergoing termination reactions, there is a spread in the molecular weight of product obtained. Thus, although one refers to the molecular weight of a polymer, such figures are averages, and the polymer concerned consists of a mixture of molecules of different size. The average molecular weights of commercial vinyl polymers normally lie in the range 10^4 to 10^6.

Chain Transfer Reactions

Chain transfer reactions, in which the kinetic chain is transferred from the growing polymer chain to another molecule, commonly occur in free-radical vinyl polymerisations, and have important effects on molecular weight, and on polymer structure and properties. Chain transfer can occur to polymer, to monomer, to solvent, or to other compounds in the reaction medium.*

In chain transfer to polymer, a growing polymer radical abstracts a hydrogen atom from a finished polymer chain, leaving an unpaired electron somewhere along the length of the polymer chain:

$$\sim\!\!\sim\!\!-CH_2\underset{X}{C}H\cdot + \underset{\substack{X \\ polymer}}{\sim\!\!\sim\!\!-CH_2CH\!\!\sim\!\!\sim} \longrightarrow \sim\!\!\sim\!\!-CH_2\underset{X}{C}H_2 + \sim\!\!\sim\!\!-CH_2\underset{X}{\overset{\bullet}{C}}\!\!\sim\!\!\sim$$

Further propagation reactions, followed eventually by a termination reaction, will lead to the formation of a branch on the original polymer chain:

$$\sim\!\!\sim\!\!-CH_2\underset{X}{\overset{\bullet}{C}}\!\!\sim\!\!\sim + CH_2\!\!=\!\!\underset{X}{C}H \longrightarrow \sim\!\!\sim\!\!-CH_2\underset{X}{\overset{\overset{\textstyle \overset{\bullet}{C}HX}{CH_2}}{C}}\!\!\sim\!\!\sim \xrightarrow{\text{etc.}} \sim\!\!\sim\!\!-CH_2\underset{X}{\overset{\overset{\textstyle \overset{\{}{CHX}}{CH_2}}{C}}\!\!\sim\!\!\sim$$

The extent to which this type of reaction occurs varies considerably from polymer to polymer. In cases where it does occur, the branching produced can have important effects on the properties of the polymer (see Chapter 8).

In chain transfer to monomer, the active centre is transferred from a growing radical to a molecule of monomer, giving a new radical which can propagate and terminate as usual. The effect of this type of reaction is to cause a reduction in molecular weight, in that each time it occurs it leads to the cessation of growth of a polymer chain.

Monomers vary greatly in their susceptibility to chain transfer, monomers containing readily abstracted hydrogen atoms being

---

* Intramolecular chain transfer, in which the active centre is transferred from one position to another on the growing radical, can also occur (see page 247).

particularly susceptible. Thus, extensive chain transfer to monomer occurs in the polymerisation of vinyl acetate, which can readily undergo abstraction of one of its methyl hydrogens to give a resonance-stabilised radical:

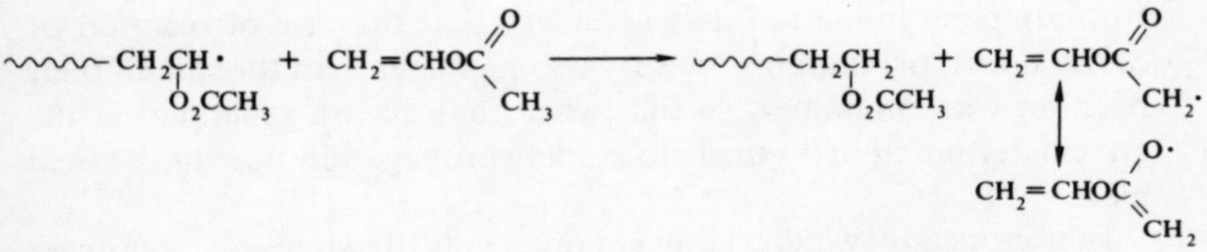

$$\text{\textasciitilde\textasciitilde CH}_2\text{CH} \cdot\ +\ \text{CH}_2\text{=CHOC}\overset{O}{\diagdown}_{CH_3} \longrightarrow \text{\textasciitilde\textasciitilde CH}_2\text{CH}_2\ +\ \text{CH}_2\text{=CHOC}\overset{O}{\diagdown}_{CH_2\cdot}$$

(with O_2CCH_3 substituents and resonance structure $CH_2\text{=CHOC}\overset{O\cdot}{\diagdown}_{CH_2}$)

In some cases, chain-transfer agents are deliberately added to the reaction mix in order to limit the molecular weight attained by the polymer. Mercaptans, for example dodecyl mercaptan, are often used for this purpose:

$$\text{\textasciitilde\textasciitilde CH}_2\underset{X}{\text{CH}} \cdot\ +\ C_{12}H_{25}\text{SH} \longrightarrow \text{\textasciitilde\textasciitilde CH}_2\underset{X}{\text{CH}}_2\ +\ C_{12}H_{25}\text{S} \cdot$$

Inhibition and Retardation

Various materials reduce the rate of polymerisation by effectively terminating reaction chains. Phenols are one class of compounds which produce this type of effect. They are presumed to suffer abstraction of the phenolic hydrogen to give a phenoxy radical which, because of resonance stabilisation, is relatively unreactive and does not add to monomer molecules:

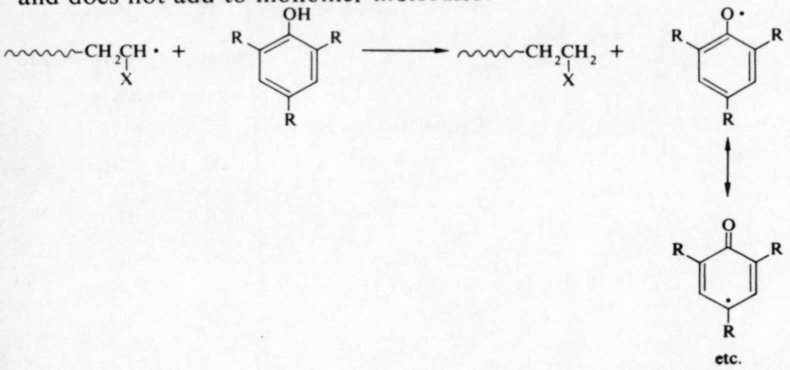

$$\text{\textasciitilde\textasciitilde CH}_2\underset{X}{\text{CH}} \cdot\ +\ \text{(phenol with OH and R substituents)} \longrightarrow \text{\textasciitilde\textasciitilde CH}_2\underset{X}{\text{CH}}_2\ +\ \text{(phenoxy radical)}$$

etc.

It is assumed that the inhibitor radical produced undergoes coupling either with another inhibitor radical or with a polymer radical. Quinones, polyhydric phenols and aromatic amines exhibit the same type of activity, although the mechanisms involved vary.

Agents of this type are called *inhibitors* or *retarders* depending on whether they completely prevent the production of high polymer for a time, or whether they merely reduce the rate of polymerisation. For complete inhibition it is necessary that the rate of reaction of radicals with the inhibitor be very much greater than the rate of their reaction with monomer, so that when radicals are generated in the system, termination occurs before many propagation steps have taken place.

Inhibitors are widely used commercially to stabilise monomers against spontaneous polymerisation in storage; typically, hydroquinone and t-butylcatechol are used for this purpose in concentrations ranging from 0·001 to 0·1%.

POLYMERISATION OF CONJUGATED DIENES

The situation when conjugated dienes are polymerised is more complicated than in the case of polymerisation of monomers containing a single double bond. Butadiene, for example, can enter the polymer chain with a 1,2- or a 1,4-structure, and the 1,4-structure can have either the *cis* or *trans* configuration:

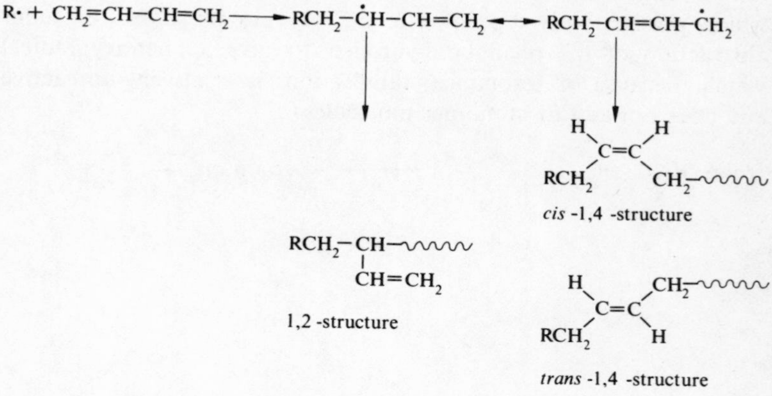
R· + CH2=CH–CH=CH2 → RCH2–ĊH–CH=CH2 ↔ RCH2–CH=CH–ĊH2
RCH2–CH–
CH=CH2
1,2 -structure
H H
C=C
RCH2 CH2–
cis -1,4 -structure
H CH2–
C=C
RCH2 H
trans -1,4 -structure

Free radical polymerisation of butadiene gives a product containing 15–20% of 1,2-structures. The proportion of *cis*- and *trans*-1,4-units in the remaining 80–85% depends on the temperature at which the polymerisation is carried out, low temperatures favouring the formation of *trans*-units.

With substituted dienes, e.g. isoprene (2-methylbutadiene) or chloroprene (2-chlorobutadiene), there is a further complication in that 3,4-entry into the chain is possible.

COPOLYMERISATION

In copolymerisation, a mixture of two (or more) monomers is caused to polymerise to give a product containing both (or all) the monomers. This procedure is technically important because by its means a very considerable degree of control over polymer properties can be exerted.

At first sight one might expect that the composition of a copolymer could be controlled at will by varying the proportion of the monomers in the polymerising mixture, but, except in a small number of cases, this is not so. Generally the composition of the polymer depends not only on the proportions of the monomers, but also on their identity. Some combinations of monomers will not copolymerise at all. In many other cases there is a tendency for one monomer to enter the chain more readily than the other(s). A full discussion of the theory of copolymerisation is beyond the scope of this book, and we shall merely consider two extreme situations. Detailed discussions of copolymerisation will be found in many books on polymer chemistry, e.g. in those by Lenz, Margerison and East, Odian, and Ravve, listed under 'Further Reading'.

In copolymerisation of two monomers A and B, four propagation reactions are possible:

(i) Reaction of a radical with a monomer A unit at the active end with monomer A:

$$\sim\!\!\sim\!\!\sim\!\!-A\cdot \; + \; A \; \longrightarrow \; \sim\!\!\sim\!\!\sim\!\!-A\!-\!A\cdot$$

(ii) Reaction of the above type of radical with monomer B:

$$\sim\!\!\sim\!\!\sim\!\!-A\cdot \; + \; B \; \longrightarrow \; \sim\!\!\sim\!\!\sim\!\!-A\!-\!B\cdot$$

(iii) Reaction of a radical with a monomer B unit at the active end with monomer B:

$$\sim\!\!\sim\!\!\sim\!\!-B\cdot \; + \; B \longrightarrow \sim\!\!\sim\!\!\sim\!\!-B-B\cdot$$

(iv) Reaction of the above type of radical with monomer A:

$$\sim\!\!\sim\!\!\sim\!\!-B\cdot \; + \; A \longrightarrow \sim\!\!\sim\!\!\sim\!\!-B-A\cdot$$

One extreme case is that in which all these reactions proceed with equal facility, i.e. they all have the same rate constant, and in this case the composition of the polymer will be that of the monomer mix from which it is made. The distribution of monomer units in the chain will be random. Another extreme is where reactions of type (i) and (iii) do not occur, i.e. the monomers will not homopolymerise. In this case an alternating copolymer containing equal amounts of both monomers will be formed no matter what the feed composition. These extreme situations do occur in practice, but only in a limited number of cases.

FREE-RADICAL POLYMERISATION PROCESSES
In polymerisation processes, in addition to the normal requirements of a chemical process that an acceptable rate and yield be achieved, there is the requirement that the product should have the correct properties, these being determined for a particular polymer essentially by the molecular weight, the molecular weight distribution, and the degree of branching.

Bulk Polymerisation
The most obvious way of carrying out a polymerisation is to place monomer and initiator in a reactor, and to subject the mix to a temperature at which polymerisation to polymer of the desired properties proceeds at a suitable rate. This apparently simple procedure, known as bulk polymerisation, presents considerable difficulties in practice, these being mainly associated with heat transfer. Concentrated solutions of polymers are extremely viscous, so that in cases where the polymer is soluble in the monomer the reaction mix becomes more and more viscous as reaction proceeds, and heat transfer becomes more and more difficult. Since vinyl polymerisations are quite strongly exothermic this leads to difficulties in temperature control. Consequently, bulk polymerisation is of rather limited applicability.

When the polymer is insoluble in the monomer a slurry of polymer in monomer is formed rather than a viscous syrup, and this can again give rise to heat transfer problems, and also to handling difficulties.

Solution Polymerisation

One obvious way round the difficulties encountered in bulk polymerisation would at first sight appear to be to carry out the reaction
in solution so as to reduce the viscosity of the reaction mix. There are,
however, two disadvantages to this procedure. Firstly, removal of
solvent after completion of the polymerisation tends to be difficult
and expensive, and secondly, many solvents can take part in chain
transfer reactions and this results in low molecular weight polymers
being produced. In general, solution polymerisation is not widely
used for free radical vinyl polymerisation, except in cases where the
polymer is required for use in solution, e.g., where it is to be used in a
surface coating or adhesive.

Suspension Polymerisation

In suspension polymerisation, the monomer is polymerised as beads
of about $0 \cdot 1$–1 mm diameter, suspended in water in a conventional
reaction vessel. Since the droplets have a very large surface to volume
ratio, heat transfer from droplet to water is very good, and heat can
be removed from the water by cooling coils in the vessel or by jacket
cooling. Suspension of the monomer as droplets is brought about by
agitation. To prevent coalescence of the droplets, a suspending agent
is used, typically either a water-soluble polymer such as poly(vinyl
alcohol), or a finely divided water-insoluble inorganic compound,
e.g. kaolin, tricalcium phosphate, aluminium hydroxide. A monomer-
soluble initiator is used, and effectively suspension polymerisation
consists of a large number of 'micro-bulk' polymerisations. The
kinetic characteristics are the same as those of bulk polymerisation.

Suspension polymerisation has some disadvantages, the most
important of which are:

(i) The necessity to use suspending agents can cause some contamination of the polymer, with a deleterious effect on clarity or
colour.

(ii) The product must be filtered and dried before sale, and this
adds to capital and operating costs.

(iii) It has not been found feasible to carry out suspension
polymerisation in a flow system, so that batch processing has to be
used.

Despite these drawbacks, suspension polymerisation is a very useful
technique, and is widely used.

Emulsion Polymerisation

Emulsion polymerisation is superficially similar to suspension poly-merisation, but its mechanism and applications are very different. It involves the polymerisation of monomer dispersed in water in the presence of a surface-active agent, typically soap. The product of an emulsion polymerisation is a stable polymer 'latex', in contrast to the easily separable beads obtaihed in a suspension polymerisation.

When soap is dissolved in water at above a certain critical concen-tration, micelles are formed. These consist of clusters of soap mole-cules with their hydrophobic hydrocarbon tails inwards and their hydrophilic carboxylate groups outwards, towards the aqueous phase [see Fig. 7-1(a)]. When these micelles are formed in the presence of monomer, monomer is absorbed, and the micelles are swollen [see Fig. 7-1(b)]. If more monomer is present than can be taken up by the micelles, which is normal at the start of an emulsion polymerisa-tion, a separate monomer phase of droplets, stabilised by soap molecules around the outside, is formed [see Fig. 7-1(c)]. The micelles have diameters of about 5 nm, and the droplets of about 1 500 nm.

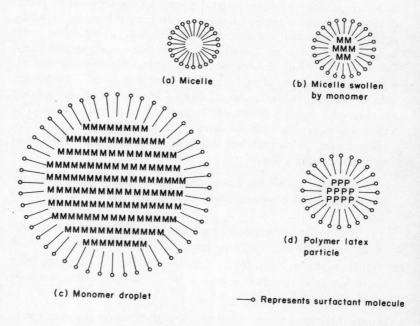

(a) Micelle

(b) Micelle swollen by monomer

(c) Monomer droplet

(d) Polymer latex particle

——o Represents surfactant molecule

Fig. 7-1. *Particles in emulsion polymerisation* (*not to scale*).

The number of micelles is very much larger than the number of droplets (about 10^{24} micelles/m^3 compared with about 10^{17} droplets/m^3).

Water-soluble initiators are used in emulsion polymerisation, and these generate free radicals in the aqueous phase. Thus, inorganic persulphates are frequently used, and are presumed to generate sulphate ion-radicals:

$$S_2O_8^{2-} \longrightarrow 2SO_4^{-} \cdot$$

When an initiator radical encounters a micelle or a droplet it diffuses into it; since the total surface area of the micelles is much greater than that of the droplets, encounters with micelles are much more frequent than with droplets, and practically all the polymerisation takes place in the micelles.

When a radical enters a micelle, it initiates polymerisation of the absorbed monomer in the normal way, and polymerisation proceeds inside the micelle. The monomer consumed by this polymerisation is replaced by diffusion of monomer through the aqueous phase from the droplets, which thus act as monomer reservoirs. The rate of generation of radicals, and the number of micelles, are such that there is, on average, a relatively long period between successive radical captures by a particular micelle (of the order of 10 seconds), so that after initiation of polymerisation in a micelle, propagation continues for an unusually long time with no possibility of termination. When the next radical enters, termination is practically instantaneous, and no further polymerisation occurs in that micelle until the next radical enters. Thus, the most important feature of emulsion polymerisation is that very high molecular weight products can be obtained at high rates of polymerisation.

As with suspension polymerisation, heat transfer presents no major problems in emulsion polymerisation.

If the polymer is required as a solid, the latex has to be coagulated, e.g. by the addition of acid, and the polymer filtered and dried, and this naturally adds to costs. The polymer obtained is contaminated by the surface-active agent, and this may be a disadvantage in some circumstances. However, if a very high molecular weight is required, e.g. in a synthetic rubber, emulsion polymerisation is generally the only feasible way of achieving this by a free-radical polymerisation.

For some applications, e.g. the manufacture of emulsion paints, the polymer is used in latex form.

Ionic Vinyl Polymerisation

Vinyl monomers can be polymerised by ionic as well as by free-radical initiators, and such ionic polymerisations have some commercial importance, though by no means so much as free-radical polymerisation. The mechanisms involved are considerably more complex than those in free-radical polymerisation, there being a much greater variation in initiation, propagation and termination steps. In view of this complexity, and the relatively limited commercial importance of ionic polymerisations, we shall consider them only briefly.

ANIONIC POLYMERISATION

In anionic polymerisation the active centre of the propagating species is negatively charged. It occurs only with monomers bearing electron-withdrawing substituents, e.g. phenyl, vinyl, nitrile and carboxyl, which are able to stabilise the propagating species. Initiation can be brought about by a number of agents, of which one important class are the alkyl-lithiums, typically n-butyl-lithium. In this case, initiation occurs as shown below:

$$\underset{X}{CH_2=CH} + C_4H_9Li \longrightarrow \underset{X}{C_4H_9CH_2CH^-}\ Li^+$$

Propagation can then occur by successive additions of monomer molecules to the active species:

$$\sim\!\!\sim\!\!\sim\underset{X}{CH_2CH^-}\ Li^+ + \underset{X}{CH_2=CH} \longrightarrow \sim\!\!\sim\!\!\sim\underset{X}{CH_2CH}\underset{X}{CH_2CH^-}\ Li^+$$

The degree of association between the active end of the propagating species and the positive counterion can vary considerably, and depends on a number of factors, including the dielectric constant and the solvating power of the polymerisation medium. The counterion may be completely dissociated from the negative site, or it may be closely associated with it as an ion-pair. Not surprisingly, the degree of association can have a major effect on the course of the reaction.

Termination by coupling or disproportionation, as occurs in free-radical polymerisation, cannot take place in ionic polymerisations, and there is no general in-built termination mechanism for anionic polymerisation. In some systems, termination by hydride ion transfer to the counterion can occur:

$$\text{\textasciitilde}\text{\textasciitilde}\text{CH}_2\underset{X}{\text{CH}}^- \text{ M}^+ \longrightarrow \text{\textasciitilde}\text{\textasciitilde}\text{CH}=\underset{X}{\text{CH}} + \text{MH}$$

In others, reactive centres can react with solvent, or with functional groups in the monomer.

Chain transfer reactions do not appear to be important in anionic polymerisation.

CATIONIC POLYMERISATION

Cationic polymerisation, which involves carbonium ion active centres, occurs essentially only with monomers containing electron-releasing substituents, e.g. phenyl, vinyl, alkoxy and 1,1-dialkyl. Initiation is by proton donation, either by protonic acids or by Lewis acids plus co-catalysts, e.g. boron trifluoride plus water:

$$\text{BF}_3 + \text{H}_2\text{O} + \text{CH}_2=\underset{X}{\text{CH}} \longrightarrow \text{CH}_3\underset{X}{\text{CH}}^+ (\text{BF}_3\text{OH})^-$$

As in anionic polymerisation, the counterion is associated with the active centre to a greater or lesser extent depending on the solvent used and other factors.

Propagation involves monomer addition at the carbonium ion site:

$$\text{\textasciitilde}\text{\textasciitilde}\text{CH}_2\underset{X}{\text{CH}}^+ (\text{BF}_3\text{OH})^- + \text{CH}_2=\underset{X}{\text{CH}} \longrightarrow \text{\textasciitilde}\text{\textasciitilde}\text{CH}_2\underset{X}{\text{CH}}\text{CH}_2\underset{X}{\text{CH}}^+ (\text{BF}_3\text{OH})^-$$

Termination of polymer chains can occur by proton transfer to the counterion:

$$\text{\textasciitilde}\text{\textasciitilde}\text{CH}_2\underset{X}{\text{CH}}^+ (\text{BF}_3\text{OH})^- \longrightarrow \text{\textasciitilde}\text{\textasciitilde}\text{CH}=\underset{X}{\text{CH}} + \text{H}^+ (\text{BF}_3\text{OH})^-$$

From the point of view of kinetics, this is a chain transfer reaction in that the initiator is now available to initiate another chain. Chain transfer to monomer, polymer and solvent also commonly occur in cationic polymerisation.

Ziegler–Natta and Related Polymerisations

In Ziegler–Natta polymerisation, named after K. Ziegler of Germany and G. Natta of Italy, reaction is carried out in the presence of a complex solid catalyst, typically formed by mixing titanium tetrachloride or trichloride with an alkylaluminium compound. Such catalyst systems have two important characteristics:

(i) They bring about the polymerisation under mild conditions of a range of monomers, some of which cannot be polymerised by free-radical or ionic initiators.

(ii) They allow a considerable degree of control over the structure of the polymer to be achieved. *Stereoregular* polymers can be made.

Polymerisation in the presence of Ziegler–Natta catalysts is of major commercial importance.

MECHANISM

The mechanism of Ziegler–Natta polymerisation appears to be extremely complex, and is still under very active investigation: it is not possible to discuss fully the present situation here. What follows is merely a brief indication of some of the major characteristics of the system.

Ziegler–Natta catalysts can be made from a wide range of transition metal compounds and organometallic compounds, but the most widely used are those prepared from alkylaluminiums and titanium tetrachloride or trichloride. Mixture of these components in an inert solvent, e.g. heptane, gives the catalyst. The catalysts thus formed are solids of very complex constitutions which have by no means been fully worked out. It is evident that some reduction of the transition metal takes place, and exchange of halogen atoms and alkyl groups occurs, and that the catalysts thus contain transition metal–alkyl compounds.

One mechanism for the polymerisation which has found quite wide acceptance is that proposed by Cossee. In this, it is proposed that polymerisation occurs at active sites at the catalyst surface, these sites involving a transition metal atom, e.g. titanium, having an octahedral configuration and with one ligand vacancy:

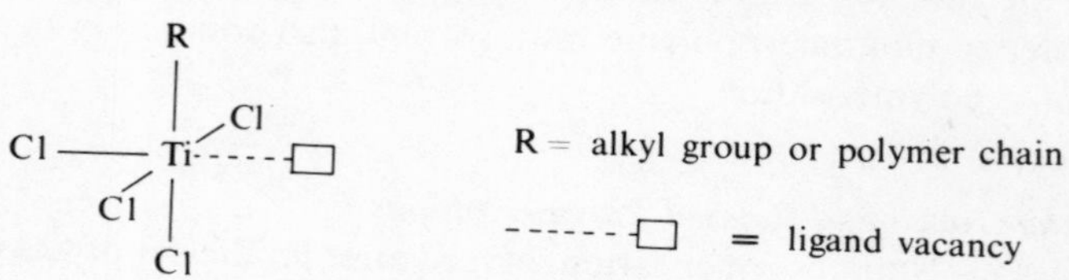

R = alkyl group or polymer chain

------ ☐ = ligand vacancy

The monomer is co-ordinated to the titanium through π-bonding, and undergoes insertion between the titanium and the alkyl group or polymer chain in the manner indicated below:

$$\text{Cl}\!-\!\overset{\overset{\displaystyle R}{|}}{\underset{\underset{\displaystyle \text{Cl}}{|}}{\overset{\diagup\text{Cl}}{\underset{\diagup}{\text{Ti}}}}}\cdots\square \; + \; \text{CH}_2\!=\!\text{CHCH}_3 \longrightarrow \text{Cl}\!-\!\overset{\overset{\displaystyle R}{|}}{\underset{\underset{\displaystyle \text{Cl}}{|}}{\overset{\diagup\text{Cl}}{\underset{\diagup}{\text{Ti}}}}}\!-\!\overset{\displaystyle \overset{\text{CH}_3}{|}\;\text{CH}}{\underset{\text{CH}_2}{\|}}$$

$$\longrightarrow \text{Cl}\!-\!\overset{\overset{\displaystyle R}{\diagdown}}{\underset{\underset{\displaystyle \text{Cl}}{|}}{\overset{\diagup\text{Cl}^{\diagdown}\text{CH}}{\underset{\diagup}{\text{Ti}}}}}\cdots\overset{\overset{\displaystyle \overset{\text{CH}_3}{|}}{}}{\underset{\text{CH}_2}{}} \longrightarrow \text{Cl}\!-\!\overset{\overset{\displaystyle \square}{\vdots}}{\underset{\underset{\displaystyle \text{Cl}}{|}}{\overset{\diagup\text{Cl}}{\underset{\diagup}{\text{Ti}}}}}\!-\!\overset{\overset{\displaystyle R}{|}}{\underset{\text{CH}_2}{\text{CHCH}_3}}$$

$$\longrightarrow \text{Cl}\!-\!\overset{\overset{\displaystyle \overset{\overset{\overset{\overset{R}{|}}{\text{CHCH}_3}}{|}}{\text{CH}_2}}{|}}{\underset{\underset{\displaystyle \text{Cl}}{|}}{\overset{\diagup\text{Cl}}{\underset{\diagup}{\text{Ti}}}}}\cdots\square$$

The polymer chain thus grows by successive insertions of monomer units at the titanium atom.

STEREOSPECIFIC POLYMERISATION

One of the important characteristics of polymerisation in the presence of Ziegler–Natta catalysts is that control over the way in which the monomer units enter the polymer chain is possible. In the polymerisation of butadiene, for instance, it is possible by a suitable choice of catalyst to prepare polymer containing predominantly *cis*-1,4-, *trans*-1,4-, or 1,2-units.

Polymers without double bonds involved in the polymer chain can also exhibit stereoregularity. Thus, if we consider the polymerisation of monomer $CH_2 = CHX$, three different types of head-to-tail polymer are possible. These are:

(i) Polymer in which all the X-substituted carbon atoms have the same configuration:

244 INTRODUCTION TO INDUSTRIAL ORGANIC CHEMISTRY

This is called an *isotactic polymer*.

(ii) Polymer in which the configuration of the X-substituted carbon atom alternates:

This is called a *syndiotactic polymer*.

(iii) Polymer in which the configurations of the monomer units are randomly distributed along the polymer chain:

This is called an *atactic polymer*.

Whereas polymers made by free-radical polymerisation generally have largely atactic structures, isotactic and syndiotactic polymers can be made from a variety of hydrocarbon monomers by polymerisation in the presence of an appropriate Ziegler–Natta catalyst. The method is not in general applicable to polar monomers such as vinyl chloride, methyl methacrylate and acrylonitrile, since these inactivate the catalyst.

Stereoregularity has important implications in terms of polymer properties (see Chapter 8).

METAL OXIDE CATALYSTS

During the 1950s, at about the same time as Ziegler–Natta catalysts were being developed, catalyst systems consisting of reduced transition metal oxides on supports such as alumina or silica were shown to be effective for the polymerisation of alkenes, and catalysts of this type are now of considerable importance for the polymerisation of ethylene. The mechanism of polymerisation in the presence of this type of catalyst is not clear; some workers suggest a mechanism similar to Ziegler–Natta polymerisation.

SYNTHESIS OF LINEAR PRIMARY ALCOHOLS AND 1-ALKENES

Ziegler–Natta catalysts were discovered by Ziegler during investigations into the reaction of ethylene with alkylaluminiums, and have, as we have seen, become of great importance. The alkylaluminium–ethylene reaction, itself discovered by Ziegler, has also found commercial application in the manufacture of alcohols and, potentially, of alkenes. Although the products of this reaction are not high polymers, it is conveniently discussed here.

If ethylene and triethylaluminium are allowed to react at about 370 K and pressures above 65 atm, the so-called 'growth reaction' occurs, in which ethylene molecules are successively inserted between the aluminium atom and the attached alkyl groups:

$$\underset{\diagdown CH_2CH_3}{\overset{\diagup CH_2CH_3}{Al - CH_2CH_3}} \ + \ (x+y+z)\ C_2H_4 \ \longrightarrow \ \underset{\diagdown (CH_2CH_2)_{z+1}H}{\overset{\diagup (CH_2CH_2)_{x+1}H}{Al - (CH_2CH_2)_{y+1}H}}$$

A reaction occurring in competition with this is loss of an alkyl group to give an alkene and dialkylaluminium hydride:

$$\underset{\diagdown (CH_2CH_2)_{z+1}H}{\overset{\diagup (CH_2CH_2)_{x+1}H}{Al - (CH_2CH_2)_{y+1}H}} \ \longrightarrow \ \underset{\diagdown H}{\overset{\diagup (CH_2CH_2)_{x+1}H}{Al - (CH_2CH_2)_{y+1}H}} \ + \ CH_2 = CH\,(CH_2CH_2)_z H$$

Since the hydride formed immediately reacts with ethylene, the net effect is the displacement of an alkene by ethylene:

$$\underset{\diagdown H}{\overset{\diagup (CH_2CH_2)_{x+1}H}{Al - (CH_2CH_2)_{y+1}H}} \ + \ CH_2{=}CH_2 \ \longrightarrow \ \underset{\diagdown CH_2CH_3}{\overset{\diagup (CH_2CH_2)_{x+1}H}{Al - (CH_2CH_2)_{y+1}H}}$$

The rate of alkene displacement is independent of ethylene pressure, while the rate of the growth reaction increases with increasing ethylene pressure, so that the formation of long alkyl chains is favoured by high pressures.

Linear alkenes can be made by carrying out the growth reaction with a controlled amount of ethylene to produce a mixture of trialkylaluminiums with the desired average length of alkyl chains, and then bringing about the displacement reaction by heating the trialkylaluminiums to 550 to 590 K in the presence of ethylene, when a mixture of 1-alkenes is produced. This process does not appear to be commercially attractive at present, compared, e.g., with the manufacture of 1-alkenes by wax cracking.

The manufacture of linear primary alcohols by this route is used commercially, e.g. in the USA and Germany. In this process, the mixture of trialkylaluminiums produced by the growth reaction is oxidised by dry air to a mixture of aluminium alkoxides, which is then hydrolysed to give a mixture of primary alcohols:

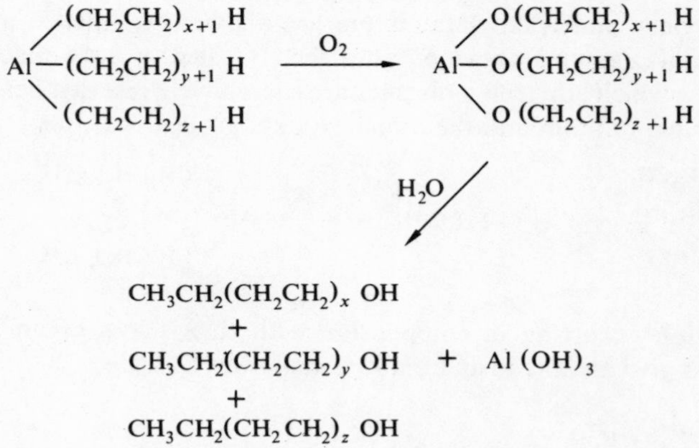

$$
Al \Bigg\langle
\begin{array}{l}
(CH_2CH_2)_{x+1}\,H \\
(CH_2CH_2)_{y+1}\,H \\
(CH_2CH_2)_{z+1}\,H
\end{array}
\quad \xrightarrow{\ O_2\ } \quad
Al \Bigg\langle
\begin{array}{l}
O\,(CH_2CH_2)_{x+1}\,H \\
O\,(CH_2CH_2)_{y+1}\,H \\
O\,(CH_2CH_2)_{z+1}\,H
\end{array}
$$

$$\Big\downarrow H_2O$$

$$
\begin{array}{l}
CH_3CH_2(CH_2CH_2)_x\ OH \\
+ \\
CH_3CH_2(CH_2CH_2)_y\ OH \quad + \quad Al\,(OH)_3 \\
+ \\
CH_3CH_2(CH_2CH_2)_z\ OH
\end{array}
$$

This process provides an alternative to the hydrogenolysis of fats and oils as a source of long-straight-chain primary alcohols.

Important Vinyl Polymers

POLYETHYLENE $\displaystyle +\!\!\left[CH_2 - CH_2 \right]\!\!_n$

Two main types of polyethylene are marketed, *low-density polyethylene*, and *high-density polyethylene*, the former being made by free-radical polymerisation, and the latter by polymerisation in the presence of Ziegler–Natta or oxide catalysts. There are marked differences in properties between the two types (see Table 8-1).

Free-radical polymerisation of ethylene requires the use of very high pressures. Typically, the polymerisation is carried out at temperatures in the range of 350 to 570 K, and at 1 000 to 3 000 atm, oxygen or an organic peroxide being added as initiator. The reaction temperatures used are above the critical temperature of ethylene, so that the polymerisation is rather unusual in that the monomer is in the gas phase. High pressures are necessary in order to favour the propagation reactions, which are strongly affected by ethylene concentration, over the termination reactions, which are not. About 6 to 25% of the ethylene is converted, the rest being recycled.

Extensive chain transfer to polymer occurs during polymerisation, and results in the formation of branched polymer molecules:

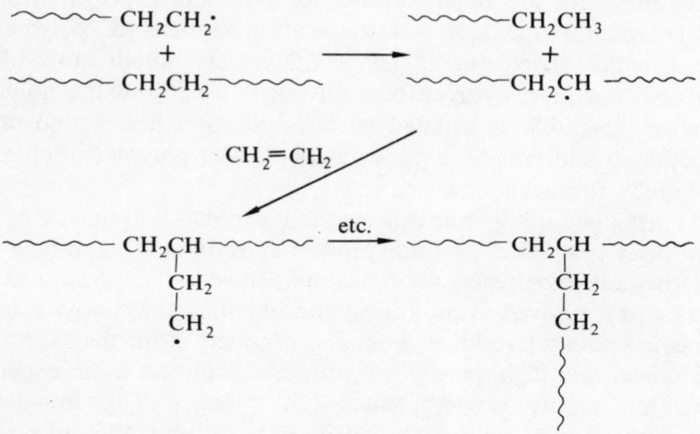

$$\sim\sim\sim CH_2CH_2\cdot$$
$$+$$
$$\sim\sim\sim CH_2CH_2\sim\sim\sim\sim$$
$$\longrightarrow$$
$$\sim\sim\sim CH_2CH_3$$
$$+$$
$$\sim\sim\sim CH_2\underset{\cdot}{C}H \sim\sim\sim$$

$$CH_2\!\!=\!\!CH_2$$

$$\sim\sim\sim CH_2CH\sim\sim\sim \xrightarrow{\text{etc.}} \sim\sim\sim CH_2CH \sim\sim\sim$$
$$\overset{|}{C}H_2 \qquad\qquad\qquad\qquad \overset{|}{C}H_2$$
$$\underset{\cdot}{C}H_2 \qquad\qquad\qquad\qquad \overset{|}{C}H_2$$
$$\qquad\qquad\qquad\qquad\qquad \{$$

In addition to the long branches that one would expect to be formed by this reaction, low-density polyethylene also contains a large number of short branches of, it is suggested, up to 5 carbon atoms. These are assumed to be formed in intramolecular chain-transfer reactions:

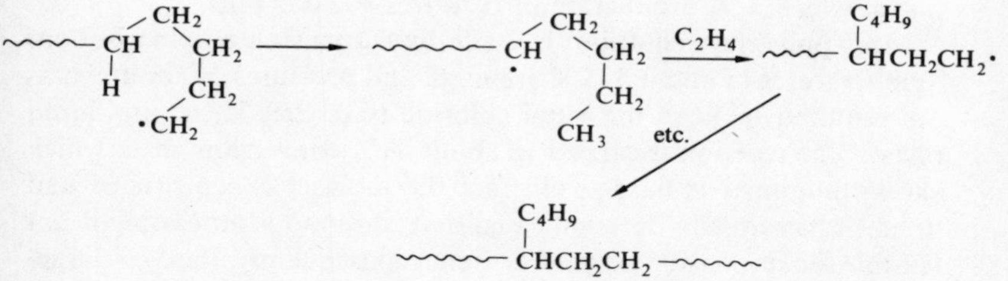

$$\sim\sim\sim \overset{\displaystyle CH_2}{\underset{\displaystyle H}{\overset{|}{C}H}}\overset{\diagup}{\underset{\diagdown}{}}\overset{CH_2}{\underset{CH_2}{\overset{\displaystyle CH_2}{}}}\cdot CH_2 \longrightarrow \sim\sim\sim \underset{\cdot}{C}H \overset{\diagup CH_2}{\underset{\diagdown CH_2}{}}\underset{CH_3}{\overset{|}{C}H_2} \xrightarrow{C_2H_4} \sim\sim\sim \underset{CHCH_2CH_2\cdot}{\overset{C_4H_9}{|}}$$

etc.

$$\sim\sim\sim \underset{CHCH_2CH_2}{\overset{C_4H_9}{|}} \sim\sim\sim$$

Typically a low-density polyethylene molecule would have a short branch about every 50 carbon atoms, and one or two long branches per molecule.

UK production of low-density polyethylene in 1970 was 306 000 tons.

High-density polyethylene is made by polymerisation in the presence of Ziegler–Natta or supported metal oxide catalysts; much lower pressures are required than for free-radical polymerisation. In the presence of a Ziegler–Natta catalyst, for instance, polymerisation can be carried out at 330 to 350 K and about atmospheric pressure. An inert hydrocarbon solvent is used, and the polymer, which is insoluble, is isolated by filtration after destruction of the catalyst by addition of, e.g., an alcohol. The polymer obtained is essentially unbranched.

At first sight, it might be expected that avoidance of the use of very high pressures would give this process an economic advantage over the free-radical polymerisation process. However, it appears that the extra costs involved in producing and handling the catalyst, and in removing catalyst residues from the product, more than outweigh this effect, and high-density polyethylene is in fact more expensive than low-density polyethylene. (UK prices, 1970, low-density polyethylene, 14·7 p/kg; high-density polyethylene, 19·7 p/kg.)

UK production of high-density polyethylene in 1970 was 61 000 tons.

POLY(VINYL CHLORIDE)
$$\left[\begin{array}{c} CH_2-CH \\ | \\ Cl \end{array} \right]_n$$

Poly(vinyl chloride), which is made only by free-radical polymerisation, is the second largest tonnage polymer made, the largest being polyethylene. UK production in 1970 was 302 000 tons.

Most poly(vinyl chloride) is made by suspension polymerisation. Temperatures of about 325 K are used, and pressures of about 9 atm are required to keep the vinyl chloride (b.p. 259 K) in the liquid phase. The reaction is carried to about 88% conversion, after which excess monomer is flashed off, and the product is centrifuged and dried. It has already been indicated that continuous processing is not feasible for suspension polymerisations; consequently, this very large-scale manufacture is carried out by batch processing.

Emulsion polymerisation is used to make some grades of polymer. In this case, except where the polymer is required as a latex, it is isolated by spray drying.

Bulk polymerisation of vinyl chloride was not used until the mid-1960s when a process was developed by Pechiney–St. Gobain in France. The problems of viscous polymer solutions do not arise in the polymerisation of vinyl chloride since the polymer is insoluble in the monomer and precipitates out as it is formed. However, the process is technically difficult since the monomer is absorbed by the polymer to such an extent that at 20% conversion there is no free liquid phase present, and the reaction mixture is in powder form. In the Pechiney–St. Gobain process polymerisation is carried to about 7% conversion in a conventional reaction vessel and then the reaction is completed in a horizontal autoclave in which agitation is provided by slowly rotating blades. Batchwise operation is used.

It is claimed that the polymer produced by this process has some advantages, e.g. in clarity, over polymer produced by suspension polymerisation, in consequence of the absence of contamination by suspending agents. The process has had a considerable degree of success, and at the time of writing the total capacity of plants throughout the world either using this process or under construction is 750 000 tons per annum.

POLYSTYRENE $\left[\text{CH}_2-\underset{\underset{\text{Ph}}{\mid}}{\text{CH}}\right]_n$

All manufacture of polystyrene is by free-radical polymerisation, the processes used being mainly bulk polymerisation and suspension polymerisation. In bulk polymerisation, the reaction is carried to about 40% conversion in a conventional reactor, up to which stage the reaction mix is still fluid enough to allow heat transfer to the reactor jacket or coils, and then polymerisation is completed adiabatically in specialised equipment.

Polystyrene is a fairly brittle polymer, and for many applications its properties are inadequate as they stand. For such applications, *high-impact* or *toughened* polystyrene is made. This consists of polystyrene with a small amount of rubber finely dispersed in it. It is normally made by dissolving rubber in styrene and then polymerising the styrene; the rubber is insoluble in the polystyrene and comes out of solution as a fine dispersion in the polymer.

Expandable polystyrene, used for example for making expanded polystyrene ceiling tiles, is produced either by suspension polymerisation of styrene containing an inert, volatile hydrocarbon, e.g. pentane, or by impregnating beads of polystyrene produced by suspension polymerisation, with such a hydrocarbon. Expanded polystyrene is made by placing the beads in a mould and heating them with steam, when the polystyrene softens and the hydrocarbon volatilises and causes the beads to expand. The softened expanded beads fuse together to give the finished article.

UK production of all grades of polystyrene in 1970 was 176 000 tons.

POLYPROPYLENE $\left[\begin{array}{c} CH_2{-}CH{-} \\ | \\ CH_3 \end{array}\right]_n$

Propene cannot be polymerised to high molecular weight polymers by free-radical initiators. The methyl hydrogens, being allylic, readily undergo abstraction by radicals, and very extensive chain transfer to monomer occurs, preventing the building up of long polymer chains:

$$\sim\!\!\sim\!\!\sim CH_2\underset{\underset{CH_3}{|}}{CH}\cdot \; + \; CH_2{=}CH{-}CH_3$$

$$\longrightarrow \;\sim\!\!\sim\!\!\sim CH_2\underset{\underset{CH_3}{|}}{CH_2} \; + \; CH_2{=}CH{-}CH_2\!\cdot \longleftarrow\!\longrightarrow \cdot CH_2{-}CH{=}CH_2$$

Commercial production of polypropylene is by polymerisation in the presence of Ziegler–Natta catalysts, the product made being isotactic polypropylene. The processes used are similar to those used for the production of high-density polyethylene, and in some cases plants can produce either polymer at choice.

At the present time, consumption of, and production capacity for, polypropylene in the UK are rising rapidly. Thus, UK consumption in 1968 was 55 000 tons, and it is estimated that by 1972 this will have risen to 135 000 tons. UK production in 1970 was 66 000 tons.

POLY(VINYL ACETATE) $\left[\begin{array}{c} CH_2{-}CH{-\!\!-} \\ | \\ O_2CCH_3 \end{array}\right]_n$

Vinyl acetate polymerisation is all carried out by free-radical methods. The major applications of poly(vinyl acetate) and of vinyl acetate copolymers are as polymer latices in emulsion paints, other surface coatings, adhesives etc., and these latices are, naturally, prepared by emulsion polymerisation. Smaller amounts of poly(vinyl acetate) are made by solution, suspension and bulk polymerisation.

It has already been indicated that the methyl hydrogens of vinyl acetate are readily abstracted, so that a substantial amount of chain transfer to monomer occurs in the polymerisation of vinyl acetate. For the same reasons, the methyl hydrogens in the polymer readily undergo abstraction, and chain transfer to polymer also occurs readily in vinyl acetate polymerisation. As would be expected, the amount of transfer to polymer depends on the conversion to which the reaction is taken, and relatively unbranched polymer can be obtained, if desired, by operating at low monomer conversion.

UK production of poly(vinyl acetate) in 1970 was 40 000 tons.

POLY(METHYL METHACRYLATE)
$$\left[CH_2-\underset{\underset{CO_2CH_3}{|}}{\overset{\overset{CH_3}{|}}{C}} \right]_n$$

Poly(methyl methacrylate) is unusual in that a substantial proportion of production is by a specialised form of bulk polymerisation in which the polymer is produced in the form of sheets ('Perspex', 'Plexiglas'). In the preparation of this material, monomer, or a prepolymerised syrup containing between 20 and 50 % of polymer, containing a free-radical initiator in solution, is poured into a mould made up of sheets of glass separated by a flexible gasket, and the whole is then placed in an oven and polymerisation allowed to proceed.

Poly(methyl methacrylate) for use in conventional plastics fabrication processes is made by suspension polymerisation, and solution and emulsion polymerisation are used for products (mainly copolymers) for use as adhesives, surface coatings, etc.

Details of the amounts of poly(methyl methacrylate) produced or consumed in the UK are not available. It is made on a substantially smaller scale than, e.g., polystyrene or polypropylene.

POLYTETRAFLUOROETHYLENE $+CF_2-CF_2\frac{}{}_n$

Polytetrafluoroethylene (PTFE) has exceptional chemical and thermal stability, in consequence of the high strength of the carbon–fluorine bond. In addition it has excellent electrical properties, a very low coefficient of friction, and good non-stick properties. It is, however, a very expensive material, so that its use is limited to applications in which its unique properties are of particular value, and consumption of PTFE is therefore small compared to the consumption of most other commercial polymers. Total world

consumption in 1969 was *ca.* 13 000 tons, of which US consumption
was 8 000 tons, and UK consumption was about 800 to 900 tons.

Tetrafluoroethylene is polymerised by passing it into water
containing a water-soluble initiator, e.g. ammonium persulphate, at
about 310 to 350 K and under about 3 to 30 atm pressure. If the
polymerisation is carried out in the presence of an emulsifying agent,
the product is an aqueous dispersion of PTFE; in the absence of an
emulsifying agent, a granular polymer is formed.

ACRYLIC FIBRES

The acrylic fibres, e.g. 'Acrilan', 'Courtelle', 'Orlon', which form one
of the major groups of synthetic fibres, are made from copolymers of
acrylonitrile with small amounts of other monomers.

Actylonitrile is appreciably soluble in water, and is most commonly
polymerised in aqueous solution, using water-soluble free-radical
initiators. Since the polymer is insoluble in water, it precipitates out
as it is formed. Polymerisation continues in the polymer particles,
and the polymerisation has some of the characteristics of emulsion
polymerisation. A number of comonomers are used in order to
produce desirable modifications in the properties of the polymer, for
instance in its dyeing characteristics (see page 311). The polymer is
essentially unbranched.

UK capacity for acrylic fibres in 1970 was about 120 000 tons per
annum. Production in 1968 was about 70 000 tons.

$$\text{POLYISOBUTENE, BUTYL RUBBER} \left[CH_2-\underset{\underset{CH_3}{|}}{\overset{\overset{CH_3}{|}}{C}} \right]_n$$

Isobutene cannot be polymerised by a free-radical mechanism, cf.
propene. However, it very readily undergoes cationic polymerisation,
in the presence of Friedel–Crafts catalysts. The molecular weight
obtained in this polymerisation is heavily dependent on the reaction
temperature, since at around ambient temperature extensive chain
transfer to monomer occurs:

$$\sim CH_2-\overset{CH_3}{\underset{CH_3}{C^+}} + CH_2=\overset{CH_3}{\underset{CH_3}{C}} \longrightarrow \sim CH=\overset{CH_3}{\underset{CH_3}{C}} + CH_3-\overset{CH_3}{\underset{CH_3}{C^+}}$$

Thus, if polymerisation is carried out in the presence of aluminium
chloride and hydrogen chloride at 280 to 320 K, materials with

molecular weights of up to about 3 000, which range from thin to viscous oils, are obtained. These find a variety of outlets, for example as lubricating oil additives.

At lower temperatures, the rate of this chain transfer reaction relative to the propagation reaction is reduced, and high molecular weight polymer may be obtained by carrying out the polymerisation at reduced temperatures. However, more important than high molecular weight homopolymers of isobutene are copolymers with 1·5 to 4·5% of isoprene, known as butyl rubber. Since these copolymers contain double bonds in the polymer chain they can be vulcanised by conventional methods (see Chapter 8); polyisobutene, being completely saturated, cannot. In the manufacture of butyl rubber, the polymerisation is carried out at about 180 K, in solution in methyl chloride, with aluminium chloride or boron trifluoride as catalyst. The polymer precipitates out as it is formed, and is isolated by running the reaction mix into hot water, when the catalyst is decomposed and the solvent and residual monomer are flashed off.

Butyl rubber, because of its low degree of unsaturation, has very good resistance to attack by chemicals, oxygen and ozone. It has very low permeability to gases, and this has led to it finding its major use in the manufacture of tyre inner tubes, and in lining tubeless tyres.

UK capacities for low molecular weight polymers of isobutene and butyl rubber respectively in 1970 were 50 000 tons per annum and 36 000 tons per annum.

POLYBUTADIENE

Polybutadiene elastomers, produced by polymerisation in the presence of sodium or potassium, were developed by I.G. Farbenindustrie in the late 1920s. These products had quite good mechanical properties, but had a number of major drawbacks, notably very poor processing properties. Despite these disadvantages, these elastomers achieved a limited commercial success, and sodium-polymerised butadiene was still being made on a limited scale in 1966. At about the same time that sodium-polymerised butadiene was being developed, it was shown by workers at I.G. Farbenindustrie that butadiene could be polymerised in emulsion systems. However, all attempts to obtain usable polymers failed, and finally attention was turned to butadiene copolymers, and elastomeric butadiene–styrene and butadiene–acrylonitrile copolymers were developed. Butadiene–styrene copolymers subsequently became the major general purpose synthetic elastomers.

Interest in polybutadiene itself revived in 1956 when the Phillips Petroleum Company announced the development of a polybutadiene with a high proportion of *cis*-1,4-units in the polymer chain, produced by polymerisation in the presence of a Ziegler–Natta catalyst system. This polymer had good technological properties, and was put into commercial production in 1960. Subsequently other catalyst systems have been found that give polymers of greater than 96% *cis*-1,4-content, and there are now a number of 'high cis' polybutadienes in production. Polymers with much lower *cis*-1,4-content, but with good properties are also now produced, by polymerisation in the presence of n-butyl-lithium, and by emulsion polymerisation under specially controlled conditions.

UK capacity for polybutadiene in 1970 was 80 000 tons per annum.

POLYCHLOROPRENE $$\left[\begin{array}{c} CH_2-C=CH-CH_2 \\ | \\ Cl \end{array}\right]_n$$

Polychloroprene, or neoprene, another elastomeric polymer, was introduced by du Pont in 1932. In addition to good mechanical properties, it has good resistance to solvents and chemicals, and to oxidation. Its relatively high price excludes it from general purpose applications, and it finds its outlets as one of the 'speciality rubbers', in uses where its resistance to solvents, chemicals and oxidative attack is of value.

Neoprene is made by emulsion polymerisation; the latex is coagulated by acidification and freezing in order to isolate the polymer. In contrast to the free-radical polymerisation of butadiene, which gives a polymer containing 1,2-, and *cis*- and *trans*-1,4-units, neoprene contains mainly *trans*-1,4-units.

UK capacity for neoprene in 1970 was 30 000 tons per annum.

STYRENE–BUTADIENE SYNTHETIC RUBBERS

Styrene–butadiene copolymers, generally known as SBR rubbers, were the first synthetic rubbers to find major use as general-purpose elastomers, and are still the largest-tonnage synthetic elastomers made. UK capacity for SBR rubbers in 1970 was 217 000 tons per annum.

SBR rubbers, which typically contain about 24% by weight of styrene, are made by emulsion polymerisation. The original products, the so-called 'hot' rubbers, are made by polymerisation at about 320 K, potassium persulphate being used as the initiator. In the late

1940s, 'redox' initiation systems, which generate radicals at low temperatures and allow the polymerisation to be carried out at 278 K, to give 'cold' rubbers, were developed. A number of redox systems are used; one involves a mixture of an organic hydroperoxide and a ferrous salt, and generates free radicals as follows:

$$\text{ROOH} + \text{Fe}^{2+} \longrightarrow \text{RO} \cdot + \text{OH}^- + \text{Fe}^{3+}$$

The cold rubbers have superior properties to the hot rubbers; however, since refrigeration is required for their manufacture, they are somewhat more expensive.

Conversion is limited to about 60% in the manufacture of SBR, since at higher conversions branching occurs, by involvement of polymer chain double bonds, in the reaction, to an extent which causes a deterioration in polymer properties. When the desired degree of conversion is attained, the reaction is arrested by adding a 'shortstop', i.e. a polymerisation inhibitor. Residual butadiene and styrene are removed by vaporisation and, except where the product is required as a latex, dilute sulphuric acid or aluminium sulphate is added to coagulate the rubber, which is then filtered and dried. Production is generally by continuous processing, in a number of stirred-flow reactors in series. Polymerisation is carried out under a few atmospheres pressure to keep the butadiene in the liquid phase.

Polymerisation of Carbonyl Compounds

Polymerisation through a carbonyl group can be brought about by both anionic and cationic mechanisms. One such polymerisation that has achieved commercial importance is that of formaldehyde, to give the so-called *acetal resins*, e.g. 'Delrin' (du Pont), and 'Celcon' (Celanese). These are polymers with very good mechanical properties, which can be used as replacements for metals in various engineering applications, and consequently are often referred to as engineering plastics.

Acetal resins may be produced by polymerisation of formaldehyde, or its cyclic trimer, trioxane. Formaldehyde can be polymerised by both anionic and cationic mechanisms, but trioxane will give high polymers only in the presence of cationic initiators. A large number of initiator systems have been suggested in patents. For example, formaldehyde may be polymerised by tertiary amines with water as co catalyst, and it is thought that in this case the initiation and propagation reactions are as follows:

$$R_3N \; + \; H_2O \; \rightleftharpoons \; (R_3NH)^+ \, OH^-$$

$$(R_3NH)^+OH^- + CH_2=O \longrightarrow HOCH_2O^- \, (R_3NH)^+$$

$$HOCH_2O^- \, (R_3NH)^+ + CH_2=O \longrightarrow HOCH_2OCH_2O^- \, (R_3NH)^+$$

$$\xrightarrow{\text{etc.}} HO\text{+}CH_2O\text{+}_n \, CH_2O^- \, (R_3NH)^+$$

There are no kinetic termination steps, but chain transfer to polar compounds readily occurs, e.g. chain transfer to the water co-catalyst:

$$HO\text{+}CH_2O\text{+}_n CH_2O^- \, (R_3NH)^+ \; + \; H_2O \longrightarrow HO\text{+}CH_2O\text{+}_n CH_2OH \; + \; (R_3NH)^+OH^-$$

The polyoxymethylenes thus obtained have poor thermal stability, in that depolymerisation by successive loss of formaldehyde molecules from the hydroxyl end groups can occur. This degradation, graphically described as an 'unzipping' reaction, is catalysed by acids and bases, and presumably involves mechanisms of the following type:

acid catalysed

$$\sim\sim\sim-O-CH_2-O-CH_2-OH \xrightarrow{H^+} \sim\sim\sim-O-CH_2-O-CH_2-OH_2^+ \xrightarrow{-H_2O}$$

$$\sim\sim\sim-O-CH_2-O-CH_2^+ \longrightarrow CH_2=O + \sim\sim\sim-O-CH_2^+ \xrightarrow{\text{etc.}}$$

base catalysed

$$\sim\sim\sim-O-CH_2-O-CH_2-OH \xrightarrow{OH^-} H_2O + \sim\sim\sim-O-CH_2-O-CH_2-O^-$$

$$\longrightarrow CH_2=O + \sim\sim\sim-O-CH_2-O^- \xrightarrow{\text{etc.}}$$

It occurs so readily that the polymer undergoes extensive degradation under normal fabrication conditions, which involve melting the polymer, and a major factor in the development of the acetal resins was finding methods of preventing the occurrence of unzipping. Two main methods have been used.

In the manufacture of Delrin, the polymer chains are 'end-capped' by acetylation with acetic anhydride:

$$\sim\sim\sim\sim\text{OCH}_2\text{OH} \xrightarrow{(\text{CH}_3\text{CO})_2\text{O}} \sim\sim\sim\sim\text{OCH}_2\overset{\displaystyle O}{\overset{\displaystyle \|}{\text{OCCH}}}_3$$

A different approach is used in the manufacture of Celcon. This is a copolymer of formaldehyde and ethylene oxide. After polymerisation it is subjected to depolymerising conditions, e.g. in the presence of basic catalysts, and the chains unzip until a —OCH₂CH₂O— unit is reached, when the degradation, of necessity, stops, e.g.,

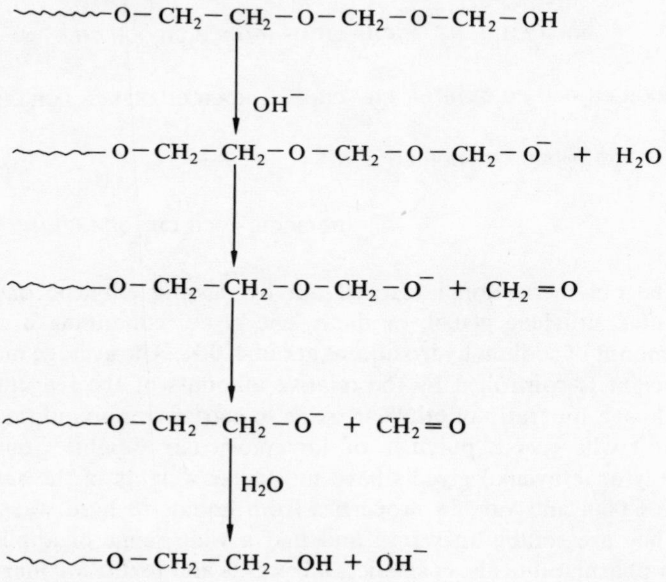

$$\sim\sim\sim\sim\sim\text{O}-\text{CH}_2-\text{CH}_2-\text{O}-\text{CH}_2-\text{O}-\text{CH}_2-\text{OH}$$

$$\downarrow \quad \text{OH}^-$$

$$\sim\sim\sim\sim\text{O}-\text{CH}_2-\text{CH}_2-\text{O}-\text{CH}_2-\text{O}-\text{CH}_2-\text{O}^- \quad + \quad \text{H}_2\text{O}$$

$$\downarrow$$

$$\sim\sim\sim\sim\text{O}-\text{CH}_2-\text{CH}_2-\text{O}-\text{CH}_2-\text{O}^- \quad + \quad \text{CH}_2=\text{O}$$

$$\downarrow$$

$$\sim\sim\sim\sim\text{O}-\text{CH}_2-\text{CH}_2-\text{O}^- \quad + \quad \text{CH}_2=\text{O}$$

$$\downarrow \quad \text{H}_2\text{O}$$

$$\sim\sim\sim\sim\text{O}-\text{CH}_2-\text{CH}_2-\text{OH} \quad + \quad \text{OH}^-$$

A stable polymer with hydroxyethyl ether end-groups is produced.

Acetal resins are relatively expensive, and are consequently used only in moderate tonnages. UK consumption in 1970 was about 4 700 tons. There is no manufacture in the UK at the time of writing, but the construction of a plant of capacity 20 000 tons per annum is under consideration.

Polymerisation of 1,2-Epoxides

1,2-Epoxides will polymerise by both cationic and anionic mechanisms. Such polymerisations are important in preparing derivatives of ethylene oxide and propylene oxide, and in the crosslinking of epoxy resins, discussed later in this chapter.

Poly(oxyethylene) glycols are an important group of derivatives of ethylene oxide. They are made by the base-catalysed polymerisation of ethylene oxide in the presence of water, ethylene glycol, or diethylene glycol, the reactions involved being of the type shown below:

$$HOCH_2CH_2OH + NaOH \rightleftharpoons HOCH_2CH_2O^- Na^+ + H_2O$$

$$HOCH_2CH_2O^- Na^+ + CH_2\underset{O}{\overset{}{-}}CH_2 \longrightarrow HOCH_2CH_2OCH_2CH_2O^- Na^+$$

$$HOCH_2CH_2OCH_2CH_2O^- Na^+ + CH_2\underset{O}{\overset{}{-}}CH_2 \longrightarrow HOCH_2CH_2OCH_2CH_2OCH_2CH_2O^- Na^+$$

$$HOCH_2CH_2\left[OCH_2CH_2\right]_n OCH_2CH_2O^- Na^+ + H_2O \rightleftharpoons$$

$$HOCH_2CH_2\left[OCH_2CH_2\right]_n OCH_2CH_2OH + NaOH$$

The polymerisation is carried out by passing ethylene oxide into water, ethylene glycol, or diethylene glycol containing a catalytic amount of sodium hydroxide, at about 400 K. The average molecular weight is controlled by the relative amounts of the reagents used. Thus, a low ratio of ethylene oxide to starter compound (i.e. water, etc.) will give a polymer of low molecular weight. Commercial poly(oxyethylene) glycols have molecular weights in the range 200 to 6 000, and vary in properties from liquids to hard waxy solids. They are soluble in water, and find a wide range of applications, in pharmaceuticals, cosmetics, lubricants and textile auxiliaries.

Advantage is taken of the hydrophilic nature of the polyoxyethylene chain in the preparation of non-ionic surface-active agents. A number of important products of this type are made by polymerising ethylene oxide onto a hydrophobic starter compound, under conditions similar to those described above. Alkylphenols are commonly used as the hydrophobic starter compound:

$$C_9H_{19}\text{—}\langle\text{—}\rangle\text{—OH} + nCH_2\underset{O}{\overset{}{-}}CH_2 \xrightarrow{NaOH} C_9H_{19}\text{—}\langle\text{—}\rangle\text{—}O\left[CH_2CH_2O\right]_n H$$

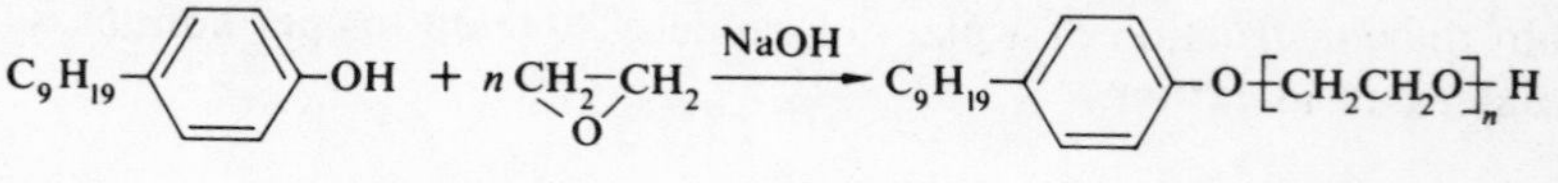

Poly(oxypropylene) glycols, triols, and higher polyols are important intermediates in the manufacture of polyurethanes (see later). They are made in a manner similar to that described for poly-

(oxyethylene) glycols, using appropriate starter compounds. Thus, triols may be obtained by using glycerol as starter:

$$
\begin{array}{l}
CH_2OH \\
| \\
CHOH \\
| \\
CH_2OH
\end{array}
\;+\;(n+n'+n'')\;
\underset{O}{CH_2\text{—}CHCH_3}
\;\longrightarrow\;
\begin{array}{l}
CH_2\!\!\left[OCH_2CH(CH_3)\right]_{\!n}\!OH \\
| \\
CH\!\!\left[OCH_2CH(CH_3)\right]_{\!n'}\!OH \\
| \\
CH_2\!\!\left[OCH_2CH(CH_3)\right]_{\!n''}\!OH
\end{array}
$$

There are two possible sites for nucleophilic attack on propylene oxide:

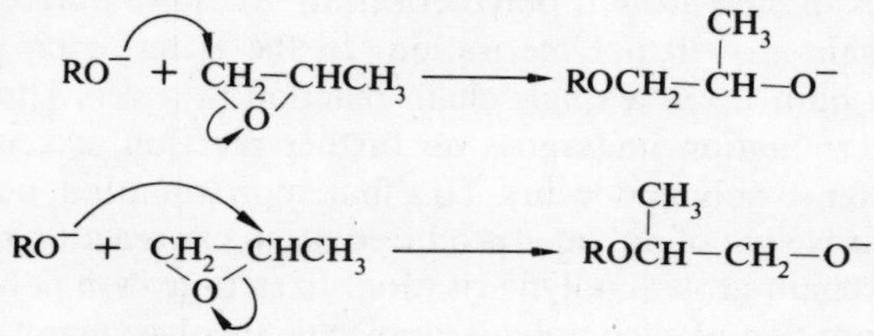

$$
RO^- + \underset{O}{CH_2\text{—}CHCH_3} \longrightarrow ROCH_2\text{—}\overset{CH_3}{\underset{}{CH}}\text{—}O^-
$$

$$
RO^- + \underset{O}{CH_2\text{—}CHCH_3} \longrightarrow RO\overset{CH_3}{\underset{}{CH}}\text{—}CH_2\text{—}O^-
$$

In fact, attack is almost entirely at the primary position, and the products are essentially head-to-tail polymers terminated by secondary alcohol groups. This has important implications in polyurethane technology because of the lower reactivity of secondary alcohols compared to primary alcohols.

A side reaction involving reaction of alkoxide or hydroxide ions with the monomer can occur, and leads to the formation of polymer chains with terminal carbon–carbon double bonds:

$$
RO^- + \underset{O}{CH_3CH\text{—}CH_2} \longrightarrow ROH + {}^-\underset{O}{CH_2CH\text{—}CH_2} \longrightarrow CH_2{=}CHCH_2O^-
$$

$$
CH_2{=}CHCH_2O^- + \underset{O}{CH_2\text{—}CHCH_3} \longrightarrow CH_2{=}CHCH_2OCH_2\overset{CH_3}{\underset{}{CHO}}{}^- \xrightarrow{\text{etc.}}
$$

This reaction is undesirable in the preparation of polyols for polyurethane manufacture, in that it leads to the production of molecules of lower functionality than the hydroxy compound used as starter for the reaction.

STEP-GROWTH POLYMERISATION

In step-growth polymerisation, polymer molecules are built up through many separate reactions of functional groups, and the chemistry of step-growth polymerisation is thus to a substantial extent the chemistry of the functional groups involved. We shall consider the general characteristics of step-growth polymerisation, and then discuss some of the more important classes of polymers which are made in this way.

General Characteristics of Step-Growth Polymerisation

The kinetics of step-growth polymerisation are quite different from those of chain-growth polymerisation. In the latter, each polymer molecule is built up in a single chain reaction in a very short time, and after termination undergoes no further reaction, except where chain transfer to polymer occurs. To a first approximation, no change in molecular weight of polymer produced occurs as reaction proceeds in a typical chain-growth polymerisation. In step-growth polymerisation the formation of each polymer molecule involves many separate reactions, and in the later stages of the polymerisation much of the reaction is between polymer molecules of intermediate size. There is thus a progressive increase in molecular weight as the polymerisation proceeds.

LINEAR STEP-GROWTH POLYMERISATION

Let us consider the situation where one starts with an exactly equimolar mixture of bifunctional monomers, using as an example the polyesterification of ethylene glycol by adipic acid. The first reaction product in this system will be the hydroxy acid formed by the reaction of adipic acid and ethylene glycol:

$$HO_2C(CH_2)_4CO_2H + HOCH_2CH_2OH \longrightarrow HO_2C(CH_2)_4CO_2CH_2CH_2OH + H_2O$$

This product can then react further, (i) with a molecule of glycol, to give a diol:

$$HO_2C(CH_2)_4CO_2CH_2CH_2OH + HOCH_2CH_2OH$$

$$\longrightarrow HOCH_2CH_2O_2C(CH_2)_4CO_2CH_2CH_2OH + H_2O$$

(ii) with a molecule of adipic acid, to give a diacid:

$$HO_2C(CH_2)_4CO_2CH_2CH_2OH + HO_2C(CH_2)_4CO_2H$$

$$\longrightarrow HO_2C(CH_2)_4CO_2CH_2CH_2O_2C(CH_2)_4CO_2H + H_2O$$

or (iii), with a further molecule of hydroxy acid, to give a new hydroxy acid:

$$2HO_2C(CH_2)_4CO_2CH_2CH_2OH$$

$$\longrightarrow HO_2C(CH_2)_4CO_2CH_2CH_2O_2C(CH_2)_4CO_2CH_2CH_2OH + H_2O$$

The products of these reactions can then undergo further, analogous reactions, and so on.

If it is assumed that no side reactions occur in the system, there will at all times be an exactly equivalent number of hydroxyl and carboxyl groups, and if the reaction could be carried to completion the final product would be a single molecule incorporating all the monomer units, with a carboxyl group at one end, and a hydroxyl group at the other. It will not surprise the reader to learn that this is never achieved in practice.

It is evident from the above discussion that the molecular weight of the polymer obtained in a step-growth polymerisation depends on the extent to which reaction between functional groups has occurred, i.e., on the conversion to which the reaction is taken. It can readily be shown* that in a polymerisation of the type under discussion, in order to achieve a degree of polymerisation of 50, i.e. a polymer containing, on average, 50 monomer units per molecule, 98% of the carboxyl and hydroxyl groups must have reacted. In the case of poly(ethylene adipate) such a polymer has a molecular weight of 4 300, which is substantially below the level at which polymers usually develop their useful mechanical properties. To attain degrees of polymerisation of 100 and 200 respectively, the reaction must proceed to 99 and 99·5% completion.

Thus, one necessary condition for the formation of high molecular weight linear polymers by step-growth polymerisation is that the reaction involved must be one in which a very high degree of reaction of functional groups can be achieved. This requires that the reaction should be fairly fast, so that the polymerisation may be completed in a reasonable time. Also, where the reaction is reversible, which is the case with the majority of commercially important step-growth polymerisations, it is essential that the other product of reaction be removed during the reaction in order that polymerisation may continue.

* See, for example, the books by Lenz, Margerison and East, or Odian, detailed in the 'Further Reading' list.

If instead of starting with an exact balance of functional groups one uses an excess of one of the monomers, then at some stage in the reaction one type of group will be completely consumed, and reaction will cease. If, for example, in the case under consideration an excess of glycol is used, at some stage all the product molecules will have a hydroxyl group at each end, and further esterification will be impossible. If the excess of glycol is large, no high polymer will be formed, and the major product will be bis(2-hydroxyethyl) adipate:

$$HOCH_2CH_2OH \text{ (excess)} \quad + \quad HO_2C (CH_2)_4 CO_2 H$$

$$\longrightarrow \quad HOCH_2CH_2O_2C (CH_2)_4 CO_2CH_2CH_2OH \quad + \quad 2H_2O$$

The extent of imbalance between the functional groups in the original monomer mix thus determines the maximum molecular weight attainable. If, for example, in the reaction of two bifunctional monomers, a 1% excess of one monomer is used, then the maximum possible degree of polymerisation attainable is 199. Thus, in linear step-growth polymerisation accurate proportioning of the monomers is essential if high molecular weight polymers are to be made, except in certain cases, where the nature of the reaction is such that other steps can be taken to ensure completion of reaction [see 'Poly(ethylene terephthalate)'].

Monomers used in step-growth polymerisations are generally required to be of a very high degree of purity, to ensure accurate proportioning of monomers, and because monofunctional impurities can limit the molecular weight by 'capping' polymer chains; e.g.:

$$\sim\sim\sim CH_2CH_2OH + RCO_2H \longrightarrow \sim\sim\sim CH_2CH_2O_2CR + H_2O$$

Even in a situation where the functional groups are originally in balance, the occurrence of side reactions can lead to an imbalance, and limit the molecular weight achievable. In a polyesterification, for instance, hydroxyl groups can be removed, e.g., by dehydration and by ether formation:

$$\sim\sim\sim CH_2CH_2OH \longrightarrow \sim\sim\sim CH=CH_2 \quad + \quad H_2O$$

$$
\begin{array}{ll}
\sim\sim\sim CH_2 CH_2OH & \sim\sim\sim CH_2CH_2 \\
+ \longrightarrow & \qquad\qquad\quad\; O \quad + \; H_2O \\
\sim\sim\sim CH_2 CH_2OH & \sim\sim\sim CH_2CH_2
\end{array}
$$

The requirements for the attainment of high molecular weight linear polymer by step-growth polymerisation are, therefore, quite demanding; there is only a limited number of reactions which have been used commercially for the production of such polymers.

NON-LINEAR STEP-GROWTH POLYMERISATION

In the case that one or both of the monomers has more than two functional groups, the polymer formed will not be linear, but will consist of a three-dimensional network. Thus, if we consider the reaction of a triol (I) with a dibasic acid (II), after a small number of reaction steps intermediates of the type shown below will be present in the reaction mix:

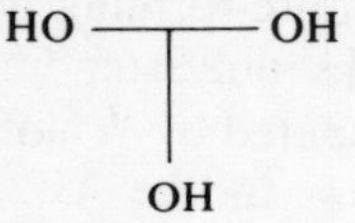

$$HO \underset{\underset{OH}{|}}{\overset{}{\rule{2cm}{0.4pt}}} OH \quad (I) \qquad\qquad HO_2C \rule{2.5cm}{1pt} CO_2H \quad (II)$$

$$HO \underset{\underset{OH}{|}}{\overset{}{\rule{1.5cm}{0.4pt}}} O_2C \rule{2cm}{1pt} CO_2 \underset{\underset{O_2C}{|}}{\overset{}{\rule{1.5cm}{0.4pt}}} O_2C \rule{2cm}{1pt} CO_2H$$

$$CO_2 \underset{\underset{OH}{|}}{\overset{OH}{\rule{1.5cm}{0.4pt}}}$$

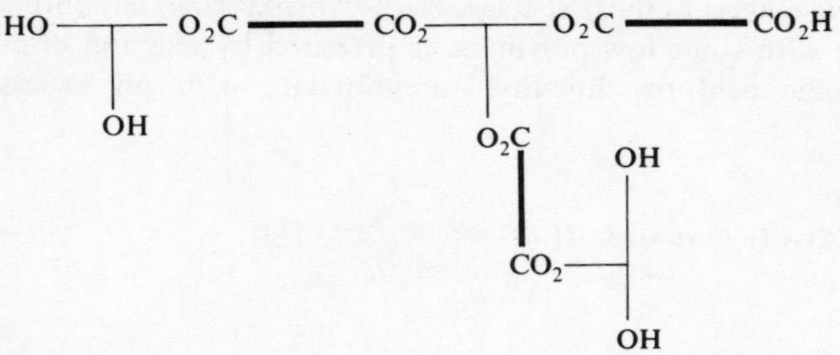

A very rapid build-up of molecular weight and functionability of the reacting entities occurs. If such a reaction is carried to completion, all the monomer is effectively incorporated into a single molecule of extremely high molecular weight, and the product is an insoluble, infusible mass. Although such a product would clearly be useless if formed as such in a reactor, since it would not be possible to remove it from the reactor other than with, e.g., a road drill, the formation of such three-dimensional networks is of importance in the application of thermosetting polymers (see page 283). Up to a certain stage in the reaction (the 'gel point'), the product remains liquid, and the manufacture of some thermosetting polymers, notably phenol-formaldehyde resins and amino resins, involves taking such a reaction to only partial completion and then completing the polymerisation during the fabrication of the finished article.

Polyesters

A number of synthetic methods can be used for making polyesters. Which is, in fact, used in a particular case depends on specific circumstances, and methods will therefore not be discussed in general terms. It is, however, worth pointing out that the manufacture of polyesters by the reaction of acid chlorides with glycols is not used except in special circumstances, since acid chlorides are normally relatively expensive sources of acyl groups.

POLY(ETHYLENE TEREPHTHALATE)

Poly(ethylene terephthalate) is one of the major fibre-forming polymers, being the polymer from which most of the polyester fibres, e.g. 'Terylene', 'Dacron', are made. It was first prepared by Whinfield and Dickson of the Calico Printers Association in the UK, in 1941.

Manufacture of poly(ethylene terephthalate) is normally carried out in two stages. In the first stage, bis(2-hydroxyethyl) terephthalate, together with some low polymers, is prepared by reaction of either terephthalic acid or dimethyl terephthalate with an excess of glycol:

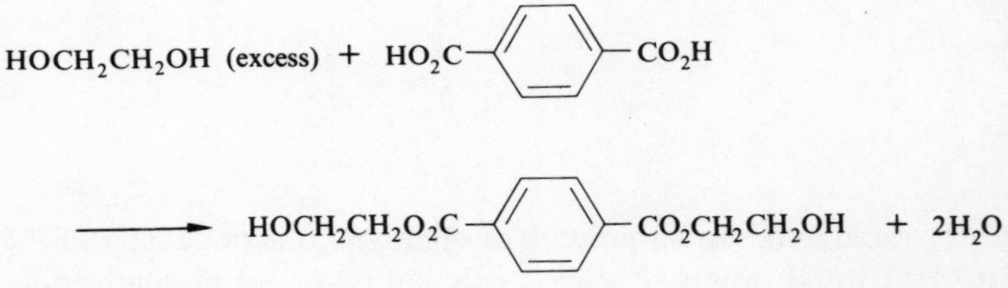

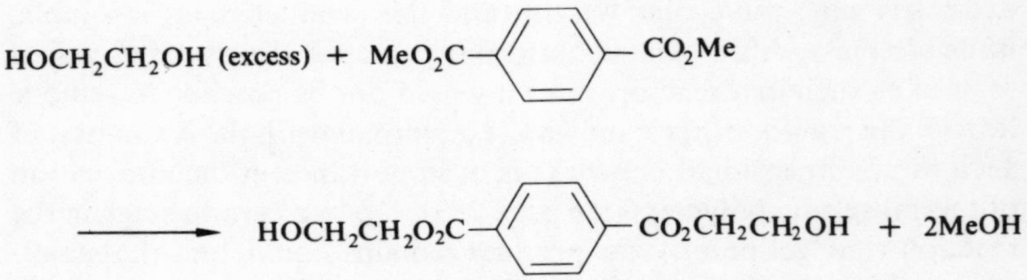

$HOCH_2CH_2OH$ (excess) $+$ HO_2C—⟨benzene ring⟩—CO_2H

$\longrightarrow$ $HOCH_2CH_2O_2C$—⟨benzene ring⟩—$CO_2CH_2CH_2OH$ $+$ $2H_2O$

$HOCH_2CH_2OH$ (excess) $+$ MeO_2C—⟨benzene ring⟩—CO_2Me

$\longrightarrow$ $HOCH_2CH_2O_2C$—⟨benzene ring⟩—$CO_2CH_2CH_2OH$ $+$ $2MeOH$

Use of dimethyl terephthalate as raw material for this stage arose because this material is easier to purify (by distillation) than terephthalic acid, which is involatile, and has a very low solubility in most solvents. The dimethyl terephthalate-based process is currently the most important method of manufacture. The reaction is carried out at 420 to 460 K in the presence of a mixed catalyst, designed to promote both the initial ester interchange reaction, and the subsequent polymerisation. For example, a mixture of zinc acetate and antimony trioxide may be used.

Some recently developed processes for terephthalic acid give a product which requires very little purification, and there is a growing tendency to adopt the acid-based route to poly(ethylene terephthalate).

The second stage of the process is carried out at about 530 K, and involves a further ester interchange, in which ethylene glycol is eliminated:

$$n\text{HOCH}_2\text{CH}_2\text{O}_2\text{C}\longrightarrow\text{CO}_2\,\text{CH}_2\text{CH}_2\text{OH}$$

$$\longrightarrow \left[\text{OC}\longrightarrow\text{CO}_2\text{CH}_2\text{CH}_2\text{O}\right]_n + n\text{HOCH}_2\text{CH}_2\text{OH}$$

The glycol is continuously removed either by carrying out the reaction under vacuum, or by passing a stream of inert gas through the reaction mixture. The molecular weight is thus controlled by the extent to which the reaction is allowed to proceed, and not, as in most step-growth polymerisations, by the accuracy of the proportioning of the original charge of monomers. In fact, some acetic acid is incorporated in the reaction mixture to limit the molecular weight attained by end-capping polymer chains, and thus prevent the formation of products with excessively high melt viscosities, which would be difficult to spin.

Although much manufacture of poly(ethylene terephthalate) has been by batch processing, a number of continuous polymerisation processes have recently been developed.

Production of poly(ethylene terephthalate) fibre in the UK in 1969 was about 67 500 tons, and about another 6 500 tons of the polymer were used for making polyester film.

POLYCARBONATES

The only polymer of this class of major commercial importance is the polycarbonate of bisphenol A. This polymer, another of the engineering plastics, has very good mechanical properties but, because of its relatively high price, is used in relatively small quantities. It is manufactured in the USA and Germany, but not in the UK.

Polycarbonates are made by condensing bisphenol A with phosgene, the reaction typically being carried out in solution in methylene chloride, with pyridine present as a hydrogen chloride acceptor:

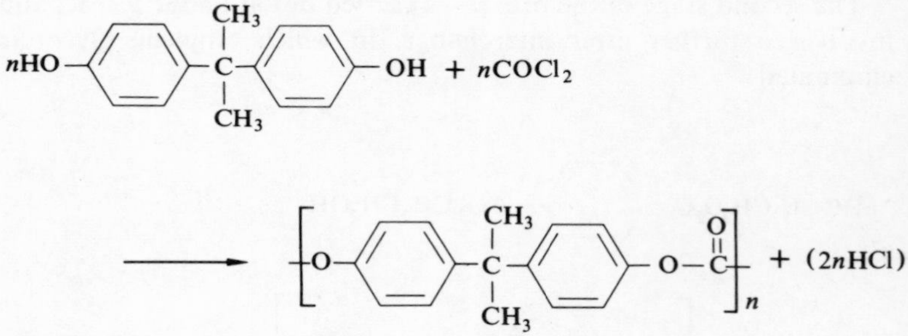

$$n\text{HO} - \text{C}_6\text{H}_4 - \overset{\displaystyle CH_3}{\underset{\displaystyle CH_3}{C}} - \text{C}_6\text{H}_4 - \text{OH} + n\text{COCl}_2$$

$$\longrightarrow \left[\text{O} - \text{C}_6\text{H}_4 - \overset{\displaystyle CH_3}{\underset{\displaystyle CH_3}{C}} - \text{C}_6\text{H}_4 - \text{O} - \overset{\displaystyle O}{\overset{\displaystyle \|}{C}} \right]_n + (2n\text{HCl})$$

In this case, preparation by esterification with the acid is, of course, impossible.

The polymer can also be made from bisphenol A and diphenyl carbonate by ester interchange.

OTHER LINEAR POLYESTERS

The polyterephthalate of 1,4-bis(hydroxymethyl)cyclohexane is also used as a synthetic fibre ('Kodel'), but is very much less important than poly(ethylene terephthalate):

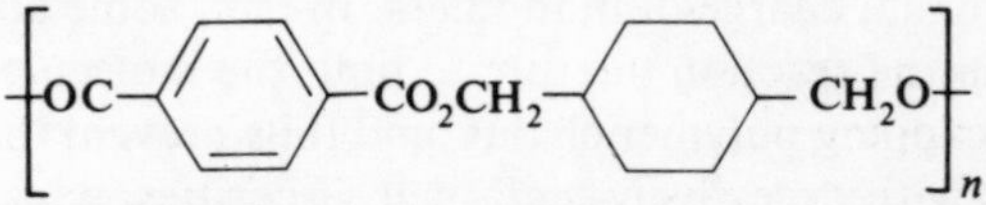

$$\left[\text{OC} - \text{C}_6\text{H}_4 - \text{CO}_2\text{CH}_2 - \text{C}_6\text{H}_{10} - \text{CH}_2\text{O} \right]_n \qquad \text{'Kodel'}$$

Hydroxyl-ended polyesters of relatively low molecular weight (*ca.* 2 000), are of importance as raw materials for polyurethane manufacture. The acids most commonly used in these products are adipic acid and phthalic acid (as the anhydride), and the glycols most widely used are ethylene-, propylene-, and diethylene-glycols.

Branched products can be produced by the use of some triol, e.g. glycerol, in conjunction with the glycol. The molecular weight is controlled by the excess of hydroxyl compound used. Since the molecular weights required are low, procedures of the type used in the manufacture of poly(ethylene terephthalate) are not necessary.

Low molecular weight unsaturated polyesters, prepared by the reaction of glycols (generally propylene glycol) with maleic and phthalic anhydrides, are of considerable importance as thermosetting resins (see page 283). UK production in 1969 was 37 000 tons.

ALKYD RESINS

Alkyd resins are branched polyesters prepared from phthalic anhydride, sometimes together with maleic anhydride, a polyhydroxy compound, typically glycerol, and fatty acids. They are of very major importance in the surface coatings industry.

Alkyd resin technology is complex and specialised, and will not be discussed here, but it should be appreciated that the manufacture of alkyd resins provides important outlets for phthalic and maleic anhydrides, and for glycerol.

Polyamides

The linear aliphatic polyamides form an extremely important group of fibre-forming polymers. Polymers of this type are called 'nylons', and a system of numbering indicates from which monomers a particular nylon is made. Thus, nylon 66 is made by the reaction of two monomers each containing six carbon atoms, i.e. hexamethylene-diamine and adipic acid, and nylon 6 is made from one monomer containing six carbon atoms, i.e. ω-aminocaproic acid or caprolactam. Nylon 6 and nylon 66 are by far the most important products, though a number of others are made.

As is the case with polyesters, the method used for manufacture of a polyamide is determined by the circumstances in that particular case.

NYLON 66

Nylon 66, poly(hexamethylene adipamide), was first prepared by Carothers, of du Pont, in the USA, in about 1930. The discovery that this polymer could be converted into fibres of outstanding properties was an event of very major importance to polymer chemistry and to industrial chemistry in general, and marked the start of the synthetic fibre industry. Nylon 66 is the most important synthetic fibre-forming polymer in the USA and the UK; in most European countries other than the UK, this position is held by nylon 6.

Production of nylon 66 fibre in the UK in 1968 was in the region of 100 000 tons.

In the manufacture of nylon 66, balance of functional groups is ensured by preparing and purifying the salt of hexamethylene diamine and adipic acid, 'nylon salt', in a separate stage, and then using this as feed for the polymerisation stage. Nylon salt is usually prepared by mixing solutions of the acid and diamine in methanol:

$$HO_2C\,(CH_2)_4\,CO_2H \; + \; H_2N(CH_2)_6\,NH_2 \longrightarrow \; {}^-O_2C\,(CH_2)_4\,CO_2^-\,{}^+H_3N\,(CH_2)_6\,NH_3^+$$

nylon salt

The salt precipitates out, and is purified by recrystallisation.

In the polymerisation stage, nylon salt is slurried in about 20–40% of its weight of water in an autoclave, and polymerisation is carried out at a gradually increasing temperature, initially under a pressure of about 18 atm. As reaction proceeds, steam is bled off from the autoclave, first at such a rate as to maintain the pressure, and finally to reduce the pressure to atmospheric. The reaction is completed at 540 K and at atmospheric pressure, or under a partial vacuum:

$$n\left[{}^-O_2C(CH_2)_4\,CO_2^-\; {}^+H_3N(CH_2)_6\,NH_3^+\right]$$

$$\longrightarrow \; \left[OC(CH_2)_4\,CONH(CH_2)_6\,NH\right]_n \; + \; 2nH_2O$$

The molecular weight is controlled within the range 12 000–20 000 by the addition of 0·6–1·2% acetic acid. Polymers with molecular weights in this range give good fibres, but at the same time have low enough melt viscosities to be readily spun.

Much manufacture of nylon 66 is by batch processing; however, a number of continuous polymerisation processes are now available.

NYLON 6

Nylon 6 is made from caprolactam rather than from the alternative monomer, ω-aminocaproic acid, because the lactam is more readily made than the amino acid. The polymerisation is not a simple step-growth polymerisation of the type that we have been discussing in this section, but involves both chain and stepwise reactions. However, because of the close structural and technological relationship between nylon 6 and nylon 66, it is conveniently considered here.

Polymerisation is carried out in the presence of catalytic amounts of water. This causes initiation by hydrolysing some caprolactam to ω-aminocaproic acid:

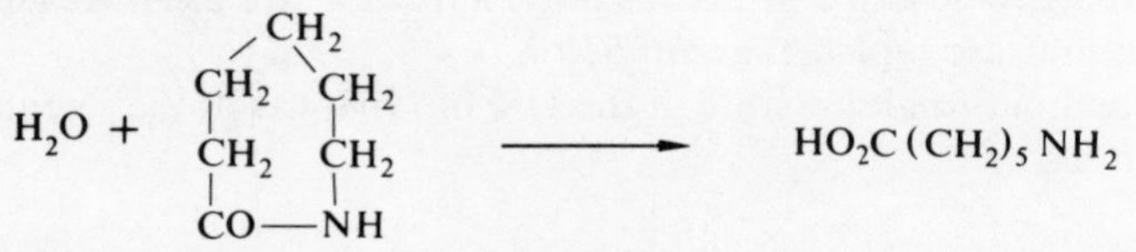

$$H_2O + \begin{array}{c} CH_2 \\ CH_2 \quad CH_2 \\ CH_2 \quad CH_2 \\ CO-NH \end{array} \longrightarrow HO_2C(CH_2)_5NH_2$$

A series of propagation reactions then occurs, in which the lactam ring is opened by attack by the amino group, in the first place of the amino acid, and subsequently, of the polymer:

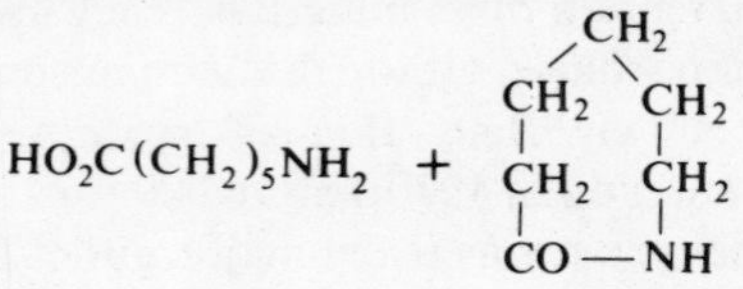

$$HO_2C(CH_2)_5NH_2 + \begin{array}{c} CH_2 \\ CH_2 \quad CH_2 \\ CH_2 \quad CH_2 \\ CO-NH \end{array}$$

$$\longrightarrow HO_2C(CH_2)_5NHCO(CH_2)_5NH_2$$

$$\xrightarrow{\text{etc.}} HO_2C(CH_2)_5NH\left[CO(CH_2)_5NH\right]_n CO(CH_2)_5NH_2$$

The molecules built up by this reaction can take part in stepwise formation of amide linkages by reaction between amino groups and carboxyl groups:

$$HO_2C(CH_2)_5NH\left[CO(CH_2)_5NH\right]_m CO(CH_2)_5NH_2 + HO_2C(CH_2)_5NH\left[CO(CH_2)_5NH\right]_n CO(CH_2)_5NH_2$$

$$\downarrow$$

$$HO_2C(CH_2)_5NH\left[CO(CH_2)_5NH\right]_{m+n+2} CO(CH_2)_5NH_2 + H_2O$$

It is thought that in the initial stage of the polymerisation, the ring opening chain reaction discussed above gives rise to polymer chains in the molecular weight range 8 000 to 14 000, and that these then undergo stepwise condensation reactions to give higher molecular weight polymer.

It is impossible to achieve complete conversion in the manufacture of nylon 6, since the polymer is in equilibrium with about 10% of monomer and low molecular weight polymer. As there is no other product of reaction, there is no practicable way of displacing the equilibrium. Unconverted monomer and low molecular weight polymers are removed from the polymer before spinning, e.g. by extraction with water.

Both batchwise and continuous polymerisation are used. Reaction temperatures are typically about 520 K.

Production of nylon 6 fibre in the UK in 1968 was in the region of 20 000 tons.

Phenol-Formaldehyde Resins

The phenol-formaldehyde resins are the oldest synthetic plastics, their commercial history having started in 1906 when Baekeland took out his first patent on these materials. They are still of considerable commercial importance, although their consumption is expanding much more slowly than that of more recently developed polymers. UK production in 1970 was 76 000 tons. The manufacture of phenol-formaldehyde resins is the major outlet for phenol.

PHENOL-FORMALDEHYDE REACTIONS

Phenol will react with aqueous formaldehyde in the presence of both acidic and basic catalysts.

Base Catalysis

Reaction in the presence of a base is considered to involve nucleophilic attack on formaldehyde by phenoxide ion:

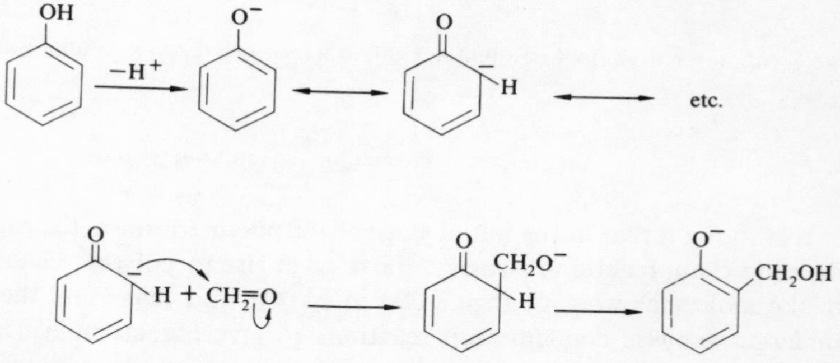

There are evidently three positions on the ring which can carry out such an attack, i.e., phenol is a trifunctional monomer, and the initial products of reaction of phenol and formaldehyde under base catalysis are mono-, di-, and trimethylolphenols.

Further reaction of these initial products occurs, resulting in the formation of methylene bridges between rings, e.g.:

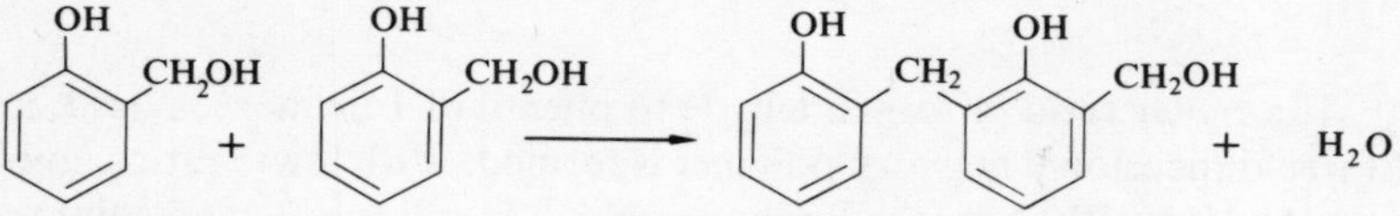

It can be seen that if this type of reaction occurs extensively a three-dimensional network polymer will be produced. However, the methylene bridge-forming reactions are substantially slower than the primary reactions with formaldehyde, and under appropriate conditions it is possible to obtain a product consisting of methylol-phenols and low molecular weight further reaction products. This type of product is called a 'resol'.

Acid Catalysis

The reactions involved in the acid-catalysed reaction of phenol with formaldehyde are of the type that has already been discussed in Chapter 5. Thus, electrophilic attack of activated rings by protonated formaldehyde occurs:

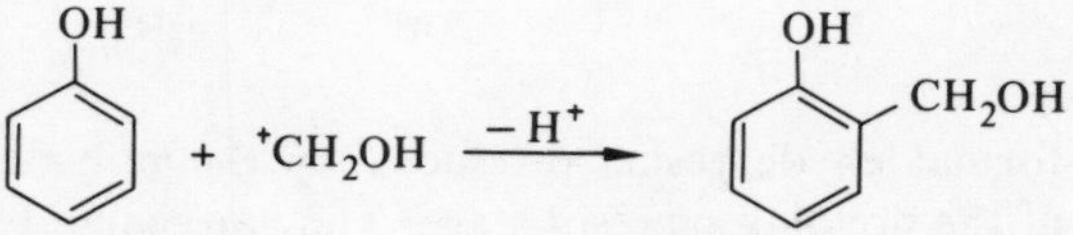

In this case, methylol derivatives are not obtained as products since they undergo rapid further reaction resulting in the formation of methylene bridges, e.g.:

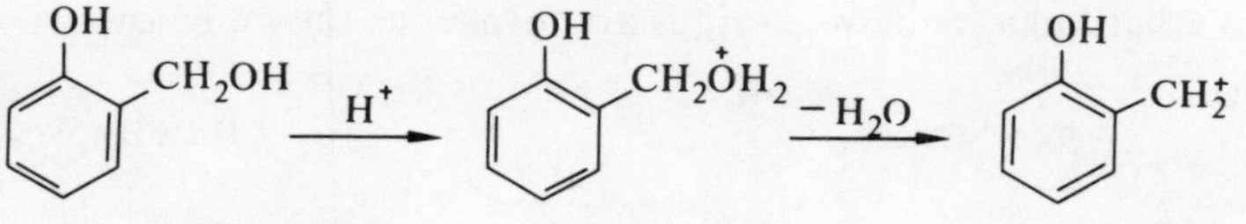

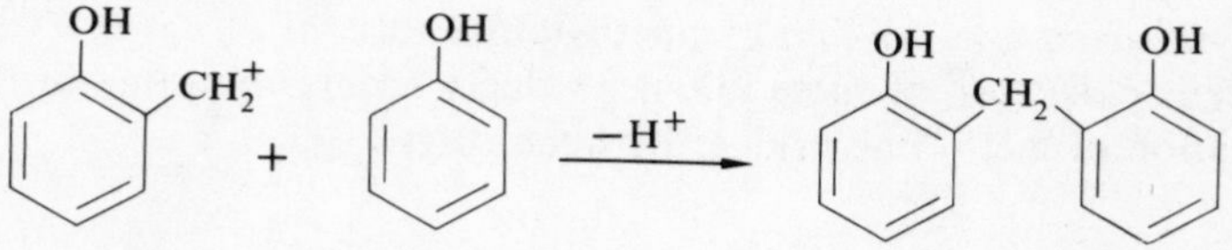

If a molar ratio of formaldehyde to phenol of 1 or more is used, a three-dimensional network polymer is formed. With lower ratios, low molecular weight branched polymer which is still fusible and soluble is obtained. This type of product is called a 'novolak'; it has a structure of the type represented below:

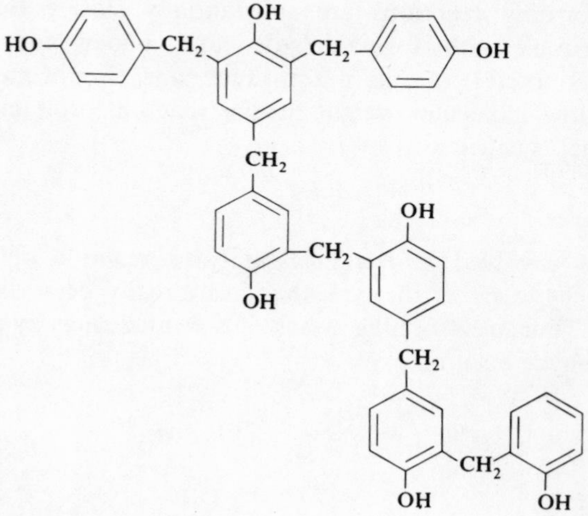

CROSSLINKING

The phenol-formaldehyde resins produced by the polymer manufacturer are at the novolak or resol stage. They are caused to cross-link to a network polymer during moulding or other fabricating or finishing processes, and are thus thermosetting resins (see Chapter 8).

Crosslinking of resol resins is brought about by the use of acid catalysts. Methylene bridges are formed from methylol groups by reactions analogous to those which occur during novolak formation, and ether linkages between rings are formed as shown below:

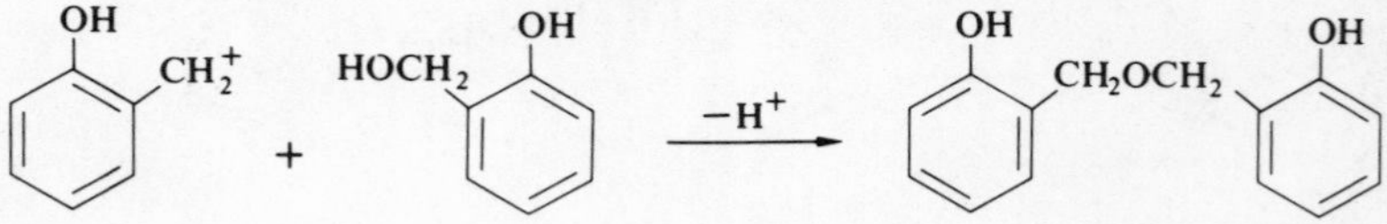

Since the methylolphenols and low molecular weight polymers are polyfunctional, these reactions rapidly lead to the formation of a crosslinked network.

Novolak resins require the addition of formaldehyde or a formaldehyde source before crosslinking can occur. Hexamethylenetetramine is widely used for this purpose:

$$(CH_2)_6N_4 \ + \ 6H_2O \ \rightleftharpoons \ 6CH_2{=}O \ + \ 4NH_3$$

hexamethylene-
tetramine

When a novolak is heated with hexamethylenetetramine, methylene bridges are formed, presumably by reactions similar to those involved in resol formation.

MANUFACTURE OF PHENOL-FORMALDEHYDE RESINS

In the manufacture of novolak resins, phenol and rather less than a one molar ratio of formaldehyde (as 37% aqueous solution) are allowed to react at reflux temperature in the presence of an acid catalyst, e.g. sulphuric acid or hydrochloric acid, until the required degree of reaction is attained. The catalyst is neutralised, and water is removed from the product, first by distillation at atmospheric pressure, and finally under vacuum at up to 430 K. If the product is to be used in solution, e.g. as an adhesive or surface coating, solvent is then added. If it is to be used as solid, it is discharged from the reaction vessel, and either flaked, or cooled and broken up into lumps, and ground to a powder. When the material is to be used in moulding applications, it is mixed with a filler, normally wood flour, i.e. finely ground wood.

Manufacture of resol resins is broadly similar to the above procedure, except that in this case control is more critical, in that the formaldehyde usage (normally between 1·5 and 3 moles of formaldehyde per mole of phenol) is such that a crosslinked polymer can be produced if the reaction is allowed to proceed too far. Catalysts typically used are sodium-, calcium-, or barium hydroxides.

Substituted or polyhydric phenols can be used to form resins, either alone, or in conjunction with phenol itself. In order that crosslinkable products be obtained, it is necessary that at least part of the phenolic monomer has a functionality greater than two. p-Cresol, for instance, cannot form a crosslinked product, since only two positions on the ring are open to attack.

Aminoplastics

This term covers a number of products of which by far the most important are urea-formaldehyde and melamine-formaldehyde resins. The technology of these materials is similar to that of phenol-formaldehyde resins, but they have the advantage over phenol-formaldehyde resins of being colourless and odourless, and of not discolouring on exposure to light. Major uses are as moulding materials, e.g. in the manufacture of electrical fittings and kitchen ware; in the manufacture of laminates, e.g. 'Formica' and related materials; as adhesives, e.g. in plywood and chipboard manufacture; in crease-resistant finishes for textiles; and in papermaking where they are used in the production of high-wet-strength papers. Total UK production of aminoplastics in 1970 was about 140 000 tons. Urea-formaldehyde resins were first produced commercially in the early 1930s, and melamine-formaldehyde resins came into use towards the end of that decade.

The chemistry of the formation of these products is complex and is not very well understood. The reactions involved are broadly analogous to those involved in phenol-formaldehyde resin manufacture, and consist of reaction between the bifunctional formaldehyde and the polyfunctional urea or melamine to give relatively low molecular weight branched, but not crosslinked, products, which can then be caused to react further to give crosslinking during the moulding, or other forming operation. The main polymerisation reaction is probably formation of a methylol derivative of an amino group, followed by condensation with another amino group to give a methylene bridge, e.g.:

$$NH_2\overset{O}{\overset{\|}{C}}NH_2 \ + \ CH_2{=}O \longrightarrow NH_2\overset{O}{\overset{\|}{C}}NHCH_2OH$$

$$NH_2\overset{O}{\overset{\|}{C}}NHCH_2OH \ + \ NH_2\overset{O}{\overset{\|}{C}}NH_2 \longrightarrow NH_2\overset{O}{\overset{\|}{C}}NHCH_2 NH\overset{O}{\overset{\|}{C}}NH_2 \ + \ H_2O$$

The resins are made by the reaction of urea or melamine with aqueous formaldehyde at temperatures from 320 to 370 K, the exact conditions used depending on the grade being made. For a urea-formaldehyde resin, typically a molar ratio of formaldehyde to urea of 1·6:1 is used, and the reaction is carried out at pH 7·5 and 320 K. In the preparation of melamine-formaldehyde resins, 2 to 3 moles of formaldehyde per mole of melamine are used, and the reaction is carried out at pH 9 and rather higher temperatures, e.g. 360 K. Precise control of the pH is important in both cases since at low pH crosslinking may occur to give a useless product.

When the product reaches the desired stage of polymerisation, the reaction mixture is cooled, and the further treatment depends on the application for which the product is intended. If it is to be used as a moulding compound, a fibrous filler, normally cellulose or wood flour, is impregnated with the aqueous solution of the resin, and the material is then dried. Unfilled resin, for use, e.g., as an adhesive, may be spray-dried to give a powder, or may be sold as an aqueous syrup.

MANUFACTURE OF UREA AND MELAMINE

Urea is made by the reaction of ammonia and carbon dioxide at about 460 K and 100 to 200 atm:

$$2NH_3 + CO_2 \longrightarrow NH_2CONH_2 + H_2O$$

Its major use is in fertiliser applications.

Until the early 1960s, all melamine was made from dicyandiamide, itself made from calcium cyanamide derived from calcium carbide:

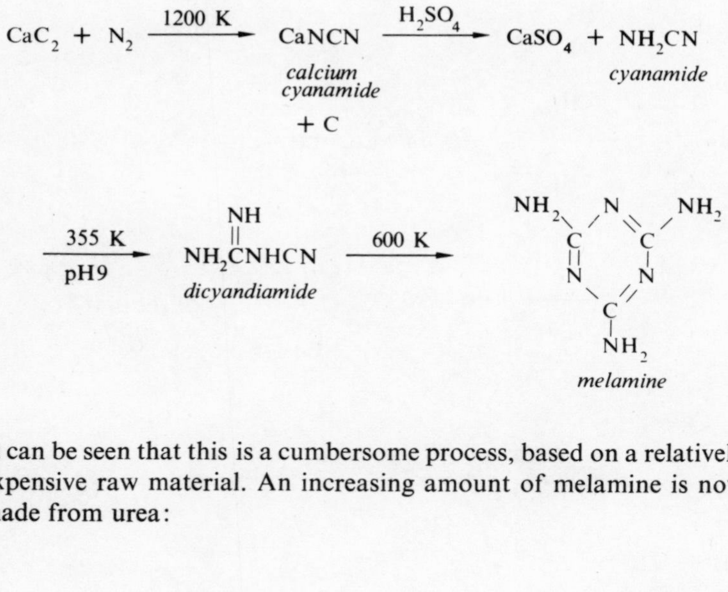

$$CaC_2 + N_2 \xrightarrow{\ 1200\ K\ } \underset{\substack{\text{calcium}\\ \text{cyanamide}}}{CaNCN} \xrightarrow{\ H_2SO_4\ } CaSO_4 + \underset{\text{cyanamide}}{NH_2CN}$$

$$+\ C$$

It can be seen that this is a cumbersome process, based on a relatively expensive raw material. An increasing amount of melamine is now made from urea:

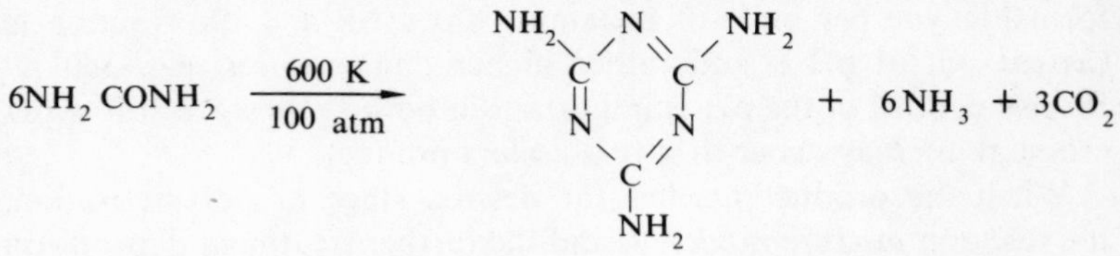

$$6NH_2 CONH_2 \xrightarrow[\text{100 atm}]{\text{600 K}}$$

$+ \quad 6NH_3 + 3CO_2$

The carbon dioxide and ammonia produced can be re-converted into urea.

The only large-scale use of melamine is in the manufacture of polymers.

Epoxy Resins

Epoxy resins form an important class of thermosetting polymers. By far the most important are those made by the reaction of epichloro-hydrin with bisphenol A, and these are the only ones we will consider.

Reaction between epichlorohydrin and bisphenol A is base catalysed. It involves as a first step addition of phenoxide ion to the epoxide ring of the epichlorohydrin, followed by elimination of hydrogen chloride to regenerate an epoxide ring:

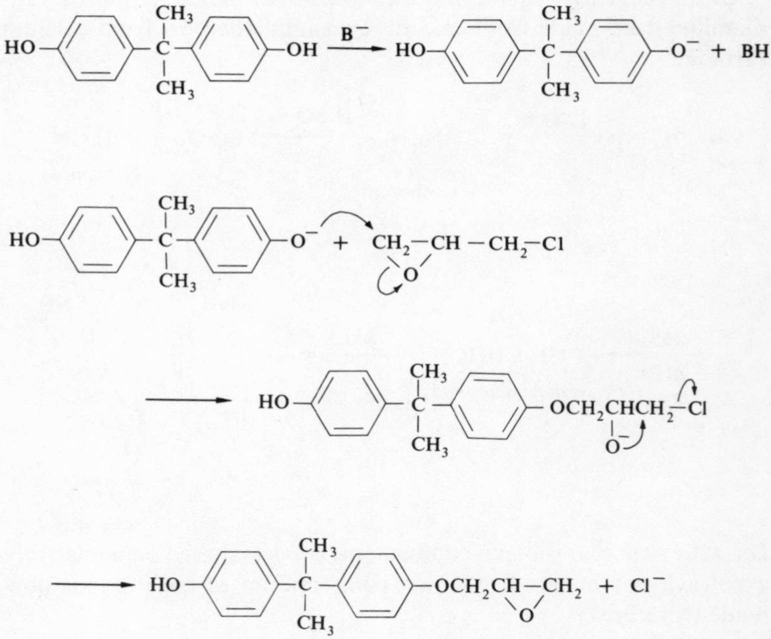

If a large excess of epichlorohydrin is used, a monomeric product, the diglycidyl ether of bisphenol A, is obtained:

$$CH_2-CHCH_2O-\langle\ \rangle-\underset{\underset{CH_3}{|}}{\overset{\overset{CH_3}{|}}{C}}-\langle\ \rangle-OCH_2CH-CH_2$$

When smaller proportions of epichlorohydrin are used, the glycidyl compounds initially formed undergo further stepwise reactions with phenoxide ions:

$$CH_2-CHCH_2O-\langle\ \rangle-\overset{\overset{CH_3}{|}}{\underset{\underset{CH_3}{|}}{C}}-\langle\ \rangle-O^- \ + \ CH_2-CHCH_2O-\langle\ \rangle-\overset{\overset{CH_3}{|}}{\underset{\underset{CH_3}{|}}{C}}-\langle\ \rangle-OH$$

$$\downarrow$$

$$CH_2-CHCH_2O-\langle\ \rangle-\overset{\overset{CH_3}{|}}{\underset{\underset{CH_3}{|}}{C}}-\langle\ \rangle-OCH_2CHCH_2O-\langle\ \rangle-\overset{\overset{CH_3}{|}}{\underset{\underset{CH_3}{|}}{C}}-\langle\ \rangle-OH$$

BH

$$CH_2-CHCH_2O-\langle\ \rangle-\overset{\overset{CH_3}{|}}{\underset{\underset{CH_3}{|}}{C}}-\langle\ \rangle-OCH_2CHCH_2O-\langle\ \rangle-\overset{\overset{CH_3}{|}}{\underset{\underset{CH_3}{|}}{C}}-\langle\ \rangle-OH \ + \ B^-$$

etc.

$$CH_2-CHCH_2-\left[O-\langle\ \rangle-\overset{\overset{CH_3}{|}}{\underset{\underset{CH_3}{|}}{C}}-\langle\ \rangle-OCH_2CHCH_2\right]_n O-\langle\ \rangle-\overset{\overset{CH_3}{|}}{\underset{\underset{CH_3}{|}}{C}}-\langle\ \rangle-CH_2CH-CH_2$$

Thus, in the absence of side reactions, and assuming that an excess of epichlorohydrin was used, the product obtained is a linear polymer with an epoxide group at each end.

Typically, epoxy resins are made by the reaction of bisphenol A and epichlorohydrin in the presence of sodium hydroxide, at temperatures in the range 320 to 380 K. The commercial products in the main have average molecular weights in the range 350 to 4 000, and range from liquids to brittle solids. They can be caused to crosslink to materials with good mechanical properties and chemical resistance by a variety of reactions involving the epoxy groups and/or the hydroxyl groups. For example, chain-growth polymerisation of the epoxy groups may be brought about by both anionic and cationic initiators, and, since there are two epoxy groups per molecule, leads to the formation of a network polymer. Crosslinking by stepwise reaction of epoxy and/or hydroxyl groups with polyfunctional reagents is also used.

Epoxy resins find a wide range of fairly small-tonnage uses, e.g. in surface coatings and adhesives, as potting compounds for electrical equipment, in laminated materials, and for flooring. UK production of epoxy resins in 1970 was 13 500 tons.

Polyurethanes

The name polyurethane is used to describe products obtained by the reaction of di-isocyanates with polyhydroxy compounds. It is not an accurate description of the nature of the polymers, in that in most cases many of the structural units in the polymer chain are not urethane units.

The polyurethanes are an extremely versatile group of products, and find a wide range of applications, as, e.g., surface coatings, fibres, elastomers and insulating foam, but their main use at the present time is as flexible foams for use in furniture, car seats, foam-backed fabrics, etc. Polyurethane chemistry and technology are very complex; only a brief discussion is given here.

Isocyanates will react with a wide range of materials containing active hydrogen; with alcohols, for instance, they give urethanes:

$$RN{=}C{=}O \ + \ R'OH \longrightarrow RNH\overset{\displaystyle O}{\overset{\displaystyle \|}{C}}OR'$$

Under carefully controlled conditions, glycols and di-isocyanates will yield linear polymers that are true polyurethanes. Thus, it was shown by Bayer in 1937 that 1,4-butylene glycol and hexamethylene di-isocyanate gave a linear polymer with good fibre-forming and mechanical properties:

$$HO(CH_2)_4OH \ + \ OCN(CH_2)_6NCO \longrightarrow \left[O(CH_2)_4O\overset{\displaystyle O}{\overset{\displaystyle \|}{C}}NH(CH_2)_6NH\overset{\displaystyle O}{\overset{\displaystyle \|}{C}}\right]_n$$

This polymer found limited application in Germany, but its high cost in comparison to other polymers with similar properties, e.g. nylons, prevented its widespread acceptance.

Development work on various other applications of polyurethanes continued in Germany during the war, and rigid foam, adhesive and surface-coating systems were developed. When information about these products became available after the war, interest was aroused outside Germany, and particularly in the USA. Polyurethane flexible foams were developed in the early 1950s, and rapidly became the most important of the polyurethane products.

FLEXIBLE FOAMS

The production of polyurethane foams is simple in concept but presents very considerable difficulties in practice. In essence, the raw materials are caused to polymerise, while at the same time a gas is generated so as to blow the polymer into a foam. The problems arise in controlling the various physical and chemical processes that are occurring, so as to obtain a foam with the desired properties. The present technology is the result of a large amount of development work by both the raw material producers and the foam manufacturers.

Most flexible foams are now made by the 'one-shot' method, in which di-isocyanate, polyhydroxy compound, and a number of other components are rapidly mixed together and the mixture is flowed either onto a conveyor belt, or into a mould. A foam is formed within a few seconds, and the product achieves its full physical properties within hours. In the case that the mixture is flowed onto a conveyor belt, the foam is obtained as 'slabstock', typically about 60 cm thick and 2 m wide, and this is converted to the required shape by cutting; moulding gives the final product direct, but is technically more difficult.

The di-isocyanate used is almost invariably a mixture of 2,4- and 2,6-tolylene di-isocyanates (TDI). The polyol is most commonly a poly(oxypropylene) triol or glycol, or, less often, a hydroxyl-ended aliphatic polyester. Water is generally present to provide carbon dioxide to blow the foam (see below), and an inert low boiling liquid, typically trichlorofluoromethane (b.p. 297 K), may also be present as a supplementary blowing agent. In addition, a number of catalysts are used to control the rates of the various reactions involved, and foam-stabilising agents, e.g. a silicone oil surfactant, are also added. There are a large number of reactions which can occur in such a system; we shall consider only those which are thought to be of major importance.

Reaction of isocyanate groups with hydroxyl groups of the polyol gives rise to urethane linkages:

$$\sim\!\!\sim\!\!\sim-\text{NCO} + \text{HO}-\!\!\sim\!\!\sim\!\!\sim \longrightarrow \sim\!\!\sim\!\!\sim-\overset{\overset{\displaystyle O}{\displaystyle \|}}{\text{NHC}}\text{O}-\!\!\sim\!\!\sim\!\!\sim$$

Reaction of water with isocyanate groups gives substituted carbamic acids, which immediately lose carbon dioxide to form primary amines:

$$\text{\textasciitilde\textasciitilde\textasciitilde—NCO} + H_2O \longrightarrow \text{\textasciitilde\textasciitilde\textasciitilde—NHCO}_2H \longrightarrow \text{\textasciitilde\textasciitilde\textasciitilde—NH}_2 + CO_2$$

The primary amines thus formed readily undergo further reaction with isocyanate to give urea linkages:

$$\text{\textasciitilde\textasciitilde\textasciitilde—NH}_2 + \text{OCN—\textasciitilde\textasciitilde\textasciitilde} \longrightarrow \underset{\displaystyle}{\overset{\displaystyle O}{\text{\textasciitilde\textasciitilde\textasciitilde—NHĈNH—\textasciitilde\textasciitilde\textasciitilde}}}$$

The amount of carbon dioxide liberated, and therefore the density of the resultant foam, is determined by the amount of water used. The supplementary blowing agent, if present, is volatilised by the heat of reaction, thus providing extra gas to blow the foam. The use of such an agent allows a reduction in the usage of the relatively expensive TDI, and also gives some extension of the range of foam properties achievable.

The urea and urethane groups formed in the reactions discussed above can undergo further attack by isocyanate-containing molecules to form substituted biuret and allophanate groups respectively:

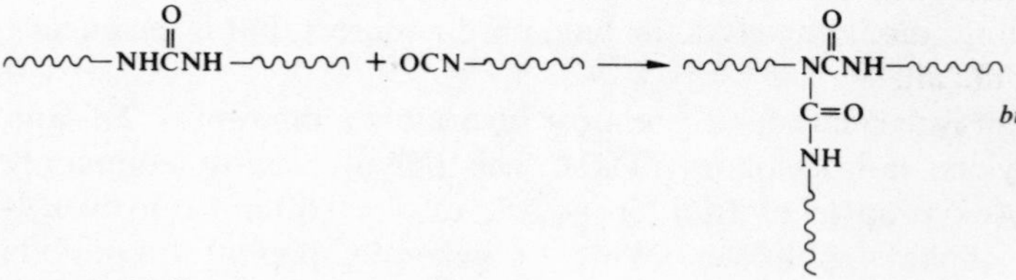

biuret unit

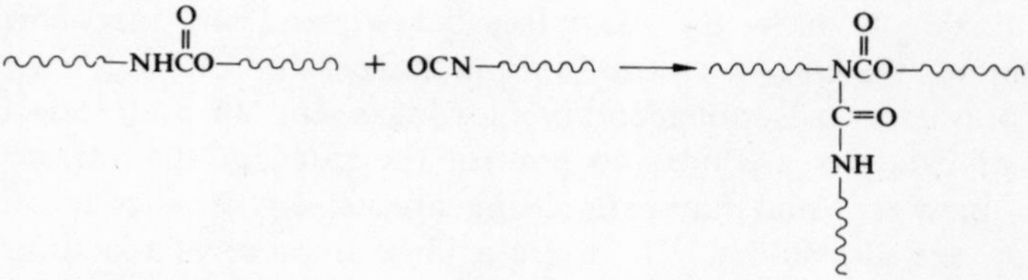

allophanate unit

The occurrence of these reactions leads to branching, and eventually to crosslinking of the polymer.

In order to obtain foam of the desired properties, control must be maintained over the relative rates of polymerisation and carbon dioxide evolution, and also over the degree of crosslinking obtained. As has been indicated above, crosslinking can result from the formation of biuret and allophanate structures. It may also be introduced by using a triol or higher polyol as the hydroxy compound, and most flexible foam manufacture is now based on poly(oxypropylene) triols of average molecular weight about 3 000. Control over the relative rates of the various reactions involved is achieved by the use of selective catalysts, mixtures of tertiary amines and tin compounds normally being used.

UK production of flexible foams in 1970 was 40 000 tons.

OTHER POLYURETHANE PRODUCTS

Rigid polyurethane foams are used e.g. in insulating and structural reinforcement applications. They are made by procedures similar to those used for flexible foams, but are more highly crosslinked. This is achieved by using a polyol of higher functionality than those used in flexible foam manufacture. One of the advantages of rigid polyurethane foams over other materials of this type, e.g. expanded polystyrene, is their ability to be foamed in place, so that cavities with complicated shapes can easily be filled. In this type of application, the relatively high vapour pressure of the toxic TDI is inconvenient, and less volatile di-isocyanates are used, the most commonly used being 4,4'-diphenylmethane di-isocyanate (MDI):

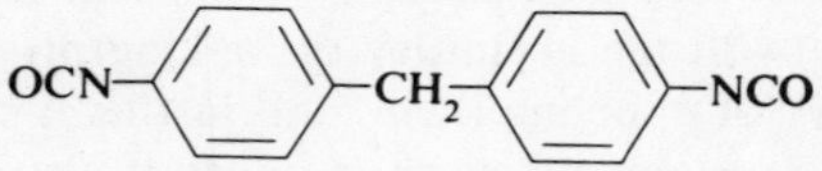

OCN—⟨◯⟩—CH₂—⟨◯⟩—NCO   *MDI*

Polyurethanes find a variety of other applications, e.g. as elastomers, surface coatings and fibres. The reactions involved in the formation of these products are of the type discussed above, except that since the products required are not foams, water is not added to the mixture of reactants. In fact, in most non-foam applications, water has to be rigorously excluded so that the evolution of carbon dioxide is avoided.

Chapter 8

Polymer Properties, Reactions and Applications

MOST of the discussion in this book has been of the chemistry involved in making chemicals. However, it should be remembered that the chemical industry exists because, at the end of various process chains, products result which can be used for some purpose other than making other chemicals. These goods then pass out of the chemical industry to be used, e.g., as solvents, antifreeze in car engines, dyestuffs and so on. Consequently, the chemical industry is concerned not only with the chemistry of making products, but also with the chemistry of applying them, and just as it carries out process research and development work, so does it also carry out product research and development work.

As the products of the industry are used in such a diverse range of applications, it would not be feasible to discuss here a major part of this applications chemistry; some aspects have been touched upon in previous chapters. In this chapter we will discuss selected aspects of the applications chemistry of the most important single group of products of the organic chemical industry, polymers.

CLASSES OF POLYMER PRODUCTS

Polymers are used in a very wide variety of applications, of which the most important can conveniently be divided into five categories:

   (i) Plastics applications
  (ii) Fibres
 (iii) Surface coatings
 (iv) Adhesives
  (v) Rubber applications.

Most of these categories are more or less self-explanatory, the least obvious being (i). The term 'plastic' is, in fact, not very well defined.

282

Possibly the easiest method of definition here is to say that plastics applications are in the main those applications of polymers which are not covered by categories (ii) to (v). Typically they involve the use of polymers in the manufacture of articles such as washing-up bowls, buckets, gramophone records, film for packaging applications, etc. Plastics applications provide the largest-tonnage outlet for synthetic polymers.

In quite a large number of cases, a polymer may be used in more than one class of application. For example, nylon 66, nylon 6 and poly(ethylene terephthalate) all find plastics applications in addition to their main use as fibres, and epoxy resins are applied as plastics, surface coatings and adhesives.

The properties required of a fibre-forming polymer are rather specific, and so only a small number of polymers are used in this type of application. The relationship between polymer structure and properties in fibre-forming polymers is interesting, and we will discuss some aspects of this later in this chapter. Similarly, the elastomeric polymers used in rubber applications form a rather coherent group, and will be discussed in a separate section.

The uses of polymers in surface coatings and in adhesives are complex and specialised fields, and it is not proposed to discuss them here.

Thermoplastic and Thermosetting Polymers

Polymers used in plastics applications are divided into two types: *thermoplastics* and *thermosetting resins*. Thermoplastic polymers are linear or branched, but not crosslinked, polymers. On heating they undergo the purely physical change of melting or softening, and on cooling they harden again. This process of melting by heating, and hardening on cooling, can, if desired, be repeated indefinitely, or until the polymer decomposes. Polyethylene, poly(vinyl chloride) and polystyrene, are typical thermoplastic polymers.

Thermosetting resins undergo chemical changes leading to crosslinking when they are heated or subjected to other curing procedures, and are thereby converted into an infusible, insoluble mass. The change is not reversible, and once the setting process has been carried out, the polymer cannot be melted.

A variety of types of reaction can be used in crosslinking thermosetting resins, and we have already met some of these in the last chapter. A good example is provided by the unsaturated polyesters which are the basis of the glass-fibre reinforced plastics. As we saw in Chapter 7, these polyesters are usually made by reaction of maleic anhydride, phthalic anhydride and propylene glycol. They are of low molecular weight, and are viscous liquids or glassy solids. Crosslinked polymer is produced from them by subjecting a mixture of the resin and a vinyl monomer, usually styrene, to free-radical polymerisation conditions, when copolymerisation of the styrene and polyester resin occurs:

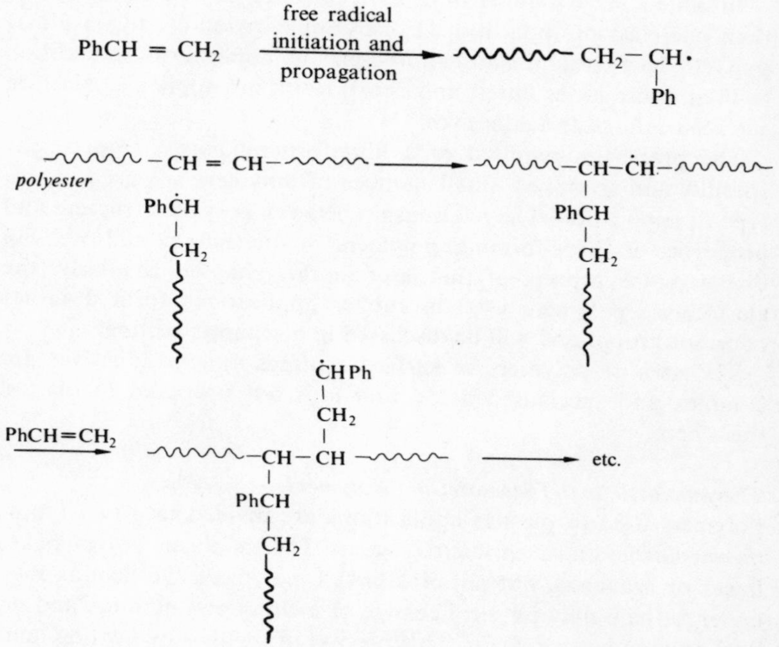

free radical
initiation and
propagation
PhCH = CH2
CH2— CH•
Ph
polyester
CH = CH
PhĊH
CH2
CH–ĊH
PhCH
CH2
PhCH=CH2
ĊHPh
CH2
CH—CH
PhCH
CH2
etc.

Since the polyester molecules each include a number of carbon–carbon double bonds, this copolymerisation rapidly leads to the formation of a three-dimensional network polymer.

Crosslinked thermosetting polymers may be made harder, stiffer, and more heat-resistant than thermoplastics. The methods used for forming finished articles from the two types of polymer differ substantially.

The most important methods of forming thermoplastics involve melting or softening the polymer by heating it, forming it into the desired shape by some means, and then allowing it to cool. Two of the major methods of carrying this out are *extrusion*, and *injection moulding*. In extrusion, polymer, in the form of 'nibs' or small pellets, is fed into what is effectively a heated, outsize version of a domestic mincing machine, i.e., a machine that consists essentially of a screw rotating inside a heated cylinder. The hot polymer is continuously forced out through a die, the shape of the die determining the cross-section of extrudate obtained. Thus, a circular die will give cylindrical rod, an annular die will give piping, a slit-shaped die will give sheeting, and so on. In injection moulding, polymer is forced from a heated cylinder, through a narrow connecting passage, into a mould of the appropriate shape. Once the article has cooled sufficiently that it can be handled without distortion, the mould can be opened, the article removed, the mould closed, and the next shot of polymer can be injected into the mould. Both processes can be highly automated, and high rates of production can be achieved.

The situation when forming thermosetting resins is different in that the article must be held in shape while a chemical reaction takes place. With some polymer systems, the forming may have to be carried out at elevated temperatures and, in some cases, under pressure; others will crosslink at ambient temperature. In general, forming thermosetting resins tends to be a slower operation, with a higher labour content, than forming thermoplastics.

Consumption of thermoplastics is about three times that of thermosetting polymers, and is growing more rapidly.

STRUCTURE AND PHYSICAL PROPERTIES

As has already been indicated, and as the reader will have observed from his own experience, polymers exhibit a wide range of properties, which are quite different from those exhibited by monomeric organic compounds. They can, for instance, be hard and rigid, soft and flexible, resilient, fibre-forming and so on. The relationship between the structures and physical properties of polymers is a matter of considerable technological importance and theoretical interest which has been the subject of much investigation, and the theory of the relationship is quite highly developed. The discussion that follows is a simplified qualitative discussion of some of the salient points of the theory.

Let us start by considering as an example the simplest organic polymer, high-density polyethylene. This is essentially a high molecular weight linear alkane (if we ignore the small number of branches that occur, and the end groups) and is therefore a higher homologue of, for example, n-hexane.

In a liquid sample of n-hexane the molecules are undergoing random thermal motion through the sample, and are essentially independent of each other except for the occurrence of collisions, and the van der Waals attractive forces which exist between the molecules. If the temperature of the sample is reduced, the average thermal energy of the molecules decreases, and the thermal motion slows. At 179 K the thermal energy of the molecules is no longer sufficient to overcome the attractive forces between them, and they pack into a crystal lattice and the sample solidifies.

Let us now consider a sample of molten polyethylene. The first point to note is that since the polyethylene molecules are some hundreds of times larger than the hexane molecules, the van der Waals forces between them will be considerably larger than the forces between hexane molecules.

Any particular molecule in the sample can exist in a very large number of conformations, of which one is that where the chain is fully extended in a regular zig-zag in which all the bonds to carbon have the tetrahedral angle:

$$\mathrm{CH_2\ CH_2\ CH_2\ CH_2\ CH_2\ CH_2\ CH_2\ CH_2\ CH_2}$$
$$\mathrm{CH_2\ CH_2\ CH_2\ CH_2\ CH_2\ CH_2\ CH_2\ CH_2}$$

Since there is free rotation about the carbon–carbon bonds, the molecule will be undergoing thermal segmental motion about these bonds, i.e., it will be undergoing continuous coiling and wriggling movements, and the above fully extended conformation will in fact be a highly improbable one. For most of the time the molecules will be more or less tightly coiled up, and since the chains are long there will be a considerable degree of entanglement of molecules with one another. This entanglement imposes a resistance to overall movement of molecules with respect to one another, which is reflected in the high viscosity of molten polymers.

If the temperature of the sample of polyethylene is now progressively reduced, the rate of segmental and overall motion of the molecules will drop, and a point will be reached at which they begin to pack into a crystal lattice. Such packing requires that the molecules be in the extended conformation, and it can readily be seen that it is highly unlikely that complete polyethylene molecules will be incorporated into a lattice in the same way that hexane molecules are. What does happen is that portions of the molecules enter into such ordered 'crystalline regions', and other portions of the same molecules remain in disordered 'amorphous regions'. The temperature at which this occurs is called the crystalline melting point (T_m) of the polymer, and at this temperature the sample changes from a viscous liquid to a solid. The solid obtained has, of course, quite different properties from those of solid n-hexane, the latter being weak and brittle, whereas polyethylene is tough and has useful mechanical properties.

The characteristic physical properties of polymers are dependent on their long chain structures and the high intermolecular forces and chain entanglement to which these give rise, and for any particular polymer there is a minimum molecular weight below which useful mechanical properties are not displayed. For polyethylene this minimum molecular weight is about 10 000–15 000, polymers with molecular weights below this ranging from oily liquids to waxes.

Crystalline and Amorphous Polymers

Polymers in which ordered crystalline regions of the type discussed above can be formed are called *crystalline polymers*. It should be emphasised that the term crystalline does not have the same meaning as when applied to monomeric materials, and that crystals of the type formed by monomeric compounds are not formed under the conditions described above.* The term implies that the polymer concerned is capable of exhibiting some degree of ordered crystal lattice-type arrangement of molecular segments in the solid state. The proportion of the whole which exists in such ordered arrangements, which can be estimated, e.g., by X-ray diffraction, is called the 'degree of crystallinity'.

The degree of crystallinity of a sample of a polymer depends on the structure of the polymer (see below), and on the history of the particular sample. Thus, if a sample of a crystallisable polymer is cooled slowly from the melt, the arrangement of molecules into the lattice is facilitated so that a product with a high degree of crystallinity tends to be obtained, and conversely rapid cooling

* In recent years, very small crystals of a number of polymers, including polyethylene, have been grown from very dilute solutions. It is not known whether these have the same structures as the crystalline regions in the corresponding bulk polymers.

tends to lead to a low degree of crystallinity, or indeed, in some cases, to an amorphous polymer.

By no means all polymers exhibit crystallinity, the necessary requirement being that the polymer chain should have a regular structure which can readily fit into a crystal lattice, or at least that long segments of it should have such a structure. High-density polyethylene evidently fulfils this requirement, and it exhibits degrees of crystallinity of up to about 90%. The molecule of low-density polyethylene has periodic irregularities in the form of branches, but crystallisation of the segments of chain between the branches occurs, and low-density polyethylene is a moderately crystalline polymer (ca. 55% crystallinity). Commercial polystyrene, on the other hand, has an irregular structure in that it is an atactic polymer (see page 244), i.e., the phenyl groups on alternate chain carbon atoms are randomly distributed in D- and L-configurations. Consequently, the molecules cannot pack into a crystal lattice, and commercial polystyrene is a non-crystalline or *amorphous polymer*. The same applies to most other polymers of monosubstituted ethylenes, except when, as is the case with polypropylene, they have been made by stereospecific polymerisation.

If we consider a sample of a polymer of irregular structure, such as polystyrene, being allowed to cool from the molten state, initially it will behave in the same way as a crystallisable polymer, i.e., as the thermal motions of the molecules slow down, the viscosity of the melt increases. However, since the molecules cannot pack into a crystal lattice, the polymer will not exhibit a crystalline melting point, and will not undergo a relatively sharp transition into a solid. Rather, as the temperature drops, the thermal segmental motions and the overall mobility of the molecules will gradually diminish, and the polymer will change from a viscous liquid, to a rubbery solid, then to a leathery solid and finally, at a stage when the thermal wriggling motions of the molecules have ceased, to a glassy solid. This latter change occurs fairly sharply, and the temperature at which it occurs is called the glass transition temperature (T_g) of the polymer. It can be seen that the properties of an amorphous polymer, and therefore its potential applications, depend on whether it is above or below its glass transition temperature at the service temperature concerned. Thus, polystyrene $(T_g = 373$ K) is a hard, stiff plastic at ambient temperature, whereas poly(ethyl acrylate) $(T_g = 203$ K) is soft and rubbery.

Crystalline polymers also exhibit glass transition temperatures associated with the amorphous regions of the polymer structure, and the properties of the polymer in service are dependent on the level at which this lies. Thus, in a crystalline polymer with a glass transition temperature well below the service temperature, the hard, stiff crystalline regions are associated with flexible amorphous regions, and the polymer tends to be tough. A crystalline polymer with a glass transition temperature above the service temperature will be more brittle. Thus, polypropylene (T_g = 298 to 308 K) is much more brittle at low ambient temperatures than high-density polyethylene (T_g = 148 K).

Since the intermolecular forces in crystalline regions are higher than in amorphous regions, then it would be expected that, other things being equal, the higher the degree of crystallinity of a polymer, the greater its mechanical strength and the higher its melting point. Further, as the crystalline regions are more closely packed, the density of a polymer would be expected to increase with the degree of crystallinity. That these generalisations are valid in the case of polyethylene is indicated by the data in Table 8-1.

TABLE 8-1
PROPERTIES OF POLYETHYLENES

Property	Low-density polyethylene	High-density polyethylene
specific gravity	0·915–0·925	0·945–0·965
approx. % crystallinity	55	80
approx. crystalline m.p. (K)	382	403
approx. softening point (K)	360	400
tensile strength (psi)	1 250–2 000	3 000–4 600

Plasticisation

Incorporation of relatively low molecular weight polymer-miscible materials into polymers can have marked effects on their physical properties. The molecules of such additives distribute themselves between, and solvate, the polymer molecules, thus reducing the attractive forces between them. The consequent increase in mobility of the polymer molecules leads to, for example, a reduction in the glass transition temperature, increased flexibility and distensibility, and reduced melt viscosity. Additives of this type, called *plasticisers*, are widely used to modify the processing and service properties of polymers.

By far the most important application of plasticisers is in

poly(vinyl chloride), and about 70% of total plasticiser output is used for this purpose. Poly(vinyl chloride) is unique amongst commercial polymers in the range of properties which it can exhibit when plasticised; depending on the amount of plasticiser used, it can cover the range from a hard, rigid plastic (unplasticised p.v.c.) to a very soft, flexible plastic (plasticiser content, *ca.* 45%). It is, in consequence, an exceptionally versatile polymer, and is used in a very wide range of applications.

A large number of plasticisers are used in poly(vinyl chloride), but the most important are the dialkyl phthalates, and in particular, the diphthalates of 2-ethylhexanol, and of 'iso-octanol' (see page 202). For many applications these have the optimum combination of the various properties required of a plasticiser, e.g. good plasticising efficiency, low volatility, good stability to heat, light and oxidation, and low cost.

Other important uses of plasticisers are in cellulose nitrate and acetate, and in polymer latices, notably poly(vinyl acetate), for use in surface coatings and adhesives. They are used in relatively small amounts in a number of other polymers, often as a processing aid rather than with the purpose of altering the service properties of the product.

Crosslinked Polymers

The above discussion of structure and properties applies to linear and branched polymers. The situation with crosslinked or network polymers is different in that overall motion of segments of the polymer network with respect to each other is not possible unless bond scission occurs. Thus, crosslinked polymers cannot be melted and will not flow. The extent of segmental motion which is possible depends on the density of crosslinking. Highly crosslinked polymers are hard and rigid; lightly crosslinked polymers may be soft and flexible (see 'Elastomeric Polymers').

Crystallinity can be exhibited by lightly crosslinked polymers.

POLYMER DEGRADATION

It is important that polymers should not undergo degradative reactions leading to major changes in their properties either during forming operations or during the service life of the product. Susceptibility to such degradation may make necessary the use of stabilising additives and/or restrict the possible applications of a polymer.

There are a variety of degradative reactions which can occur with commercial polymers. We will consider two which are of major importance, oxidation of polymers, and thermal and photolytic degradation of poly(vinyl chloride).

Oxidation of Polymers

In most applications polymers are exposed to air during their service life, and in many cases contact between hot polymer and air occurs during processing. With a number of polymers, e.g. natural and some synthetic rubbers, polyethylene and polypropylene, oxidative degradation can occur to such an extent during forming and/or service that serious deterioration in polymer properties results. Methods of preventing such oxidative degradation are of great technological importance.

SATURATED POLYMERS

Polyethylene is rather readily attacked by oxygen, both at fabrication temperatures, and at ambient temperatures under the influence of light. Polypropylene is even more readily attacked and, for example, outdoor exposure of unstabilised polypropylene fibre for 2 months leads to a complete loss of useful mechanical properties.

Oxidation of polyethylene and polypropylene in air is thought to proceed by reactions closely analogous to those involved in the liquid-phase free-radical oxidation of alkanes (see Chapter 3). Thus, the initial product is a hydroperoxide:

initiation

$$\sim\!\!\wedge\!\!\wedge\!\!\wedge\!\!- CH_2CH_2CH_2 \sim\!\!\wedge\!\!\wedge\!\!\wedge + X \longrightarrow \sim\!\!\wedge\!\!\wedge\!\!\wedge\, CH_2\overset{\bullet}{C}HCH_2 \sim\!\!\wedge\!\!\wedge$$

propagation

$$\sim\!\!\wedge\!\!\wedge\!\!\wedge\, CH_2\overset{\bullet}{C}HCH_2 \sim\!\!\wedge\!\!\wedge + O_2 \longrightarrow \sim\!\!\wedge\!\!\wedge\, CH_2\underset{\underset{OO\cdot}{|}}{C}HCH_2 \sim\!\!\wedge\!\!\wedge$$

$$\sim\!\!\wedge\!\!\wedge\, CH_2\underset{\underset{OO\cdot}{|}}{C}HCH_2 \sim\!\!\wedge\!\!\wedge + \underset{(polymer)}{RH} \longrightarrow \sim\!\!\wedge\!\!\wedge\, CH_2\underset{\underset{OOH}{|}}{C}HCH_2 \sim\!\!\wedge\!\!\wedge + R\cdot$$

Cleavage of the hydroperoxide gives an alkoxy radical:

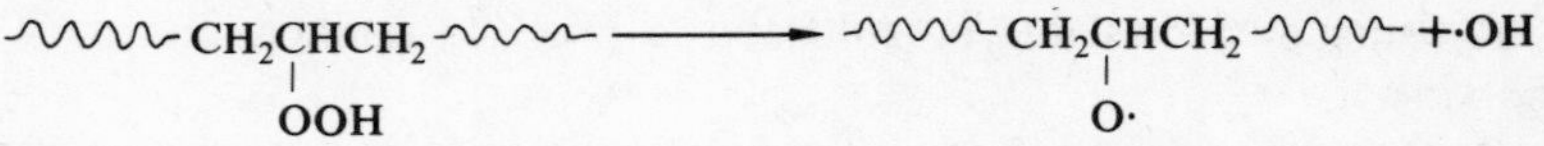

~~~~ CH$_2$CHCH$_2$ ~~~~ ⟶ ~~~~ CH$_2$CHCH$_2$ ~~~~ + ·OH
     | |
    OOH   O·

Further reactions of this radical can result in the formation of a carbonyl group or a hydroxyl group on the polymer chain:

~~~~ CH$_2$CHCH$_2$ ~~~~ + R'· ⟶ ~~~~ CH$_2$CCH$_2$ ~~~~ + R'H
        | ‖
        O·   O

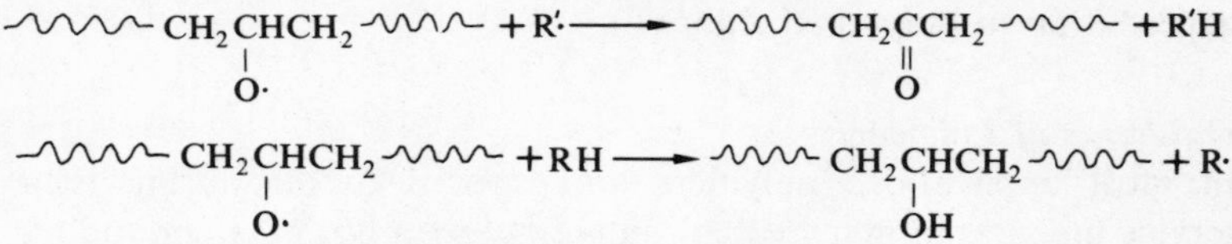

~~~~ CH$_2$CHCH$_2$ ~~~~ + RH ⟶ ~~~~ CH$_2$CHCH$_2$ ~~~~ + R·
      | |
     O·   OH

These reactions would not be expected to have much effect on the mechanical properties of the polymer except at high extents of reaction. However, the alkoxy radical can also undergo *beta*-cleavage which results in polymer chain scission:

~~~~ CH$_2$CH—CH$_2$ ~~~~ ⟶ ~~~~ CH$_2$CH + ·CH$_2$ ~~~~
              | ‖
            O·   O

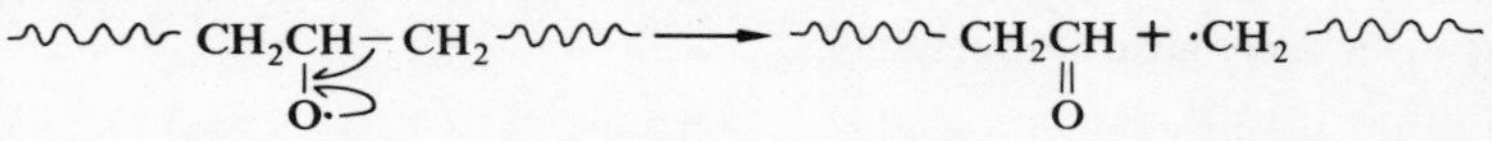

Since the useful mechanical properties of a polymer are only developed above a certain minimum molecular weight, quite a small amount of chain scission can cause a dramatic deterioration in properties. Formation of crosslinks also occurs, and results in changes in properties.

The effects of oxidative degradation become evident at only small extents of reaction. For example, low density polyethylene which has been oxidised to the extent where its oxygen content is about 1 % has lost most of its original mechanical properties.

It has already been indicated that polypropylene is much more readily oxidised than polyethylene. This is attributed to the high proportion of relatively easily abstracted tertiary hydrogen atoms on the polymer chain:

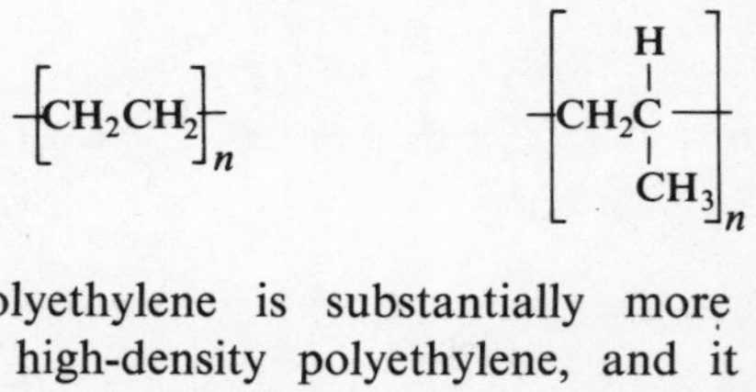

$$\left[\text{CH}_2\text{CH}_2\right]_n \qquad \left[\begin{array}{c} \text{H} \\ | \\ \text{CH}_2\text{C}— \\ | \\ \text{CH}_3 \end{array}\right]_n$$

Low-density polyethylene is substantially more susceptible to oxidation than high-density polyethylene, and it was originally thought that this was due to the tertiary hydrogen atoms at the branches on the low-density polyethylene chains. However, it now appears that the difference is mainly due to the lower degree of crystallinity of low-density polyethylene: above the melting point, both polymers oxidise at approximately the same rate. An amorphous polymer structure probably allows easier diffusion of oxygen, and presents less steric hindrance to attack than a crystalline structure.

A number of other saturated polymers undergo oxidative degradation of the type discussed above. Nylons, for instance, are fairly readily oxidised, both thermally and photolytically. On the other hand, polystyrene is rather resistant to oxidation. This is surprising, since it contains hydrogen atoms that are both tertiary and on a carbon atom attached to a benzene ring, and which would consequently be expected to be readily abstracted (cf. cumene oxidation). It is presumed that the bulky phenyl groups shield the chain hydrogens from attack.

UNSATURATED POLYMERS

Polymers containing carbon–carbon double bonds, typified by natural rubber, tend to be highly susceptible to oxidative degradation in consequence of the occurrence of allylic hydrogen atoms on the polymer chain. Abstraction of such hydrogen atoms, which gives a resonance-stabilised radical, occurs readily:

$$\sim\!\!\!\sim\!\!\!\sim CH_2 - \underset{\underset{CH_3}{|}}{C} = CH - CH_2 \sim\!\!\!\sim\!\!\!\sim$$

$$\Big\downarrow -H\cdot$$

$$\sim\!\!\!\sim\!\!\!\sim - CH_2 - \underset{\underset{CH_3}{|}}{C} = CH - \overset{\cdot}{C}H \sim\!\!\!\sim \longleftrightarrow \sim\!\!\!\sim CH_2 - \underset{\underset{CH_3}{|}}{\overset{\cdot}{C}} - CH = CH \sim\!\!\!\sim$$

Unless protected by additives, such polymers rapidly deteriorate in air at ambient temperatures even in the dark.

With polymers of this type the oxidative reactions are considerably more complicated than those of saturated polymers, since the double bonds can become involved, and there is still a great deal of uncertainty about the details of the reactions which occur. Overall, oxidation of natural rubber results in chain scission, with consequent loss of strength and softening of the polymer. With butadiene-containing rubbers crosslinking reactions predominate, and stiffening of the polymer is caused.

Antioxidants

The rate of oxidative degradation of polymers, and also of other organic products, e.g. gasoline, lubricating oils, foodstuffs, can be markedly reduced by the addition of relatively small quantities of various types of organic compounds to the material in question, and the use of such *antioxidants* is of very major technological importance. Antioxidants fall into two main groups in terms of mode of action.

CHAIN-BREAKING ANTIOXIDANTS

Chain-breaking antioxidants, which are usually phenols or aromatic amines, act by terminating kinetic reaction chains. Since many polymer molecules may undergo oxidation in one kinetic chain, a small amount of antioxidant can produce a large reduction in the extent of oxidation.

The activity of phenols as antioxidants is a consequence of the ease with which the phenolic hydrogen may be abstracted, to give a resonance-stabilised phenoxy radical (cf. polymerisation inhibitors, page 233). In an oxidising polymer, abstraction of phenolic hydrogen by intermediate radicals, usually polymer peroxy radicals, occurs:

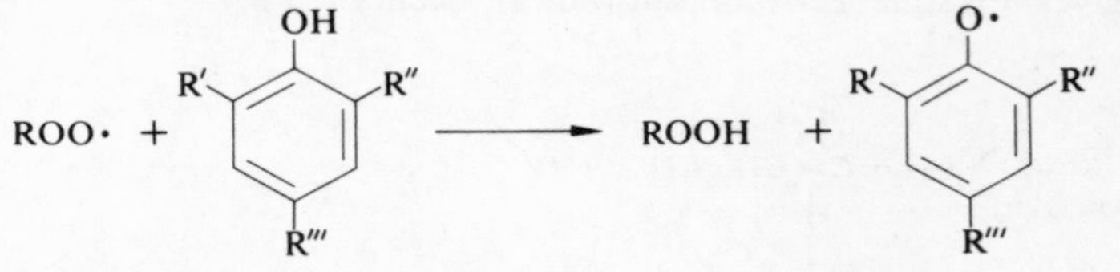

ROO· = *polymer peroxy radical*

The effect on the overall reaction depends on the subsequent reactions of the phenoxy radical. If, for example, it abstracts a hydrogen atom from a polymer molecule, then the phenol is merely acting as a chain transfer agent.

In order for kinetic chains to be terminated, the phenoxy radicals must undergo further reactions leading to inactive products. The details of such reactions are not known with certainty, but they are supposed to be coupling reactions of the following type (A· = phenoxy radical):

$$2A· \longrightarrow AA$$

$$A· + ROO· \longrightarrow AOOR$$

These reactions do not necessarily involve the phenoxy oxygen since the unpaired electron is delocalised. Thus, the following reaction of 2,6-di-t-butyl-4-methylphenol with peroxy radicals has been demonstrated:

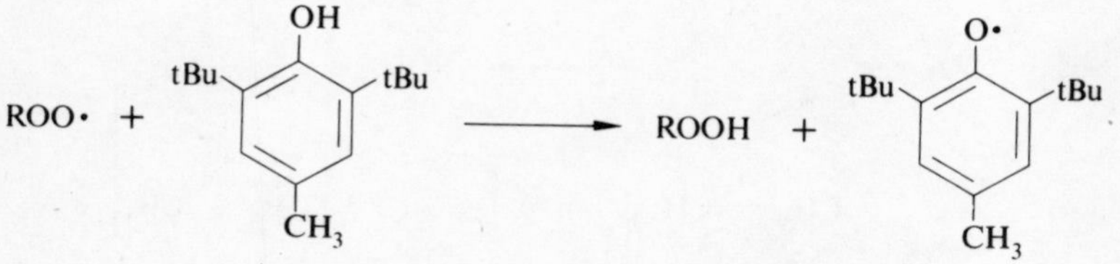

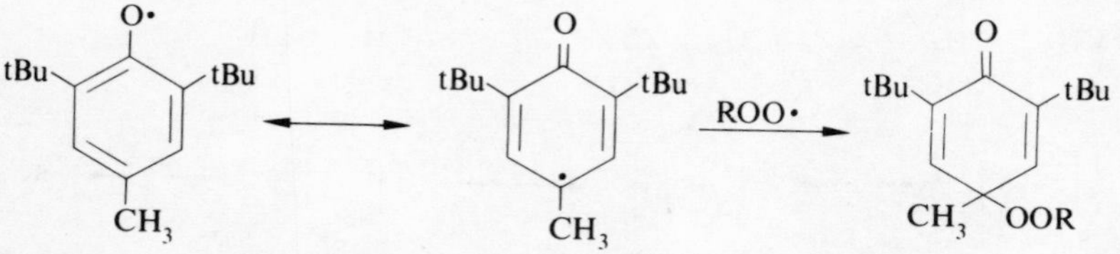

The effectiveness of a phenol as a chain-breaking antioxidant depends on the facility with which it undergoes the various possible reactions in the system in question. Thus, it should react readily with polymer peroxy radicals, and the phenoxy radicals formed should undergo reactions leading to inactive products rather than abstract hydrogen from polymer chains. It should not undergo hydrogen atom abstraction by oxygen, since this results in the initiation of reaction chains:

$$AH + O_2 \longrightarrow A\cdot + HOO\cdot$$

The relationship between phenol structure and antioxidant activity is complex and not fully understood. In general terms, antioxidant activity is increased by electron-donating groups in the positions *ortho-* and *para-* to the hydroxyl group(s), and by steric hindrance of the hydroxyl group(s) by bulky groups in the *ortho*-positions. 2,4,6-trisubstituted phenols are very commonly used as antioxidants.

The main mode of action of amine antioxidants is probably analogous to that of phenols. However, the antioxidant activity of amines is not completely destroyed by total *N*-alkylation, and, for example, *N,N'*-tetramethyl-*p*-phenylenediamine is quite an effective antioxidant. It is therefore suggested that the formation of electron-transfer complexes between antioxidant and peroxy radicals may be of some importance with amine antioxidants:

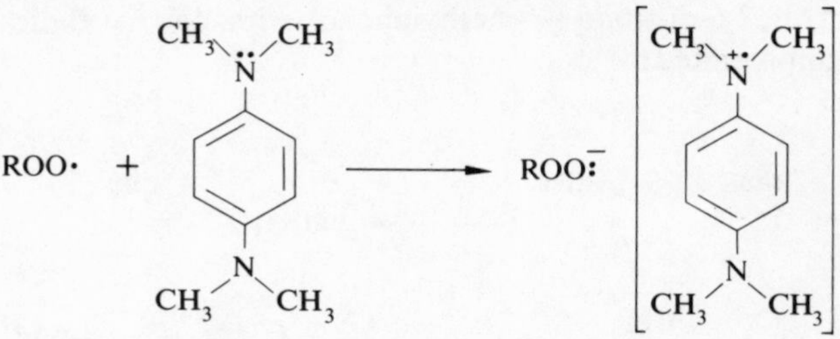

Radical ions of the type shown above are heavily resonance-stabilised:

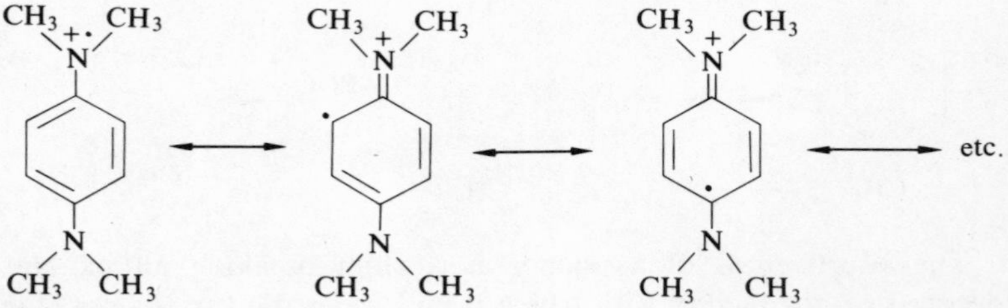

PREVENTATIVE ANTIOXIDANTS

An alternative method of reducing the rate of oxidation to that discussed above is to prevent the initiation of reaction chains rather than interrupt them once started.

Once a free-radical air-oxidation has started, most of the initiation of chains results from the cleavage of hydroperoxide (see page 76). One class of preventative antioxidants acts by bringing about the decomposition of hydroperoxides to non-free-radical products, and thus preventing their participation in initiation reactions. A variety of sulphur compounds are used for this purpose; the details of the manner in which they bring about decomposition of hydroperoxides is not at present well understood.

As was discussed in Chapter 3, a number of metal ions induce the cleavage of hydroperoxides to radicals, and consequently have profound catalytic effects on free-radical air-oxidations, e.g.:

$$ROOH \ + \ Co^{2+} \longrightarrow RO\cdot + \ Co^{3+} \ + \ OH^-$$

$$ROOH \ + \ Co^{3+} \longrightarrow ROO\cdot + \ Co^{2+} \ + \ H^+$$

Addition of compounds which strongly complex the ions to their maximum co-ordination number, or which stabilise one valency state at the expense of others, prevents this effect, and such materials act as antioxidants for materials containing traces of metals such as cobalt, iron and copper. A number of amine chain-breaking antioxidants, e.g. *N,N'*-diphenyl-*p*-phenylenediamine (I), have important copper-deactivating activity.

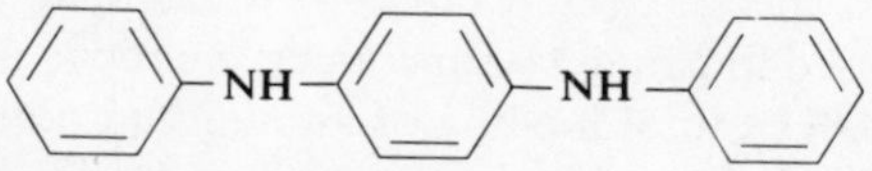

(I)

It has already been indicated that exposure to light has important effects in promoting the oxidation of polymers. Irradiation leads to the production of radicals by the photolysis of, e.g., hydroperoxide or carbonyl groups in the polymer:

$$\text{ROOH} \xrightarrow{h\nu} \text{RO·} + \text{·OH}$$

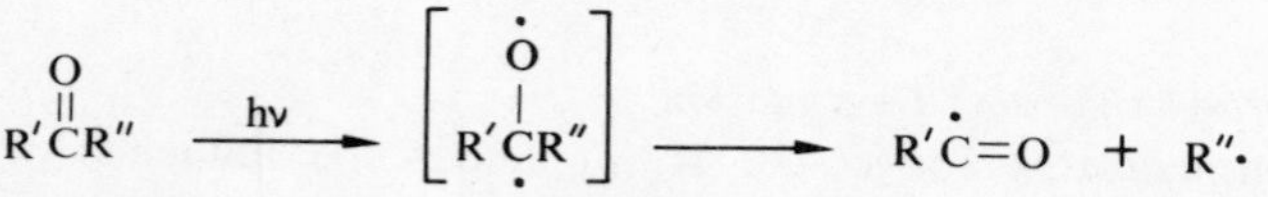

The radicals formed can initiate reaction chains in the normal way.

This effect may be prevented by adding to the polymer materials which absorb the light and dissipate the energy without producing radicals capable of starting reaction chains. The cheapest and most effective such agent is carbon black, and this is very often used where a black product can be tolerated. Where this is not the case, the most widely used ultraviolet deactivators are 2-hydroxybenzophenone derivatives. It is suggested that these absorb light with the production of a diradical, as indicated above, but that rather than undergo cleavage to two radicals, the diradical isomerises to an enolic quinone:

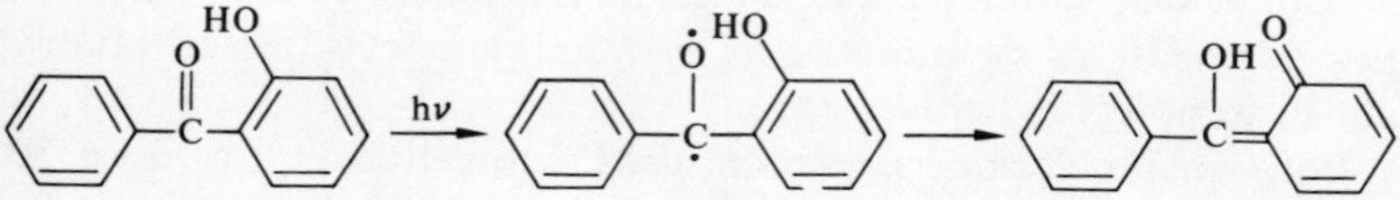

ANTIOXIDANT APPLICATIONS

Antioxidant technology is complex, and the development of antioxidants and the design of antioxidant systems remains something of an art. Since there are a number of different modes of anti-oxidant activity, synergistic effects are often observed, i.e. better effects are obtained with mixtures of antioxidants than can be produced by the use of the individual components, and mixtures of antioxidants are often used. Polypropylene, for example, is typically stabilised with a chain-breaking antioxidant, a peroxide-decomposing antioxidant, and an ultraviolet deactivator. Other additives may be necessary in special circumstances; for example, if the polypropylene is to be used in contact with copper, a copper-deactivating agent is also added. The total loading of antioxidant additives is up to about 0·5%. Polyethylene, which is much less readily oxidised, is adequately protected for most applications by the addition of 0·02–0·1% of a chain-breaking antioxidant.

Unsaturated rubbers require rather high loadings of antioxidants, typically of the order of 1%. As a consequence of these high loadings, and of the large tonnage of rubber used (UK consumption 1969, 447 000 tons), the rubber industry is by far the largest consumer of antioxidants for polymer applications.

Poly(vinyl chloride) Degradation

At temperatures above 420 K, or under the influence of light, poly(vinyl chloride) undergoes decomposition involving the evolution of hydrogen chloride. The first effects on the polymer that are noted are the development of discoloration, and this can reach a level at which the polymer is rendered technologically unacceptable for many applications after the loss of only a very small amount (0·1% or less) of hydrogen chloride. In air the colour develops through the sequence yellow $\longrightarrow$ orange $\longrightarrow$ brown $\longrightarrow$ black; in an inert atmosphere, the sequence is pink $\longrightarrow$ red $\longrightarrow$ brown $\longrightarrow$ black. As the decomposition proceeds, the mechanical properties of the polymer deteriorate, and it eventually becomes stiff and brittle. The degradation occurs at such a rate that if it were not possible to suppress it, the applications of poly(vinyl chloride) would be very severely restricted. In fact, stabilisers which minimise this degradation, or minimise its undesirable effects, are invariably added to poly(vinyl chloride).

Poly(vinyl chloride) has been used commercially for over 30 years, and the problem posed by its degradation has existed for the same time, so that at first sight it is somewhat surprising to find that the mechanism by which it degrades is not known with any degree of certainty. There appear to be two main reasons for this state of affairs. Firstly, the problem is complex and difficult to investigate, for one thing because of the difficulties in characterising both the original and the degraded polymer. Secondly, the subject has not received a great deal of attention from academic chemists, and most of the work on poly(vinyl chloride) degradation has been carried out by workers in industry and in research institutes. Quite naturally, such workers have not put their major effort into long-term fundamental investigations of the problem, but have in the main concentrated on developing, largely empirically, systems of stabilisers which extend the usefulness of the polymer. Their success in this is evident from the fact that poly(vinyl chloride) is the second largest tonnage polymer made. At the same time a substantial amount of investigation into the mechanism has been carried out, but many of the results are inconclusive or contradictory, as will become evident in the following discussion.

MECHANISM OF DEGRADATION

The colour that develops during degradation is generally supposed to be due to the production of conjugated polyene sequences, $+ CH = CH +_n$, by loss of hydrogen chloride. Structures of this type, which contain highly delocalised systems of π-electrons, begin to exhibit colour when there are about 5 to 7 double bonds in conjugation. Estimates of the length of the sequences in degraded poly(vinyl chloride) generally are in the range 5 to 20. The formation of such long sequences at very low extents of dehydrochlorination indicates that loss of hydrogen chloride from the polymer is not a random process, in which case long sequences of conjugated double bonds would only develop at high extents of degradation, but that loss occurs from successive monomer units:

$$\sim\!\!\sim\!\!\sim CH_2 - CH - CH_2 - CH - CH_2 - CH - CH_2 - CH - CH_2 - CH \sim\!\!\sim\!\!\sim$$
$$\qquad\quad | \qquad\qquad | \qquad\qquad | \qquad\qquad | \qquad\qquad |$$
$$\qquad\quad Cl \qquad\qquad Cl \qquad\qquad Cl \qquad\qquad Cl \qquad\qquad Cl$$

$$\downarrow \quad -HCl$$

$$\sim\!\!\sim\!\!\sim CH_2 - CH - CH_2 - CH - CH_2 - CH - CH_2 - CH - CH = CH \sim\!\!\sim\!\!\sim$$
$$\qquad\quad | \qquad\qquad | \qquad\qquad | \qquad\qquad |$$
$$\qquad\quad Cl \qquad\qquad Cl \qquad\qquad Cl \qquad\qquad Cl$$

$$\downarrow \quad -HCl$$

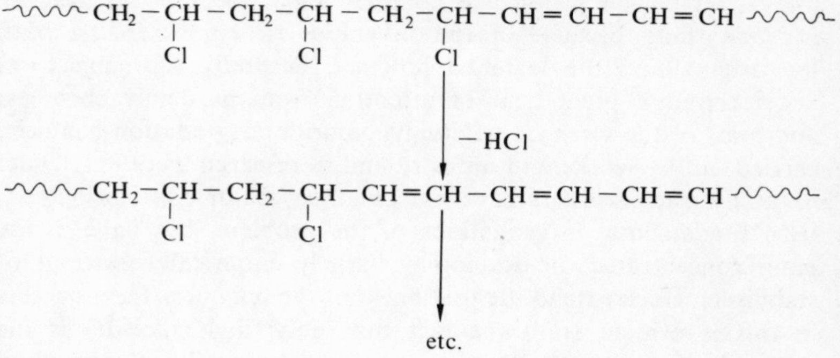

$\sim\!\!\sim\!\!\sim\mathrm{CH_2-CH-CH_2-CH-CH_2-CH-CH=CH-CH=CH}\sim\!\!\sim\!\!\sim$

with Cl, Cl, Cl substituents

$-\,\mathrm{HCl}$

$\sim\!\!\sim\!\!\sim\mathrm{CH_2-CH-CH_2-CH-CH=CH-CH=CH-CH=CH}\sim\!\!\sim\!\!\sim$

with Cl, Cl substituents

etc.

The reaction is termed a 'zipper' reaction.

From its idealised structure, $\{\mathrm{CH_2CHCl}\}_n$, poly(vinyl chloride) would not be expected to be particularly thermally labile. Simple secondary alkyl chlorides, for example, do not decompose thermally below about 620 K. Consequently, it has been necessary to postulate that the zipper reaction starts either at some structural irregularity introduced into the polymer chain during polymerisation, or alternatively that it is initiated by extraneous impurities. Amongst the structural irregularities commonly postulated as being important are structures involving allylic chlorine atoms, produced, e.g., by termination by disproportionation:

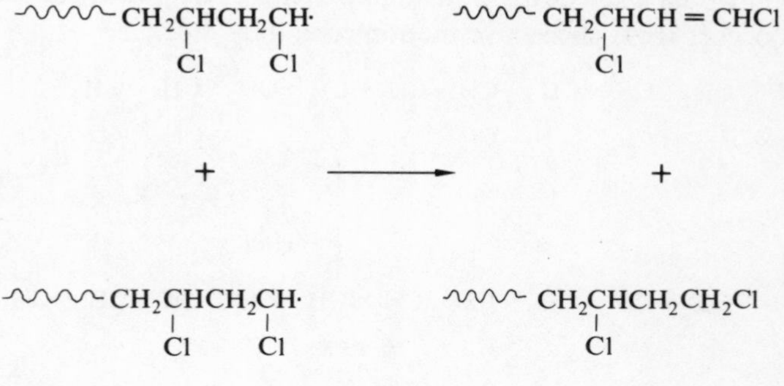

$\sim\!\!\sim\!\!\sim\mathrm{CH_2CHCH_2CH\cdot}$ (Cl, Cl) $\sim\!\!\sim\!\!\sim\mathrm{CH_2CHCH=CHCl}$ (Cl)

$+$ $\longrightarrow$ $+$

$\sim\!\!\sim\!\!\sim\mathrm{CH_2CHCH_2CH\cdot}$ (Cl, Cl) $\sim\!\!\sim\!\!\sim\mathrm{CH_2CHCH_2CH_2Cl}$ (Cl)

and tertiary chlorine atoms at branch points formed by chain transfer to polymer:

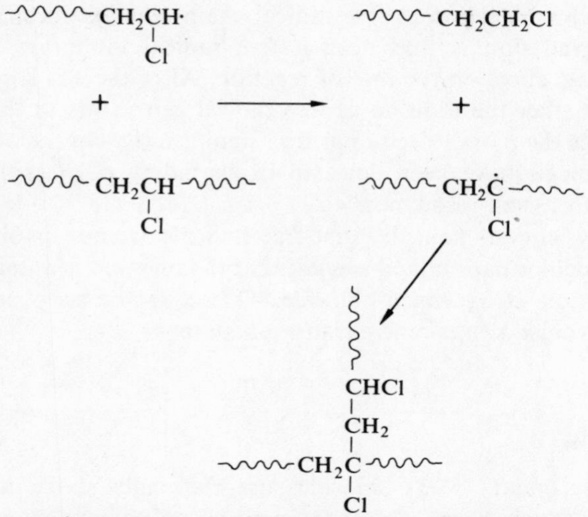

$$\sim\!\!\sim\!\!\sim CH_2\overset{\displaystyle \cdot}{C}H \qquad\qquad \sim\!\!\sim\!\!\sim CH_2CH_2Cl$$
$$\underset{\displaystyle Cl}{|}$$

$$+ \qquad\longrightarrow\qquad +$$

$$\sim\!\!\sim\!\!\sim CH_2CH\sim\!\!\sim\!\!\sim \qquad\qquad \sim\!\!\sim\!\!\sim CH_2\overset{\displaystyle \cdot}{C}\sim\!\!\sim\!\!\sim$$
$$\underset{\displaystyle Cl}{|} \qquad\qquad\qquad\qquad \underset{\displaystyle Cl}{|}$$

$$CHCl$$
$$|$$
$$CH_2$$
$$|$$
$$\sim\!\!\sim\!\!\sim CH_2\overset{\displaystyle }{C}\sim\!\!\sim\!\!\sim$$
$$\underset{\displaystyle Cl}{|}$$

A variety of mechanisms involving free-radical intermediates have been suggested for the dehydrochlorination. Thus, one suggestion is a free-radical chain-reaction of the following type:

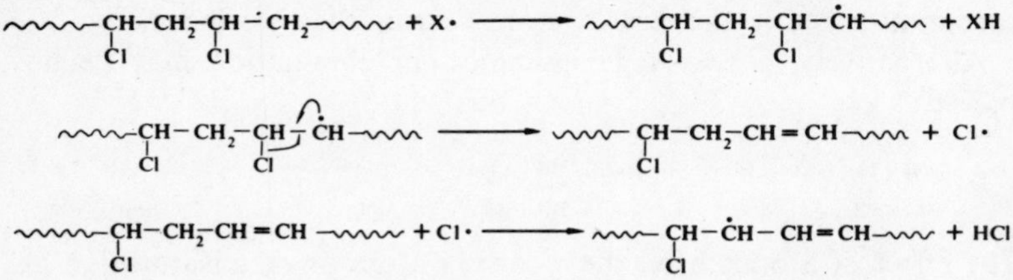

$$\sim\!\!\sim\!\!\sim\underset{Cl}{\overset{|}{C}}H{-}CH_2{-}\underset{Cl}{\overset{|}{C}}H{-}CH_2\sim\!\!\sim\!\!\sim + X\cdot \longrightarrow \sim\!\!\sim\!\!\sim\underset{Cl}{\overset{|}{C}}H{-}CH_2{-}\underset{Cl}{\overset{|}{C}}H{-}\overset{\cdot}{C}H\sim\!\!\sim\!\!\sim + XH$$

$$\sim\!\!\sim\!\!\sim\underset{Cl}{\overset{|}{C}}H{-}CH_2{-}\underset{Cl}{\overset{|}{C}}H{-}\overset{\cdot}{C}H\sim\!\!\sim\!\!\sim \longrightarrow \sim\!\!\sim\!\!\sim\underset{Cl}{\overset{|}{C}}H{-}CH_2{-}CH{=}CH\sim\!\!\sim\!\!\sim + Cl\cdot$$

$$\sim\!\!\sim\!\!\sim\underset{Cl}{\overset{|}{C}}H{-}CH_2{-}CH{=}CH\sim\!\!\sim\!\!\sim + Cl\cdot \longrightarrow \sim\!\!\sim\!\!\sim\underset{Cl}{\overset{|}{C}}H{-}\overset{\cdot}{C}H{-}CH{=}CH\sim\!\!\sim\!\!\sim + HCl$$

The zipper nature of the reaction is accounted for by postulating that hydrogen atom abstraction by chlorine atoms occurs predominantly in the allylic position, as shown above, since this results in the formation of a resonance stabilised radical.

Various proposals have been put forward as to the source of the free radicals necessary to initiate this and other proposed free-radical chain-reactions. One suggestion is that they arise by homolysis of allylic or tertiary carbon–chlorine bonds in the polymer, e.g.,

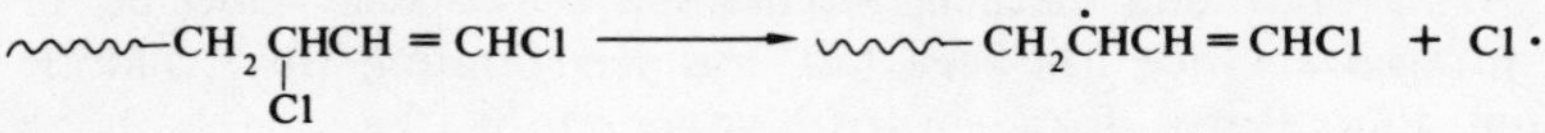

$$\sim\!\!\sim\!\!\sim CH_2\underset{Cl}{\overset{|}{C}}HCH{=}CHCl \longrightarrow \sim\!\!\sim\!\!\sim CH_2\overset{\cdot}{C}HCH{=}CHCl + Cl\cdot$$

Another is that they are formed by the cleavage of peroxides formed by the oxidation of polymer, or by the copolymerisation of oxygen with monomer during the preparation of the polymer.

A major objection to free radical chain reaction mechanisms for the degradation is that normal free radical inhibitors appear to have little effect on the rate of reaction. Also, there is some dispute as to whether the addition of free radical generators to the polymer increases the rate of decomposition significantly. On the other hand, free radicals have been detected in degrading poly(vinyl chloride) by electron spin resonance.

Many workers consider that free radicals are not involved in the dehydrochlorination, and suggest that the reaction is a unimolecular elimination of hydrogen chloride. This could be supposed to proceed through a four-centre transition state:

$$\text{~~}\overset{\text{H}}{\underset{\text{H}}{\text{C}}}-\overset{\text{H}}{\underset{\text{Cl}}{\text{C}}}\text{~~} \longrightarrow \left[\text{~~}\overset{\text{H}}{\text{C}}=\overset{\text{H}}{\text{C}}\text{~~}\atop \text{H---Cl}\right] \longrightarrow \text{~~}\overset{\text{H}}{\text{C}}=\overset{\text{H}}{\text{C}}\text{~~}\atop \overset{+}{\text{H}}\overset{-}{\text{Cl}}$$

Since secondary alkyl chlorides are thermally stable under conditions which cause decomposition of poly(vinyl chloride), it is suggested that elimination occurs initially at a structural abnormality such as a tertiary or allylic chlorine atom, and that the transition state is stabilised by the branch or the doubly-bonded carbon attached to the carbon atom bearing the chlorine.

Alternatively, a two-stage unimolecular elimination may be involved:

$$\text{~~}\underset{\text{H}}{\text{CH}}-\underset{\text{Cl}}{\text{CH}}\text{~~} \xrightarrow{slow} \text{~~}\underset{\text{H}}{\text{CH}}-\overset{+}{\text{CH}}\text{~~}\;\text{Cl}^- \xrightarrow{fast} \text{~~}\text{CH}=\text{CH}\text{~~}\atop \overset{+}{\text{HCl}}$$

The effect of a branch on the α-carbon atom or of unsaturation at the β-carbon atom would again be to stabilise the transition state in the rate-determining step. Thus, using the carbonium ion as a model for the transition state in dehydrochlorination involving an allylic chlorine:

$$\text{~~}\text{CH}_2-\overset{+}{\text{CH}}-\text{CH}=\text{CHCl} \atop \text{Cl}^- \longleftrightarrow \text{~~}\text{CH}_2-\text{CH}=\text{CH}-\overset{+}{\text{CHCl}} \atop \text{Cl}^-$$

With both unimolecular mechanisms, once one molecule of hydrogen chloride has been lost, the neighbouring vinyl chloride unit is activated for dehydrochlorination by the double bond formed, so that this type of mechanism is also consistent with the occurrence of a zipper reaction. However, it does not account for the presence of free radicals in the degrading polymer, nor is it easy to reconcile it with the accelerating effect which oxygen appears to have on the dehydrochlorination.

The role of hydrogen chloride in the degradation has been the subject of some controversy. Until the mid-1950s it was generally considered that the hydrogen chloride liberated had a catalytic effect on the decomposition. However, work published in 1953 and 1954 indicated that this was not so, and the view that there was no catalytic effect became fairly generally held. In the last few years work has been published indicating that hydrogen chloride *does* have a catalytic effect on the dehydrochlorination. As with many other aspects of this reaction, uncertainty prevails at present.

STABILISATION

A large number of compounds are used as stabilisers in poly(vinyl chloride). They can be divided into five main groups, and these are summarised in Table 8-2.

TABLE 8-2

POLY(VINYL CHLORIDE) STABILISERS

Class	Examples
Lead salts and soaps	Basic lead carbonate $2PbCO_3 . Pb(OH)_2$
	Dibasic lead stearate $(C_{17}H_{35}CO_2)_2Pb . 2PbO$
Alkaline earth salts and soaps	Barium stearate $(C_{17}H_{35}CO_2)_2Ba$
Zinc and cadmium salts and soaps	Cadmium laurate $(C_{11}H_{23}CO_2)_2Cd$
	Zinc stearate $(C_{17}H_{35}CO_2)_2Zn$
Tin compounds	Dibutyltin dilaurate $(C_4H_9)_2Sn(O_2CC_{11}H_{23})_2$
	Dibutyltin dilaurylmercaptide $(C_4H_9)_2Sn(SC_{12}H_{25})_2$
Epoxy compounds	Epoxidised soya bean oil

In view of the fact that so little is known of the mechanism of poly(vinyl chloride) degradation, it is not surprising that knowledge of the mode of action of stabilisers is meagre. It was originally thought that they acted by combining with the hydrogen chloride liberated in the degradation, thus preventing autocatalysis taking place. This theory to some extent fell into disrepute when doubt was cast on the catalytic effect of hydrogen chloride, and is, in any case, almost certainly an oversimplification. However, practically all poly(vinyl chloride) stabilisers do react with hydrogen chloride, this being possibly the only linking factor in a rather diverse group of compounds. Also, as we have seen, the catalytic effect of hydrogen chloride has to some extent been reinstated, and overall it seems likely that reactivity with hydrogen chloride is one factor in stabiliser activity.

Many other suggestions have been made as to the mode of action of stabilisers. For instance, it has been suggested that they eliminate labile sites in the polymer, that they react with polyene sequences, and so disrupt the conjugation which is the cause of coloration, and that they act as antioxidants. However, there is very little firm supporting evidence for most of these theories.

By the correct choice of stabilising system, poly(vinyl chloride) can be effectively stabilised against degradation during processing, and in service in most applications. Individual stabilisers sometimes show marked differences in effectiveness in stabilisation against heat and light. Thus, basic lead carbonate imparts good heat stability, but poor light stability, whereas cadmium laurate gives better light than heat stability. Pronounced synergistic effects are often shown, and mixtures of stabilisers are frequently used. Loadings vary from up to 10 parts per hundred parts of polymer (phr) for lead compounds in some applications, to about 0·5–2 phr of the highly effective but expensive tin compounds.

SYNTHETIC FIBRE-FORMING POLYMERS

A rather specific combination of properties is required for a polymer to be useful for the manufacture of synthetic fibres, and consequently only a small number of polymers have become important in this type of application. Since the cross-sectional area of the polymer is very small in fibre applications, a prime requirement of a fibre-forming polymer is that it should be capable of exhibiting a very high tensile strength, in order that fibres of acceptable strength be obtained.

Polymers of major importance in synthetic fibre applications are nylon 66, nylon 6, poly(ethylene terephthalate), acrylonitrile copolymers (acrylic fibres) and, to a lesser extent, polypropylene. A number of other nylons have some commercial importance, as do two other linear polyesters, and poly(vinyl alcohol) is used widely in Japan. High-density polyethylene finds some fibre applications.

Fibre-Forming Properties and Polymer Structure
Let us consider a fibre, indicated schematically in Fig. 8-1, in which all the polymer molecules are in the fully extended conformation,

and are aligned with the axis of the fibre, with a random distribution along the length of the fibre. If such a fibre is put under tension, then in order that it may be permanently deformed or broken, polymer molecules must slip over each other, or must undergo chain scission. If there are high intermolecular forces between polymer molecules these will cause resistance to slippage, and the

——— = extended polymer molecule

Fig. 8-1. *Idealised fibre structure.*

polymer will have a high tensile strength. Commercial synthetic fibres approximate to this idealised model, alignment of the molecules being brought about by physical treatment of the fibres. We will discuss the manufacture of poly(ethylene terephthalate) fibre as an example of the procedures involved.

Poly(ethylene terephthalate) fibre is made, as also are nylon, polypropylene and polyethylene fibres, by *melt spinning*. The molten polyester at 545 to 555 K ($T_m = 538$ K) is passed through a number of fine holes. The fine streams of polymer cool rapidly, and solidify to the amorphous state, rather than to the crystalline polymer that would be obtained on slow cooling. In this state the polymer molecules are more or less randomly arranged with respect to the fibre geometry, and the fibres have very poor mechanical properties. They are now subjected to the procedure of *cold drawing* in which they are stretched at a temperature somewhat above the glass transition temperature of the amorphous polymer ($T_g = ca.$ 343 K). During this process, in which the length of the fibre increases by several times, and its diameter decreases, the polymer molecules are aligned along the length of the fibres, and crystallisation occurs. After cold drawing, the fibre is held at about 490 K to allow crystallisation to be completed, and to release internal molecular strains. The fibre obtained has a high tensile strength. The glass transition temperature of the drawn fibre is over 373 K; its melting point is the crystalline melting point of the polymer, 538 K.

Broadly the same procedure is used for melt spinning other polymers, though there are differences in detail. For instance, some polymers crystallise direct from the melt, and cold drawing then causes alignment of the crystalline and amorphous regions. Other polymers do not crystallise even on cold drawing.

For melt spinning to be practicable, the polymer must be stable at the temperatures required to keep it molten, and in some cases this requirement is not met. For example, polyacrylonitrile decomposes before it melts. Such polymers are spun from solution, either by extruding a solution in a volatile solvent into a stream of hot air, *dry spinning*, or by extruding a solution into a bath of liquid which causes the polymer to come out of solution, called *wet spinning*. In either case, the fibre is normally subjected to a drawing procedure.

A balance of a number of structural features is necessary in order that strong fibres may be formed from a polymer. The polymer should be linear. Crosslinked polymers cannot be made to align, and branching of the polymer chain tends to interfere with alignment. Thus, low-density polyethylene, a highly branched polymer, gives only weak fibres, while the practically unbranched high-density polyethylene gives strong ones.

The flexibility of the polymer chain is of importance since, other things being equal, the more flexible the chain, the more readily will the polymer molecules revert to random, coiled conformations. This effect is illustrated by comparing high-density polyethylene, which is fibre-forming, with poly(ethylene oxide), which is not. Introduction of the ether linkage, which has a very low energy barrier to rotation, into the polymer causes a considerable increase in chain flexibility. Conversely, the introduction of stiffening structural features, e.g. rings, into polymers may give products with improved fibre-forming characteristics. A notable example of a fibre-forming polymer containing ring structures is poly(ethylene terephthalate).

Alignment is favoured by axial symmetry and the absence of 'kinks' in the polymer molecules. This is well exemplified by a comparison of poly(ethylene terephthalate) with the non-fibre-forming poly(ethylene phthalate):

poly (ethylene terephthalate)   *poly (ethylene phthalate)*

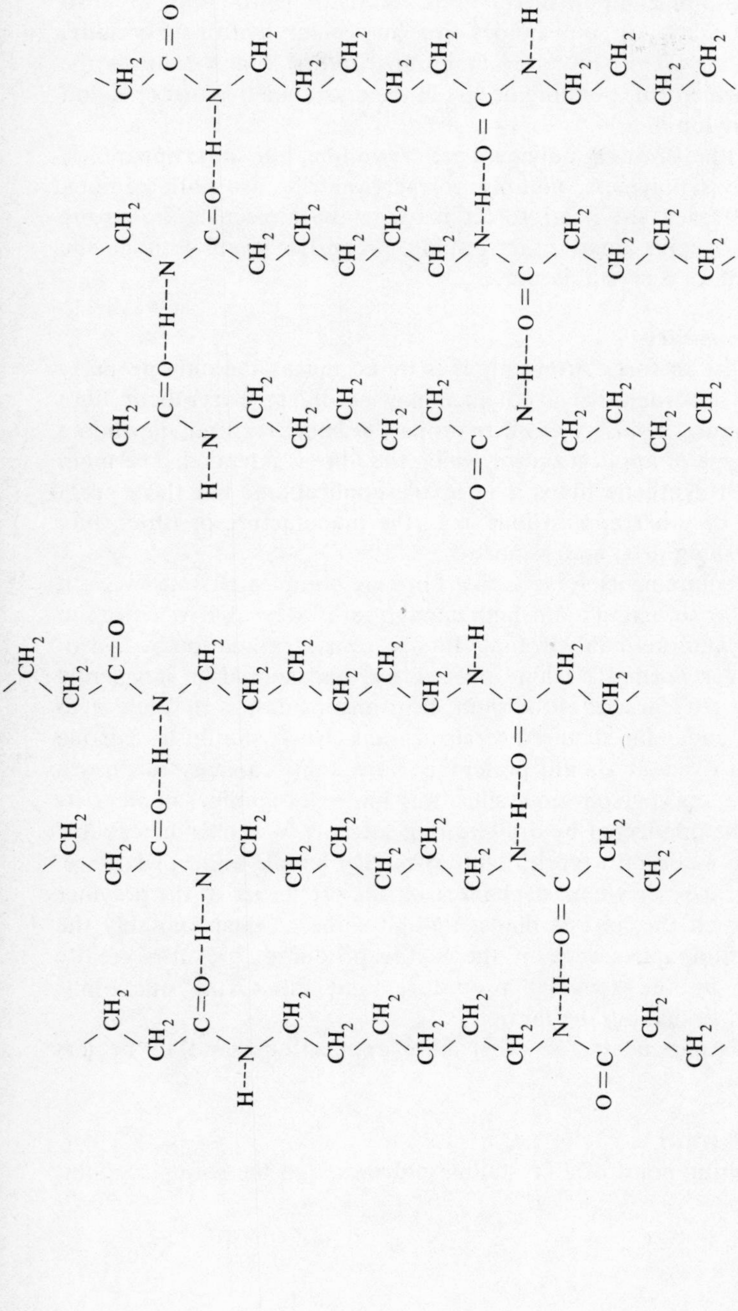

Nylon 6

Nylon 66

Fig. 8-2. *Hydrogen bonding in crystalline regions of nylon 66 and nylon 6.*

The cohesive forces between polymer chains may be simply van der Waals forces, as in the cases of polyethylene and polypropylene, but often dipole–dipole interactions and hydrogen bonding are also involved. Thus, in polyamides, intermolecular hydrogen bonding between —NH— and —C=O groups occurs. Fig. 8-2 shows the way in which this bonding occurs in the crystalline regions of nylon 66 and nylon 6.

Most fibre-forming polymers are crystalline, but some apparently amorphous polymers, notably polyacrylonitrile, are fibre-forming. In these cases, the polar forces between the molecules are strong enough to give good fibre-forming properties even without the formation of a crystal lattice.

Fibre Properties

The ability to form strong fibres is by no means the only property required in order that a polymer may be of use in synthetic fibre manufacture, a variety of other properties being required depending on the type of application for which the fibre is intended. The main outlet for synthetic fibres is in textile applications, but there are a number of other applications, e.g., the manufacture of ropes, tyre cords, fishing nets, and so on.

The requirements for a textile fibre are complex. It must have a melting or softening point high enough for it to be able to withstand washing and preferably ironing. It must resist degradation by hydrolysis under normal washing procedures, and should be sufficiently resistant to degradation by light, heat and oxidation to allow it to be used under the required service conditions. It should be capable of being dyed. It should preferably have some capacity to absorb moisture, e.g. perspiration, since this improves comfort in wear. It should be unaffected by dry-cleaning solvents. It should be capable of being made into fabrics with attractive handle and appearance. These properties are all dependent on the properties of the polymer from which the fibre is made, though some of them, notably the handle and appearance of the textile produced, are also greatly affected by the spinning procedure, and the textile operations involved in making the fabric.

The requirements for other fibre applications tend to be less stringent.

MELTING POINT

The melting point of a crystalline polymer, and the softening point of an amorphous polymer, are largely determined by the flexibility of the polymer molecules and by the intermolecular forces of attraction.

A comparison of poly(ethylene adipate) and poly(ethylene terephthalate) illustrates the effect of chain flexibility on melting point. Poly(ethylene adipate), $\leftarrow OC(CH_2)_4CO_2CH_2CH_2O \rightarrow_n$, is a fibre-forming polymer, but is of no commercial value in fibre applications since its melting point is only 323 K. Replacement of the relatively flexible $\leftarrow CH_2 \rightarrow_4$ group by the rigid p-phenylene group raises the melting point to 538 K.

The effect of intermolecular forces is indicated by a comparison of polyethylene with nylon 66. High-density polyethylene has a melting point of about 403 K. Nylon 66, which can be considered as being polyethylene in which some of the methylene groups have been replaced by $-NH-$ and $-C=O$ groups, and which, as we have seen, is heavily hydrogen bonded in the crystalline regions, has a melting point of 537 K.

CHEMICAL STABILITY

The relationship between chemical structure and resistance to degradation is fairly obvious. For example, polypropylene fibre will evidently be perfectly stable to hydrolysis, but will be liable to oxidative attack, particularly under the influence of light. On the other hand, poly(ethylene terephthalate) and nylon fibres would be expected to be susceptible to hydrolytic degradation, and under some conditions they are. However, the highly oriented structure of the fibre affords a considerable resistance to penetration by reagents, and these fibres are less readily hydrolysed than their monomeric analogues. Nylons are highly resistant to attack by alkalis, but are fairly readily hydrolysed by acids, whereas poly(ethylene terephthalate) is highly resistant to acids, but is attacked by alkalis.

DYEABILITY

Dyeability, i.e., the capacity to be permanently coloured by treatment with an aqueous solution or dispersion of a dyestuff, is of extreme importance in a textile fibre, and lack of this capacity may severely restrict the application of an otherwise attractive fibre.

The essential requirements for the successful dyeing of fibres are (i) that the dyestuff (or its precursors, in some dyeing systems) should be able to diffuse into the fibre, and (ii) that having diffused into the fibre, the dyestuff molecules should be in some way firmly held in place. Attachment to the fibre may involve the formation of covalent or ionic bonds between dyestuff and polymer molecules, or may be a result of hydrogen bonding and van der Waals forces. In general, a 'faster', i.e., more permanent, dyeing is obtained when ionic or covalent bonds are formed than when the dye is held by hydrogen bonds or van der Waals forces. The naturally occurring protein and cellulose fibres are well endowed with polar and reactive sites, and are readily dyed with a wide range of dyestuffs. Synthetic fibres present a more difficult problem, and it has been necessary to develop new dyeing systems, and in some cases to modify the fibre-forming polymers, to allow satisfactory dyeing to be carried out.

Nylons are the most readily dyed of the major synthetic fibres. The glass transition temperatures of nylons are low (around 320 K), so that at the dyeing temperature (ca. 373 K) the polymer chains in the amorphous regions are mobile, and dyestuff molecules can readily diffuse into the fibres. Amino groups at chain ends provide sites for salt formation with dyestuffs containing sulphonic acid groups:

$$Dye-SO_3H \ + \ H_2N \text{~~~} \longrightarrow Dye-SO_3^- \ H_3\overset{+}{N} \text{~~~}$$

Also, the recurring amide groups provide sites for hydrogen bonding with dyestuff molecules.

When poly(ethylene terephthalate) was first introduced, its dyeing presented difficulties, one of the reasons being that since poly(ethylene terephthalate) fibre has a glass transition temperature of over 373 K the amorphous regions of the fibre are in the glassy state at normal dyeing temperatures, and diffusion of dyestuffs molecules into the fibres cannot readily occur. This problem can be overcome either by carrying out the dyeing at above the glass transition temperature of the polymer, e.g. at 390 to 400 K, which involves dyeing under pressure, or by carrying it out at normal dyeing temperatures in the presence of a 'carrier'. Carriers are organic compounds which dissolve in the polymer and act as a temporary plasticiser for it, reducing its glass transition temperature and increasing the mobility of the polymer chains in the amorphous regions. Biphenyl and methyl salicylate are examples of materials used as carriers in the dyeing of poly(ethylene terephthalate).

Poly(ethylene terephthalate) does not contain sites which can be used for the attachment of dyestuffs to the fibre by ionic or covalent bonds.

Polyacrylonitrile fibre presents similar dyeing problems to those presented by poly(ethylene terephthalate). It has a glass transition temperature of 373 to 383 K, and carries no group which can readily be utilised for bond formation with dyestuffs molecules. In this case, however, the problems have been solved by modifying the polymer properties by copolymerisation, and acrylic fibres are usually made from terpolymers containing 85–94% acrylonitrile, together with a monomer added to provide dyeing sites, and one added to reduce the glass transition temperature of the polymer. Examples of the comonomers which may be used to depress T_g are methyl methacrylate, methyl acrylate, and vinyl acetate. 2-Vinyl-pyridine (I), and itaconic acid (II) are among the comonomers that have been mentioned in patents as being useful in providing dyeing sites:

$$\text{(I)} \qquad CH_2=C\begin{array}{l} ^{\nearrow CO_2H} \\ _{\searrow CH_2CO_2H} \end{array} \quad \text{(II)}$$

Polymers containing vinylpyridine or other basic comonomers give fibres that may be dyed with acidic dyestuffs, whereas those made from polymers containing acidic comonomers such as itaconic acid may be dyed by basic dyestuffs.

Polypropylene, being completely non-polar, has very little affinity for dyestuffs, and at the present time cannot be satisfactorily dyed. It can be coloured by mass pigmentation, i.e., by mixing a finely divided pigment with the polymer before spinning, but this is a much less flexible method of coloration than dyeing. The lack of dyeability of polypropylene has considerably restricted its use in textile applications.

MOISTURE ABSORPTION

As has been pointed out, the capacity of a fibre to absorb moisture has a considerable effect on the wearing comfort of fabrics made from the fibre, fabrics made from fibres with low moisture absorbing capacities tending to have a clammy feel. Moisture absorption also helps to reduce the development of static electricity. Probably the major disadvantage of the synthetic fibres compared with the natural fibres is that they have lower moisture absorbencies (see Table 8-3).

TABLE 8-3
MOISTURE ABSORPTION BY FIBRES

Fibre	Moisture Absorption Wt % (at 65 % relative humidity and 294 K)
cotton	8·5
wool	16
silk	11
nylon 6	4·5
nylon 66	4·5
acrylic fibres	1·3–2·5
poly(ethylene terephthalate)	0·4–0·8
polypropylene	0

Source of data: 'Synthetic Fibre-Forming Polymers', I. Goodman, R.I.C. Lecture Series 1967, No. 3. By permission of Dr. Goodman.

The capacity of a fibre to absorb water depends largely on the ability of the polymer to form hydrogen bonds with water. Thus nylons, which contain highly polar and readily hydrogen-bonded amide groups, are relatively highly absorbent; polypropylene absorbs no water at all.

ELASTOMERIC POLYMERS

Elastomeric polymers, typified by natural rubber, have the property of undergoing long-range reversible extension under relatively small applied loads. Their extensibility and resilience make them of value in a wide variety of applications.

The exhibition of elastomeric properties by polymers depends on the tendency, already discussed, of flexible long-chain molecules to exist in randomly coiled conformations in which the average distance between the ends of molecules is very much less than that in the fully extended conformation.

Consider a single such molecule undergoing random thermal segmental motion. If it were possible to take hold of the ends of this molecule, then by the application of some force, they could be moved apart until the molecule was in the fully extended conformation. On releasing the ends, the thermal motion of the molecular segments would cause the molecule to revert to its coiled conformations, and the average distance between its ends to revert to its former value. Now consider a strip of polymer made up of flexible molecules linked together into a network by a relatively small number of crosslinks, with long segments of polymer chain between the crosslinks. If the ends of this strip are pulled, then the segments of polymer chains between the crosslinks will uncoil, and approach the fully-extended conformation as described above, and the strip will stretch. When the strip is released, the thermal motion of the polymer chains will cause them to return to their coiled conformations and the piece of polymer will revert to its original shape. Overall displacement of polymer molecules, which would result in permanent distortion of the piece of polymer, is prevented by the crosslinks.

There are, therefore, two basic requirements for an elastomeric polymer. Firstly, it should have flexible chains which readily undergo thermal segmental motion at the service temperature. This implies that the polymer must not be crystalline,* and that its glass transition temperature should be substantially below the service temperature. Secondly, the structure of the polymer should be such that a lightly crosslinked network, with long segments of chain between the crosslinks, can be produced.

Major Elastomers

A wide variety of polymers are used commercially as elastomers. The more important ones, in terms of amounts used, are shown in Table 8-4.

Natural rubber, *cis*-1,4-polyisoprene, has been used commercially in significant amounts since the early part of the nineteenth century, and still accounts for a substantial proportion of rubber consumption. By the 1930s a well-developed processing technology based on natural rubber was in existence and, in the main, development work on synthetic rubbers has been aimed at producing materials which fitted in with this existing technology. Thus, most large-tonnage synthetic elastomers contain carbon–carbon double bonds, and consequently can be crosslinked by the methods used for natural rubber.

The first general-purpose synthetic rubbers made on the large scale were styrene–butadiene copolymers, generally called SBR

* A number of elastomers, including natural rubber, crystallise on stretching, the formation of a crystal lattice being favoured by the extended conformations of the chains. The crystalline regions melt when tension is released.

rubbers; as can be seen from Table 8-4, these are still the most important synthetic rubbers made.

TABLE 8-4
MAJOR COMMERCIAL ELASTOMERS

Elastomer	Structure	Date introduced	Capacity 1970 ('000 tons)	
			USA	UK
natural rubber	$\begin{bmatrix} & CH_3 & \\ & \vert & \\ & C=CH & \\ CH_2 & & CH_2 \end{bmatrix}_n$ $cis-1,4-$polyisoprene		608 (consumption 1969)	191
neoprene	$\begin{bmatrix} & Cl & \\ & \vert & \\ CH_2-C=CH-CH_2 & \end{bmatrix}_n$ polychloroprene	1931	180	30
nitrile rubbers	butadiene–acrylonitrile random copolymers	1937	117	20
SBR rubbers	butadiene–styrene random copolymers	1937	1 690	217
butyl rubber	isobutene–isoprene random copolymers, isoprene content 0·6–3·5%	1943	165	36
'High cis' polybutadiene	$\begin{bmatrix} CH=CH & \\ CH_2 & CH_2 \end{bmatrix}_n$ $cis-1,4-$polybutadiene	1960	386	80
Diene rubber	polybutadiene (*cis*-1,4–, *trans*–1,4– and 1,2–)	1961		
polyisoprene	*cis*–1,4–polyisoprene (synthetic)	1961	140	nil
ethylene–propylene rubbers	ethylene–propylene random copolymer	1962	110	3
	ethylene–propylene–diene random terpolymer	1963		

SBR rubbers were developed after early attempts to prepare useful polymers from butadiene alone by emulsion polymerisation had failed. However, in recent years, methods of preparing homopolymers of butadiene with very good properties have been developed, and such elastomers are growing rapidly in importance.

It is worth noting that whereas *cis*-1,4-polybutadiene is an elastomer, the more readily aligned *trans*-1,4-polybutadiene is a crystalline polymer with a melting point of 418 K. Similarly, *trans*-1,4-polyisoprene (gutta percha) is a crystalline polymer, whereas *cis*-1,4-polyisoprene (natural rubber) is an elastomer.

Neoprene and nitrile rubbers, which carry polar substituents, have much better solvent resistance than the hydrocarbon rubbers, and this property forms the basis of many of their applications.

The recently developed ethylene–propylene rubbers provide an interesting example of the effect of structure on the physical properties of polymers. Both high-density polyethylene and polypropylene are crystalline polymers, in which the polymer chains are tightly held in crystal lattices for a substantial proportion of their lengths, and consequently they do not have elastomeric properties. Copolymerisation of approximately equal quantities of ethylene and propene gives a polymer of highly irregular structure, which cannot crystallise. Segmental motion of the polymer chains is therefore possible, and the polymer is an elastomer. Polymers of this type cannot be crosslinked by conventional methods because of the absence of carbon–carbon double bonds. However, incorporation of some double bonds into the polymer by the use of a diene as a second comonomer gives a polymer which can be processed by conventional means.

Vulcanisation

For reasons which have already been indicated, some degree of crosslinking is usually necessary in order that a polymer may exhibit good elastomeric properties, and this is brought about by *vulcanisation*.* The extent of crosslinking required in an elastomer is much less than in a thermosetting resin, a typical vulcanisate having of the order of 100 monomer units between crosslinks. A substantially higher degree of crosslinking than this converts rubbers into hard, rigid materials.

Vulcanisation is carried out after the polymer has been formed into the required shape.

SULPHUR VULCANISATION

Sulphur vulcanisation, the main method used, was discovered by Goodyear, in the USA, in 1839, and by Hancock, in the UK, in

---

* Vulcanisation is a term which applies specifically to elastomers; it is not used to describe crosslinking of other polymers.

1842. Originally, vulcanisation was brought about simply by heating a mixture of rubber and sulphur, but this process is slow and, from very early in the history of vulcanisation technology, additives have been used to increase the rate of vulcanisation. The first such

TABLE 8-5

SOME IMPORTANT CLASSES OF VULCANISATION ACCELERATORS

Class	Examples	
Benzothiazole accelerators	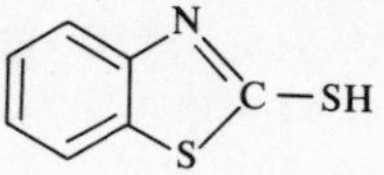	2-mercaptobenzothiazole (MBT)
	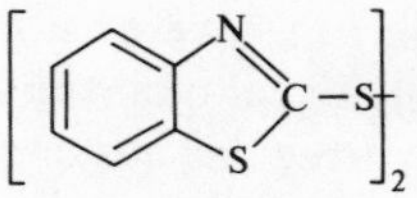	2,2'-dithiobisbenzothiazole (MBTS)
	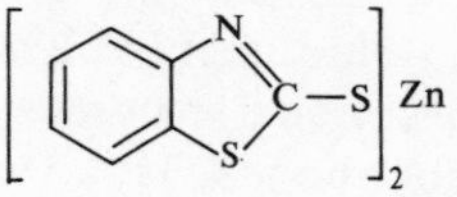	zinc benzothiazolyl mercaptide (ZMBT)
	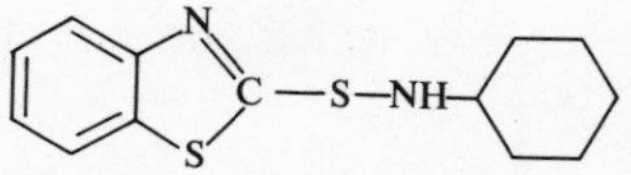	N-cyclohexylbenzothiazole-2-sulphenamide
Dithiocarbamates	$[(C_2H_5)_2 N-\overset{\overset{S}{\|}}{C}-S]_2 Zn$	zinc diethyldithiocarbamate (ZDC)
Thiurams	$[(C_2H_5)_2 N-\overset{\overset{S}{\|}}{C}-S]_2$	tetraethylthiuram disulphide (TET)
Diarylguanidines	$PhNH-\overset{\overset{NH}{\|}}{C}-NHPh$	diphenylguanidine (DPG)

additives were inorganic compounds such as lead carbonate, but modern rubber technology was made possible by the discovery, early this century, that various organic compounds had a profound effect on the vulcanisation reaction. This discovery led to the development of a group of compounds specifically designed to increase the rate of vulcanisation, the *vulcanisation accelerators*. Some of the important classes of accelerators are shown in Table 8-5.

Accelerators are now almost invariably used in sulphur vulcanisation. In addition to increasing the rate of vulcanisation, they allow a greater degree of control over vulcanisate properties than could be achieved by unaccelerated vulcanisation. There are at least 50 accelerators in commercial use; those derived from benzothiazole are the most important, accounting for more than 65% of total consumption.

Most accelerators require the presence of zinc oxide in the mix for effective action, and generally a fatty acid, e.g. stearic acid, also has to be added. The vulcanising system for a natural rubber tyre tread might typically be as follows:

sulphur	2·5 phr*
stearic acid	3·0 phr
zinc oxide	5·0 phr
accelerator(s)	0·5 phr

Mixtures of accelerators are often used in order to produce desired effects in processing characteristics and/or in vulcanisate properties. Vulcanisation is typically carried out at temperatures in the range 370 to 430 K, depending on the rubber used and the type of application.

The chemistry of both accelerated and unaccelerated sulphur vulcanisation is complex and difficult to investigate, and it is by no means fully understood. The major overall effect is to introduce mono- and polysulphide crosslinks into the polymer at carbon atoms adjacent to the double bonds, e.g.:

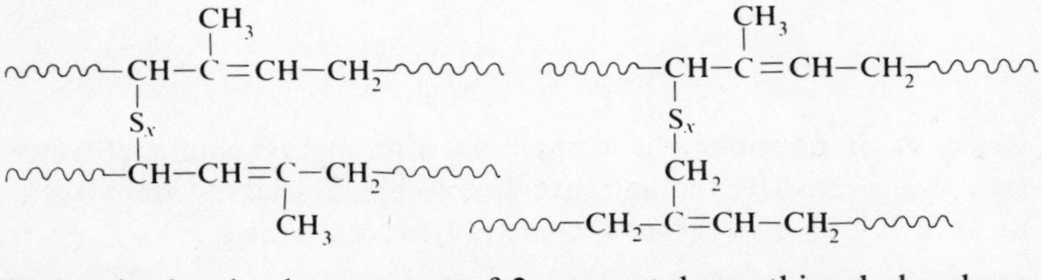

Vulcanisation in the presence of 2-mercaptobenzothiazole has been most extensively investigated, and a number of reaction schemes have been suggested. One scheme for the vulcanisation of natural rubber, which is favoured by a number of workers, is discussed briefly below.

* Parts per hundred parts of rubber.

It is proposed that reaction of 2-mercaptobenzothiazole and zinc oxide gives the zinc mercaptide, and that this forms a rubber-soluble complex with the fatty acid:

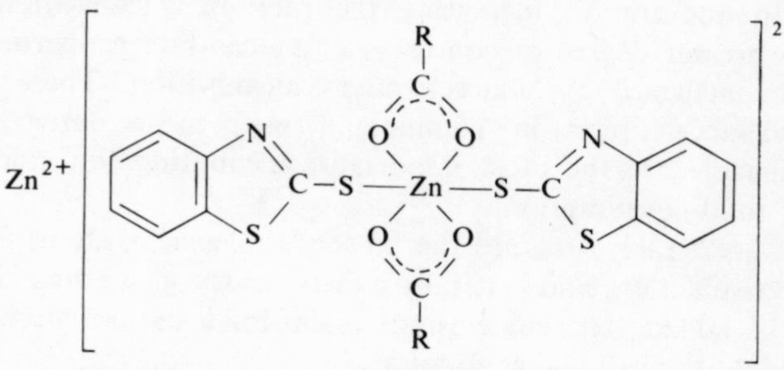

$$
\left[\text{benzothiazole–C–S—Zn—S—C–benzothiazole, with } \begin{array}{c} R \\ C \\ O \quad O \\ Zn \\ O \quad O \\ C \\ R \end{array} \right]^{2-} \quad Zn^{2+}
$$

Reaction of this with S_8 molecules gives a persulphidic complex (in the following discussion, the carboxylate ligands are omitted, and X— represents the benzothiazole residue):

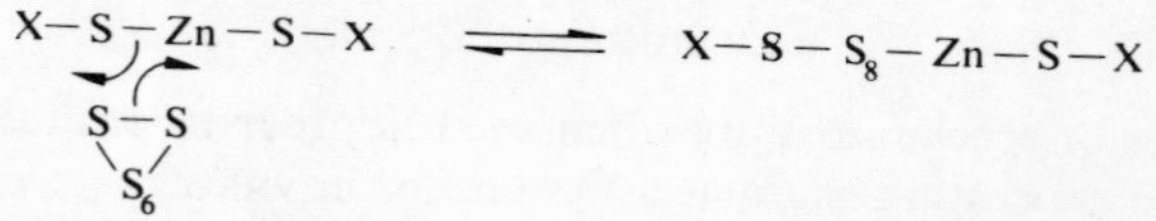

$$
\begin{array}{c} X–S \\ \quad \quad Zn–S–X \\ X–S \\ S–S \\ S_6 \end{array} \rightleftharpoons X–S–S_8–Zn–S–X
$$

Interchange with the original complex leads to the formation of a mixture of polysulphidic complexes, which are considered to be the active sulphurating species:

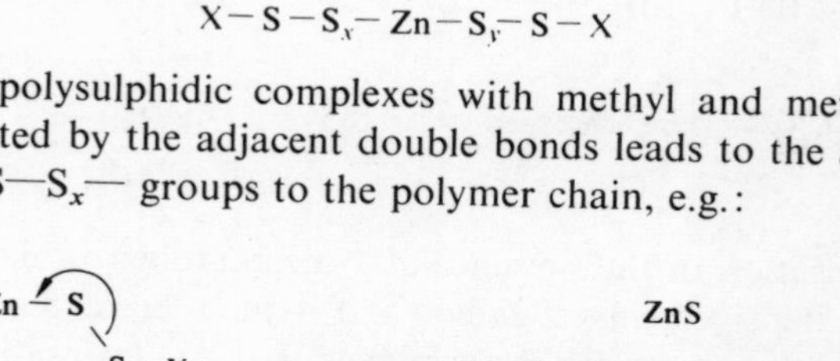

$$
X–S–S_8–Zn–S–X \; + \; X–S–Zn–S–X
$$
$$
\updownarrow
$$
$$
X–S–S_n–Zn–S–X \; + \; X–S–S_{\overline{8-n}} Zn–S–X
$$
$$
\downarrow \text{ etc.}
$$
$$
X–S–S_x–Zn–S_y–S–X
$$

Reaction of polysulphidic complexes with methyl and methylene groups activated by the adjacent double bonds leads to the attachment of X—S—S$_x$— groups to the polymer chain, e.g.:

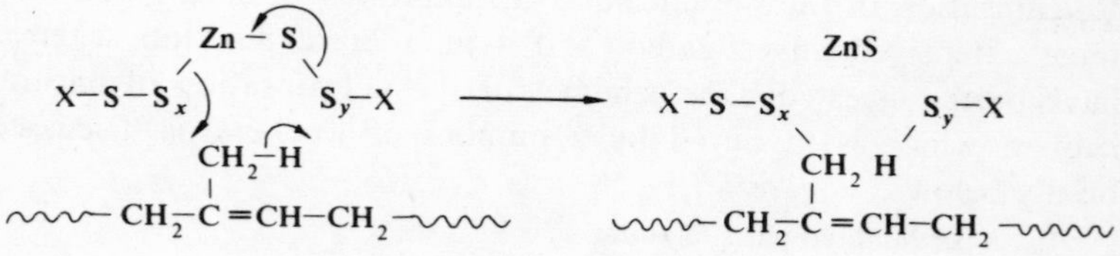

$$
\begin{array}{c}
Zn \quad S \\
X–S–S_x \quad \quad S_y–X \\
CH_2–H \\
\sim\sim CH_2–C=CH–CH_2\sim\sim
\end{array}
\longrightarrow
\begin{array}{c}
ZnS \\
X–S–S_x \quad \quad S_y–X \\
CH_2 \; H \\
\sim\sim CH_2–C=CH–CH_2\sim\sim
\end{array}
$$

A polysulphidic crosslink between polymer chains is then produced by reactions of the following type (R— = polymer chain):

$$R-S_x-S-X \quad + \quad X-S^- \longrightarrow \quad R-S_z^- \quad + \quad X-S-S_{\overline{(x-z)}}S-X$$

$$R-S_z^- \quad + \quad R-S_x-S-X \longrightarrow \quad R-S_w-R \quad + \quad X-S-S_{\overline{(x+z-w)}}$$

Further reactions occur leading to progressive shortening of the crosslink, and under some conditions mainly mono- and di-sulphide crosslinks are finally produced.

Although accelerated sulphur vulcanisation is by far the most important method of vulcanisation, a number of other systems find some use, mainly in specialist applications. Some examples of these are briefly discussed below.

The most generally applicable non-sulphur system is vulcanisation with peroxides. An organic peroxide, e.g. dicumyl peroxide, is added to the rubber. At the vulcanisation temperature it decomposes into radicals, which abstract hydrogen atoms from polymer molecules. Thus, in vulcanisation of natural rubber:

$$ROOR \longrightarrow 2RO\cdot$$

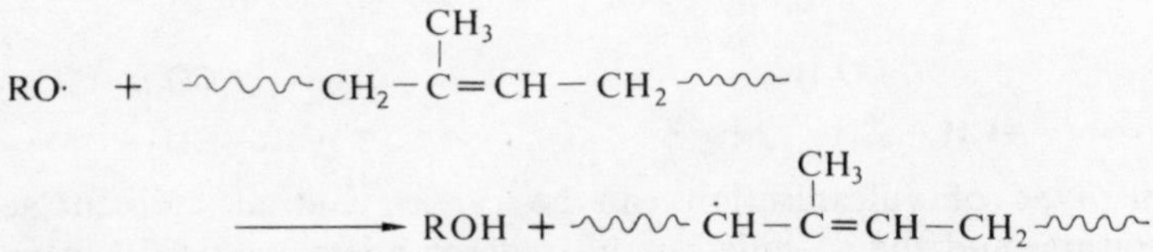

$$RO\cdot \quad + \quad \sim\!\!\sim\!\!\sim CH_2-\underset{\underset{CH_3}{|}}{C}=CH-CH_2\sim\!\!\sim\!\!\sim$$

$$\longrightarrow ROH + \sim\!\!\sim\!\!\sim \dot{C}H-\underset{\underset{CH_3}{|}}{C}=CH-CH_2\sim\!\!\sim\!\!\sim$$

The polymer radicals thus formed undergo coupling reactions to give carbon–carbon crosslinks:

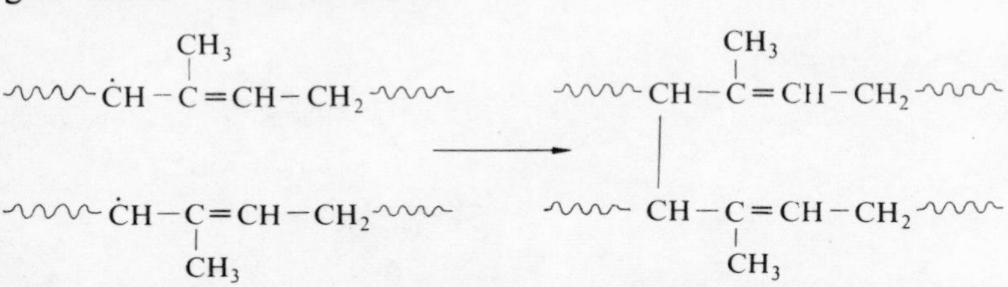

$$\sim\!\!\sim\!\!\sim \dot{C}H-\underset{\underset{CH_3}{|}}{C}=CH-CH_2\sim\!\!\sim\!\!\sim \qquad \sim\!\!\sim\!\!\sim CH-\overset{\overset{CH_3}{|}}{C}=CH-CH_2\sim\!\!\sim\!\!\sim$$

$$\longrightarrow$$

$$\sim\!\!\sim\!\!\sim \dot{C}H-\underset{\underset{CH_3}{|}}{C}=CH-CH_2\sim\!\!\sim\!\!\sim \qquad \sim\!\!\sim\!\!\sim CH-\underset{\underset{CH_3}{|}}{C}=CH-CH_2\sim\!\!\sim\!\!\sim$$

This method is more expensive, and less convenient, than sulphur vulcanisation, and is not very widely used for unsaturated rubbers.

However, it is not dependent on the presence of unsaturation in the polymer, and it thus provides a method of vulcanising saturated polymers such as ethylene–propylene rubbers.

Other non-sulphur systems are used with particular polymers. For example, vulcanisation of neoprene involves reaction of allylic chlorine atoms in the small proportion of 1,2-monomer units in the polymer:

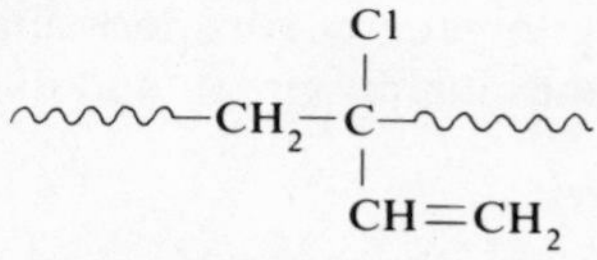

$$\sim\!\!\sim\!\!\sim\!\!-CH_2-\underset{\underset{CH=CH_2}{|}}{\overset{\overset{Cl}{|}}{C}}-\sim\!\!\sim\!\!\sim \qquad \text{1,2-chloroprene unit}$$

Vulcanising agents used are zinc oxide and magnesium oxide, together with some organic additives. The nature of the reactions involved is not fully established.

A number of speciality rubbers bearing carboxyl groups on the polymer chain are made, by the incorporation of an acidic comonomer, e.g. acrylic acid, into the polymer. These can be vulcanised by treatment with metal oxides, which leads to the formation of metal salt crosslinks:

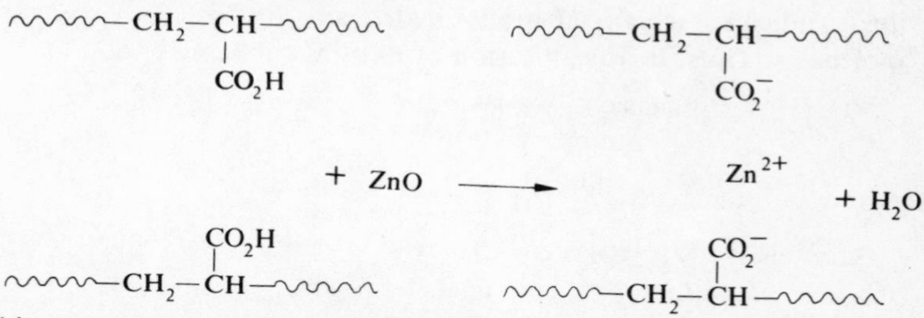

$$\sim\!\!\sim\!\!\sim\!\!-CH_2-\underset{\underset{CO_2H}{|}}{CH}-\sim\!\!\sim\!\!\sim \qquad\qquad \sim\!\!\sim\!\!\sim\!\!-CH_2-\underset{\underset{CO_2^-}{|}}{CH}-\sim\!\!\sim\!\!\sim$$

$$+ \; ZnO \longrightarrow \qquad Zn^{2+} \qquad + \; H_2O$$

$$\sim\!\!\sim\!\!\sim\!\!-CH_2-\underset{}{\overset{\overset{CO_2H}{|}}{CH}}-\sim\!\!\sim\!\!\sim \qquad\qquad \sim\!\!\sim\!\!\sim\!\!-CH_2-\underset{}{\overset{\overset{CO_2^-}{|}}{CH}}-\sim\!\!\sim\!\!\sim$$

This type of vulcanisation can be carried out at ambient temperature, and the so-called carboxylate rubbers are useful in cold-vulcanising latex adhesives, e.g. for carpet backing.

Further Reading and Sources of Information

DETAILS are given below of selected general texts, and of journals, of relevance to industrial organic chemistry and, under chapter headings, of some books and articles dealing with specific topics.

GENERAL TEXTS

Encyclopaedia of Chemical Technology, R.E. Kirk and D.F. Othmer, 2nd edn., Interscience, New York. Vol. 1 (1963) to Vol. 21 (1970) available at the time of writing.

(An invaluable, large and comprehensive work which provides a first port of call for information on any topic.)

Industrial Chemicals, W.L. Faith, D.B. Keyes and R.L. Clark, 3rd edn., John Wiley, New York, 1965.

(This gives details of processes, applications and US manufacture of most of the major organic chemicals.)

The Petroleum Chemicals Industry, R.F. Goldstein and A.L. Waddams, 3rd edn., E. & F.N. Spon, London, 1967.

(A modern, comprehensive account of the technological and economic make-up of the industry.)

Introduction to the Chemical Process Industries, R.M. Stephenson, Reinhold, New York, 1966.

(A clear, readable account, giving much technological detail. Covers the inorganic chemical industry.)

The Petrochemical Industry, A.V.G. Hahn, McGraw-Hill, New York, 1970.

(This places more emphasis on economic factors, and substantially less on technological factors, than Goldstein and Waddams.)

Chemicals From Petroleum, A.L. Waddams, 2nd edn., John Murray, London, 1968.

(A short book, which gives a very good account, at its chosen level, of the petroleum chemicals industry.)

Heavy Organic Chemicals, A.J. Gait, Pergamon Press, Oxford, 1967.

(Covers similar ground to Waddams. Good on details of manufacturers in the UK.)

322   INTRODUCTION TO INDUSTRIAL ORGANIC CHEMISTRY

Unit Processes in Organic Synthesis, P.H.Groggins, McGraw-Hill, New York, 1958.

(Becoming distinctly out of date, but still useful on the slower moving parts of the industry.)

JOURNALS

Listed below are some of the journals which provide information on processes, products and general technological and economic developments in the chemical industry.

> *British Chemical Engineering*
> *Chemical Age*
> *Chemical Engineering*
> *Chemical and Engineering News*
> *Chemical Engineering Progress*
> *Chemical and Process Engineering*
> *Chemical Technology*
> *Chemistry in Britain*
> *Chemistry and Industry*
> *Chemical Week*
> *European Chemical News*
> *Hydrocarbon Processing and Petroleum Refiner*
> *Industrial and Engineering Chemistry*
> *Oil and Gas Journal*

Of these journals, *European Chemical News* is particularly recommended as regular reading for the student who wishes to remain abreast of current developments in the chemical industry, but with the minimum expenditure of time. Despite its title, it does not confine itself to European activities.

SPECIFIC TOPICS

Chapter 1.   Introduction
Introduction to Technological Economics, D. Davies and C. McCarthy, John Wiley and Sons, London, 1967.

Research in the Chemical Industry, A. Baines, F.R. Bradbury and C.W. Suckling, Elsevier, London, 1969.

Chemistry and Industry, D.G. Jones, Chs. 4, 5 and 6, Clarendon Press, Oxford, 1967.

Chapter 2.   Reactions of Alkanes and Cycloalkanes
The Chemistry of Petroleum Hydrocarbons, B.T. Brooks, C.E. Boord, S.S. Kurtz, Jr. and L. Schmerling, Reinhold, New York, 1955. Vol. 2—Thermal Cracking, Catalytic Cracking, Catalytic Reforming, Acetylene. Vol. 3—Isomerisation, Alkylation.

Advances in Petroleum Chemistry and Refining, K.A. Kobe and J.J. McKetta, Jr., Interscience, New York. Vol. 1, 1958—Alkylation and Catalytic Reforming. Vol. 3, 1960—Isomerisation. Vol. 5, 1962—Catalytic Cracking. Vol. 6, 1962—Thermal Cracking. Vol. 8, 1964—Hydrocracking. Vol. 9, 1964—Thermal Cracking, Steam Reforming.

Modern Petroleum Technology, 3rd edn., The Institute of Petroleum, London, 1962.

Acetylene: Its Properties, Manufacture and Uses, S.A. Miller, Benn, London, 1965.

'Ethylene', R.A. Duckworth, *Chemical and Process Engineering*, **49**(2), 67 (1968).

'A chemical view of refining', L.F. Hatch, *Hydrocarbon Processing and Petroleum Refiner*, **48**(2), 77 (1969).

Catalyst Handbook, Imperial Chemical Industries Ltd., Wolfe Scientific Books, London, 1970—Steam Reforming.

Chapter 3. Oxidation

Advances in Petroleum Chemistry and Refining, K.A. Kobe and J.J. McKetta, Jr., Vol. 3, Interscience, New York, 1960.

Catalysis, P.H. Emmett, Vol. 3, Reinhold, New York, 1960.

'Oxidation of olefins with palladium chloride catalysts', J. Smidt *et al.*, *Chemistry and Industry*, **81**, 54 (1962).

Atmospheric Oxidation and Antioxidants, G. Scott, Elsevier, Amsterdam, 1965.

Liquid-Phase Oxidation of Hydrocarbons, N.M. Emanuel, E.T. Denisov and Z.K. Maizus, Plenum Press, New York, 1967.

Chemistry and Industry, D.G. Jones, Ch. 2, Clarendon Press, Oxford, 1967.

Homogeneous Catalysis, Advances in Chemistry Series No. 70, American Chemical Society, Washington, D.C., 1968.

Oxidation of Organic Compounds, Vols. 1 and 2, Advances in Chemistry Series Nos. 75 and 76, American Chemical Society, Washington, D.C., 1968.

'Make petrochemicals by liquid-phase oxidation', H.W. Prengle and N. Barona, *Hydrocarbon Processing and Petroleum Refiner*, **49**(3), 106 (1970).

Autoxidation of Hydrocarbons and Polyolefins, L. Reich and S.S. Stivala, Dekker, New York, 1969.

Chapter 4.   Halogen Compounds
Chlorine—Its Manufacture, Properties and Uses, J.S. Sconse, Reinhold, New York, 1962.

Acetylene: Its Properties, Manufacture and Uses, S.A. Miller, Vol. 2, Chs. 1 and 2, Benn, London, 1966.

'Vinyl chloride processes', L.F. Albright, *Chemical Engineering*, **74**(7), 123 (1967).

'Manufacture of vinyl chloride', L.F. Albright, *Chemical Engineering*, **74**(8), 219 (1967).

Ethylene and its Industrial Derivatives, S.A. Miller, Ch. 10, Benn, London, 1969.

The Manufacture and Uses of Fluorine and its Compounds, A.J. Rudge, Oxford University Press, London, 1962.

Bromine and its Compounds, Z.E. Jolles, Benn, London, 1966.

Chapter 5.   Aromatic Substitution and Related Reactions
The Chemistry of Petroleum Hydrocarbons, B.T. Brooks, S.S. Kurtz Jr., C.E. Boord and L. Schmerling, Vol. 3, Reinhold, New York, 1955.

Unit Processes in Organic Synthesis—see GENERAL TEXTS.

Friedel-Crafts and Related Reactions, G.A. Olah, Vol. 2, Interscience, New York, 1964.

Sulphonation and Related Reactions, E.E. Gilbert, Interscience, New York, 1965.

'Chemistry of aromatic nitrations', L.F. Albright, *Chemical Engineering*, **73**(8), 169 (1966).

'Processes for nitration of aromatic hydrocarbons', L.F. Albright, *Chemical Engineering*, **73**(9), 161 (1966).

Mechanistic Aspects of Aromatic Sulphonation and Desulphonation, H. Cerfontain, Interscience, New York, 1968.

Chapter 6.   Miscellaneous Reactions
HYDRATION
Ethylene and its Industrial Derivatives, S.A. Miller, Ch. 9, Benn, London, 1969.

HYDROGENATION AND DEHYDROGENATION
Catalysis, P.H. Emmett, Vol. 3, Reinhold, New York, 1955.

'High purity cyclohexane', F.A. Dufau, F. Eschard, A.C. Haddad and C.H. Thonon, *Chemical and Engineering Progress*, **60**(9), 43 (1964).

'The mechanism of heterogeneous catalysis', R.L. Burwell, *Chemical and Engineering News*, **44**(34), 56 (1966).

'Theory and chemistry for the hydrogenation of fatty oils', L.F. Albright, *Chemical Engineering*, **74**(19), 197 (1967).

'Commercial processes for hydrogenating fatty oils', L.F. Albright, *Chemical Engineering*, **74**(21), 249 (1967).

'Continuous processes for reducing nitroaromatics to aromatic amines', L.F. Albright, F.H. Van Munster and J.C. Forman, *Chemical Engineering*, **74**(23), 251 (1967).

HYDROFORMYLATION
Advances in Petroleum Chemistry and Refining, K.A. Kobe and J.J. McKetta, Jr., Vol. 1 Interscience, New York, 1958.

'Le Procédé Oxo', H. Lemke, *L'Industrie Chimique*, **52**, 169 (1965).

Transition Metal Intermediates in Organic Synthesis, C.W. Bird, Logos Press, London, 1967.

Chapter 7.   Polymerisation
Encyclopaedia of Polymer Science and Technology, Interscience, New York. Vols. 1 (1964) to 10 (1969) available at the time of writing.

Introduction to Polymer Chemistry, D. Margerison and G.C. East, Pergamon Press, Oxford, 1967.

Organic Chemistry of Synthetic High Polymers, R.W. Lenz, Interscience, New York, 1967.

Organic Chemistry of Macromolecules, A. Ravve, Dekker, New York, 1967.

Plastics Materials, J.A. Brydson, 2nd edn., Iliffe, London, 1969.

Principles of Polymerisation, G. Odian, McGraw-Hill, New York, 1970.

Polyurethanes: Chemistry and Technology, J.H. Saunders and K.C. Frisch, Vols. 1 and 2, Interscience, New York, 1961.

Chapter 8.   Polymer Properties, Reactions and Applications
'The Mechanism of Antioxidant Action', H.C. Bailey, *Industrial Chemist*, **38**, 215 (1962).

'Antioxidants', G. Scott, *Chemistry and Industry*, **82**, 271 (1963).

The Chemistry and Physics of Rubber-like Substances, L. Bateman, Elsevier, London, 1963.

The Stabilisation of Poly(vinyl chloride), F. Chevassus and R. de Broutelles, Edward Arnold, London, 1963.

Plasticisation and Plasticiser Properties, Advances in Chemistry Series No. 48, American Chemical Society, Washington, D.C. 1968.

Chemical Reactions of Polymers, E.M. Fettes, Interscience, New York, 1964.

Polymer Technology, D.C. Miles and J.H. Briston, Temple Press Books, London, 1956.

Atmospheric Oxidation and Antioxidants, G. Scott, Elsevier, Amsterdam, 1965.

Polymers: Structure and Bulk Properties, P. Meares, Van Nostrand, London, 1965.

Synthetic Fibre-forming Polymers, I. Goodman, Royal Institute of Chemistry Lecture Series 1967, No. 3, Royal Institute of Chemistry, London.

Organic Polymers, T. Alfrey and E.F. Gurnee, Prentice-Hall, Englewood Cliffs, 1967.

Poly(vinyl chloride), J.C. Koleske and L.H. Wartman, MacDonald, London, 1969.

Index

Acetal resins, 255–257
Acetaldehyde
  by hydrocarbon oxidation, 96
  by-product in vinyl acetate manufacture, 102
  from acetylene, 13, 194
  from ethanol, 218
  from ethylene, by Wacker Chemie process, 16, 97–101
  oxidation of, 81–83
  uses of, 101, 137, 218
Acetic
  acid
    from acetaldehyde, 9, 13, 81–83
    from butane, 89–90
    from naphtha, 19, 90
    uses of, 90, 101–103, 140, 198–202, 265, 268
  anhydride
    from acetaldehyde, 82–83
    from acetic acid, 90–91
    use of, 201–202
Acetone
  by fermentation, 9
  by-product in acetic acid manufacture, 90
  by-product in phenol manufacture, 17, 77, 81
  from isopropanol, 93, 218
  uses of, 116, 174, 219
Acetophenone, by-product in phenol manufacture, 80
Acetylene
  chemicals from, 13
  chloroprene from, 132–133
  dimerisation of, 132–133

Acetylene–*contd.*
  from calcium carbide, 12–13
  from hydrocarbons, by pyrolysis, 18–19, 20, 60–64
  hydration of, 194
  manufacturing costs, comparison with ethylene, 63–64
  perchloroethylene from, 129
  trichloroethylene from, 126
  vinyl acetate from, 103
  vinyl chloride from, 118–119, 121–122
  vinyl fluoride from, 146
'Acrilan', 252
Acrylic
  acid, in carboxylate rubbers, 320
  fibres, 252
    dyeing of, 311
    moisture absorption by, 312
    spinning of, 306
Acrylonitrile
  from acetylene, 13, 110
  from propene, 17, 109–110
  polymerisation of, 252
  uses of, 110
Addition polymerisation, definition of, 228
Adipic acid
  adiponitrile from, 208
  formation in cyclohexane oxidation, 85–86, 88
  in nylon '66' manufacture, 268
  manufacture of, 87–88
Adiponitrile, 208
Aerosol propellants, 144
Alcohols
  by Oxo process, 219–227

327

Octylphenol, 170
Orange II, 177
'Orlon', 252
Oxidation, 73–110, *see also* individual compounds, e.g. Cyclohexane, Ethylene, Naphthalene
degradation of polymers by, 291–298
Oxo process, 219–227, *see also* Hydroformylation
Oxychlorination of
benzene, 185
ethylene, 122–125

Palm oil, 6
Para Red, 178
Pent-1-ene, hydroformylation of, 220
Peracetic acid, 81
Perchloroethylene, 13, 129–130, 144, 146
Peroxides, in rubber vulcanisation, 319
'Perspex', 251
Petrol, *see* Gasoline
Petroleum
as raw material for chemicals, 5, 6, 15–19
distillation of, 29–30
refining of, 15, 27–51
uses of, 27, 30
Phenol
by cumene process, 77–81
by Dow toluene process, 103–106
by monochlorobenzene process, 3, 175, 183–184
by Raschig process, 185
by sulphonation process, 160, 182–183
from coal, 10, 11, 12
uses of, 81, 140, 174, 270–273
Phenol-formaldehyde resins, 270–273
Phenols
alkylation of, 169–171
as antioxidants, 294–295
diazonium coupling of, 177–178

1-Phenylethanol, 94
Phenylphenol, 184
2-Phenylpropan-2-ol, 79, 80
Phosgene, 128, 130, 155, 265
Phthalic
acid, isomerisation of, 179–181
anhydride, 108–109, 202, 267
Phthalocyanine pigments, 141
Plasticisers, 202–203, 289–290
Platforming, *see* Catalytic reforming
'Plexiglas', 251
Polyamides, 267–270, *see also* Nylons
Polybutadiene, 253–254, 314, 315
Polycaprolactam, *see* Nylon '6'
Polycarbonates, 265
Polychloroprene, *see* Neoprene
Polychlorotrifluoroethylene, 145
Polyesters, 264–267, 279
Poly(ethyl acrylate), 288
Polyethylene, 16, 246–248
antioxidants in, 298
glycols, 196
oxidation of, 291–293
properties and structure of, 288–289, 306
Poly(ethylene adipate), 261, 309
Poly(ethylene terephthalate), 228, 264–266
dyeing of, fibres, 310
hydrolysis of, 309
spinning of, 305
Poly(hexamethylene adipamide), *see* Nylon '66'
Polyisobutene, 252–253
Polyisoprene, 314, *see also* Natural rubber
Polymerisation, 228–281, *see also* individual monomers, e.g. Ethylene, Propene, Vinyl chloride
addition, 228
bulk, 236
chain-growth, 229–259
condensation, 228
degree of, 261
emulsion, 238–239
free-radical vinyl, 229–239

RENEWALS: 691-4574

DATE DUE

MAY 0 3			
GAYLORD			PRINTED IN U.S.A